Smart Mobility

Smart Mobility

Using Technology to Improve Transportation in Smart Cities

Bob McQueen
Bob McQueen and Associates
Perpignan, France

Ammar Safi
Senior Transportation Engineer
Ministry of Energy and Infrastructure
Abu-Dhabi

Shafia Alkheyaili
CEO, Osool Engineering Consultancy
Abu-Dhabi

WILEY

Library of Congress Cataloging-in-Publication Data applied for:
Hardback ISBN: 9781119847137

Cover Design: Wiley
Cover Image: © Blue Planet Studio/Shutterstock

Set in 9.5/12.5pt STIXTwoText by Straive, Pondicherry, India

Contents

About the Authors

Bob McQueen is an internationally recognized subject matter expert in smart mobility and the application of advanced technology to transportation. He has more than 40 years of experience in helping public and private sector organizations harness the full power of advanced technologies and has provided expert-level consulting services in Asia-Pacific, Middle East, North American, and European markets.

He has extensive experience in areas related to connected and automated vehicles and artificial intelligence. This includes work for both the public and the private sectors on the definition of use cases, system architectures, value propositions, and project concepts for the application of advanced technology to smart cities. His expertise also spans standardization, professional development program delivery, and partnerships.

He has also conducted extensive research and developed reports on connected and automated vehicles, addressing field testing and policy aspects. He also has in-depth practical knowledge of the electronic toll collection systems marketplace, having worked for both public and private sector organizations in this area. This is an important application of connected vehicle technology.

While originally from Scotland, he is now based in Reynes, France. He is currently focused on understanding the effects of investment in smart mobility and transportation management and the practical application of AI to transportation management. His work includes strategic business support for large corporations and startups, and he has had the privilege of conducting assignments in more than 100 cities around the world.

Ammar Safi stands as a luminary in the realm of engineering, boasting an illustrious career spanning over 12 years across a multitude of disciplines, including telecommunication, electrical engineering, mobility, road safety, and infrastructure. His journey has seen him traverse the corridors of power in the Federal and local governments, as well as the private sector, where he has honed his prowess in navigating intricate stakeholder landscapes and shaping pivotal decision-making processes.

Equipped with a formidable academic background, Ammar holds a bachelor's degree in telecommunication and electrical engineering, complemented by dual master's degrees in urban planning and transportation planning. Undeterred by the frontiers of knowledge, he is currently on a quest for enlightenment as a PhD candidate in the groundbreaking field of ITS and autonomous vehicle communication, pushing the boundaries of innovation and technological advancement.

At the Ministry of Energy and Infrastructure, Ammar serves as the vanguard of progress in his capacity as a senior smart mobility and smart cities engineer. His multifaceted role sees him spearheading initiatives across various transformative teams, including Artificial Intelligence, Innovation, and Smart Cities. As a luminary of scientific inquiry, Ammar's research pursuits transcend conventional boundaries, delving deep into the realms of AI and cutting-edge sciences, where he champions the relentless pursuit of excellence and progress.

Beyond his professional endeavors, Ammar's influence reverberates across the global engineering landscape as a distinguished member of prestigious organizations such as the International Association of Automation and Robotics. Closer to home, he lends his expertise to the Smart Construction Center at the esteemed American University in Sharjah, while also enriching the academic discourse as an esteemed member of the IEEE.

Not content with merely shaping the present, Ammar is actively involved in shaping the future of engineering education and practices, serving as a trusted advisor on the consulting council at the College of Engineering – Fujairah University. His indomitable spirit, relentless pursuit of knowledge, and unwavering commitment to excellence serve as a beacon of inspiration for aspiring engineers and seasoned professionals alike, as he continues to chart the course toward a future defined by innovation and progress.

Shafia Alkheyaili embodies over a decade of unparalleled expertise across civil engineering, mobility, road safety, and infrastructure. With an illustrious career enriched by her tenure within the Federal government, Shafia's leadership prowess has been finely honed, allowing her to adeptly navigate intricate stakeholder landscapes and drive impactful decision-making. Armed with a distinguished academic background – comprising a BSc in civil and environmental engineering, dual master's degrees in mobility and road safety, and her ongoing pursuit of a PhD in ITS and autonomous vehicle safety – Shafia stands at the forefront of innovation in engineering.

During her tenure as a smart cities and assets management engineer, Shafia's visionary leadership propelled development initiatives within the Federal Roads Network at the Ministry of Energy and Infrastructure, leaving an indelible mark on the infrastructure ecosystem. Her unwavering commitment to pioneering research at the nexus of AI and advanced sciences further underscores her status as a trailblazer in the field.

Beyond her professional accomplishments, Shafia is a revered figure in the global engineering community, holding esteemed memberships in a myriad of local and international boards, including the prestigious IRF, PIARC, IAARC, and the UAE Engineering Society. As the Chief Executive Officer of Osool Engineering Consultancy – based in the bustling hub of Sharjah, UAE – Shafia continues to spearhead transformative initiatives, shaping the future of engineering consultancy. Additionally, her oversight of operations at Soul Decoration Design and Implementation in Dubai Design District, UAE, underscores her commitment to fostering innovation and design excellence.

Shafia Alkheyaili's unwavering dedication, visionary leadership, and boundless passion for engineering excellence serve as an inspiring testament to her enduring impact on the industry landscape.

Preface

This is a book about people and mobility. It is about how effective application of technology can significantly increase mobility and accessibility options to all travelers, while providing new possibilities for avoiding or managing undesirable side effects. A smart mobility revolution is underway and involves sudden and dramatic change as emerging technologies make new things possible.

Smart mobility lies at the nexus between technology and business; therefore, this book deals with both. Technology is addressed and described at a level of sufficient depth that enables the reader to understand capabilities and constraints; enabling effective choices.

Business approaches and models are also important elements in this book because it is important to focus beyond technology to discover the best way to apply it. The authors have witnessed the evolution of smart mobility with a significant shift from an emphasis on making technology work to identifying and applying the best way to use it. The authors believe that the smart mobility revolution is well underway. This involves the application of emerging technologies, new business practices, and new business models to transportation.

This book is designed to enable the reader to get the most from the smart mobility revolution by providing the following:

- A set of tools and knowledge (not recipes) to enable the most effective application of smart mobility revolution elements;
- A focus on what can be done, and why, not a detailed "how-to" manual;
- The raw materials to enable practitioners to develop customized approaches to smart mobility implementation worldwide;
- An entry point to the world of smart mobility; a textbook for students;
- A collection of diverse subject materials that are relevant to smart mobility;
- A global view of progress, lessons learned, and practical experiences;
- Information on the availability of innovative data sets and emerging data science;
- An exploration of opportunities and challenges, explaining the best way to approach these through benefit–cost analysis, planning, policy, and organizational arrangements.

The materials are drawn from more than 70 years of combined experience in the application of advanced technology to transportation around the globe. The book describes both public and private sector perspectives to provide a complete picture of the best way to approach the smart mobility revolution.

The authors will consider the book successful if it:

- Ensures that the reader enjoys reading it;
- Provides new information in a compelling way;
- Delivers practical advice that can be applied in the reader's role;
- Makes a direct impact on the effectiveness of smart mobility initiatives worldwide.

We thank you for your valuable time invested in reading the book and hope that it will be repaid many times over in better smart mobility initiatives worldwide.

Acknowledgments by Bob McQueen

While the people who have provided knowledge, inspiration, and influenced the contents of this book are too numerous to mention, I give a special thanks to Ammar and Shafia, my co-authors, and the following people:

Abbas Mohaddes
Adam Lyons
Alex Sahyoun
Andrew Hall
Andy Palanisamy
Ayman El Ansary
Benoit Robinet
Bob Williams
Brett Boston
Clay Packard
Darren Capes
David Prais
Derek Turner
Donna Bergan
Doug Jamison
Dov Ganor
Dwayne Ingram
Dwight Jordan
Eli Sherer
Eric Hill
Gadi Sneh
Gary Carlin
Gavin Jackman
Glenn Cook
Greg Krueger
Harriet Chen-Pyle
Hossein Sadiq
Irene McAleese
James Garner
Janneke van der Zee
Jason JonMichael
Jay Reddy
Jeremy Dilmore
Jeris White
Jerry Quandt
Jesus Martinez

Jillian Kowalchuk
JJ Williams
John Linsley
John Slot
Kevin Borras
Konstantin Zakharov
Kyle Connor
Laetitia Parisi
Lima Saft
Marian Thompson
Mark Knellinger
Mark Lowe
Marshall Elizer
Mary Lege
Matthew Juckes
Moe Zarean
Mohamed Abdel-Aty
Mohamed Zaki Hussein
Monalisa Mohapatra
Nicola Liquori
Patrick Son
Paul Cooper
Pete Costello
Pete Kavanagh
Petros Xanthopoulus
Phil Minassian
Professor Essam Radwan
Professor Kan Chen
Professor Moshe Benbassat
Ratib Aloui
Rich Taylor
Richard Easley
Rick Schuman
Rish Malhotra
Rob Hubbard
Samiul Hasan
Sagy Amit
Scott Carlson
Steve Morello
Todd Kreter
Tony Geara
Tushar Patel
Vern Herr

Above all, I thank my wife Sonia for her love and constant support, and for all the time she gifted me to research and write. It would not have been possible without you. Thank you for being my advisor and sounding board, but most of all, thank you for being my love and best friend. Together we walk with God.

Acknowledgments by Ammar Safi

The journey of writing this book has been deeply enriched by the unwavering support of those closest to me. First and foremost, I want to express my deepest gratitude to my co-authors, Robert and Shafia, for their invaluable collaboration and dedication.

To my mother, Seba, I express my heartfelt thanks for your constant encouragement and unwavering belief. Your love has been a wellspring of strength throughout this endeavor. To my father, Majed, thank you for instilling in me the curiosity and dedication that fueled this project. Your legacy of lifelong learning is a constant source of inspiration.

My heartfelt thanks extend to my brothers, Eddy, Yousef, Dima, and Rabab. Their unwavering support and their willingness to lend a listening ear throughout this journey have been invaluable. Their understanding and encouragement have helped me navigate the challenges and celebrate the triumphs of writing this book.

I am incredibly grateful to the following individuals who provided invaluable guidance and expertise:

- Dr. Anwaar Alshimmari: Your insights and expertise were instrumental in shaping this book. Your dedication to knowledge is truly admirable.
- Professor Fernando and Teresa Avila: Their unwavering support has been instrumental throughout this journey. I am incredibly grateful for their combined guidance and encouragement.
- Professor Ali Abou-Elnour: Your guidance played a significant role in this project. Your contributions are deeply appreciated.

A special thanks to my friend and brother, Ahmad AL Dalu, for his unwavering support and encouragement during the writing process. Your camaraderie and belief in this project have been a source of immense strength.

Additionally, I extend my deepest gratitude to all my friends and colleagues who offered their help and support throughout the development of this book. Their encouragement has fueled my passion for this project and helped me bring it to fruition.

Acknowledgments by Shafia Alkheyaili

I express my deepest gratitude to my co-authors, Robert and Ammar, whose expertise, dedication, and collaboration were instrumental in bringing this book to fruition. Your insights and hard work have been invaluable.

A heartfelt thank you to my mother Amna, whose unwavering support and encouragement have been a constant source of strength and inspiration. Your belief in me has been a driving force behind this endeavor.

To my father Ali, who sadly passed away while I was writing this book, your memory has been a guiding light. I dedicate this work to you, with immense love and respect. Your wisdom and teachings continue to influence my life and work.

To my work colleagues and employees, thank you for your unwavering support and commitment. Your dedication and hard work have allowed me to pursue this project, and I am deeply grateful for your contributions.

To the esteemed professionals: Dr. Abdullah Belhaif Alnuaimi, Professor Fernando Varela, Teresa Avila, Professor Khalid Hamad, Dr. Daniel Llort, David Berrocal, Aljawhara AlMansour, Maryam AlHazami, thank you for your invaluable support, guidance, and collaboration. Each of you has played a significant role in this journey, and I am profoundly grateful for your contributions.

To my family, thank you for your patience, love, and understanding throughout this journey. Your support has meant the world to me, and I am truly grateful for each one of you.

1

Introduction

1.1 Informational Objectives

After reading this chapter, you should be able to answer the following questions:

1) Why this book? Why now?
2) What is smart mobility revolution?
3) What is smart mobility?
4) What subjects are addressed and where can they be found in the book?
5) Why were the subjects addressed in each chapter of the book chosen?
6) Who is the intended readership of the book?
7) What can the intended reader groups expect to gain by reading this book?

1.2 Introduction

According to the needs, you might decide to read the entire book front to back or dive into specific chapters that address subjects that are important at the current time. The authors hope that every chapter in the book will have something useful to specific segments of our intended readership; however, the authors will also be delighted if reading one or more chapters of the book delivers the knowledge needed to enable the delivery of better results and higher value.

1.3 What Is Smart Mobility?

At the heart of the smart mobility revolution lies "smart mobility." Smart mobility is the application of advanced information and telecommunications technologies to improve the full spectrum of activities associated with transportation. An important feature of the smart mobility revolution is the ability to take a complete approach. At the highest level, there are five categories of activities that must be supported to deliver smart mobility services:

1) Planning
2) Design

Smart Mobility: Using Technology to Improve Transportation in Smart Cities, First Edition. Bob McQueen, Ammar Safi, and Shafia Alkheyaili.
© 2024 The Institute of Electrical and Electronics Engineers, Inc. Published 2024 by John Wiley & Sons, Inc.

3) Project delivery
4) Operations
5) Maintenance

These are illustrated and summarized in Figure 1.1. This also defines some of the challenges associated with each stage of the transportation service delivery process that can be addressed by the application of advanced technology.

Note that Figure 1.1 also represents a high-level business process view of transportation, one that will be built on later in the book. One of the interesting aspects of the application of technology is that if it is delivered as effectively as it should be, it will become invisible, and the focus will be placed on the services that can be delivered because of the technology. For example, when watching television, it is highly unlikely that the viewer will be trying to understand the technology that is at work. In fact, it is highly likely that the viewer is not even aware of the technology at work and has an appropriate focus on the content, rather than the delivery technology. This represents an important measure of the maturity of a technology: if it is not obvious and the focus is on service delivery rather than making the technology work, then the technology is working well and has matured to the point where transportation practitioners do not need to be technology experts to operate it. This focus on content can be described as an outcome. In the early years of the intelligent transportation systems programs around the world, there was particular emphasis on how to make technologies work. While the authors believe that this is still the case for emerging and new technologies, a new level of maturity has been achieved for robust, well-proven, and well-understood technologies. For these mature technologies, the focus should be placed on what can be done with them to deliver results and outcomes.

This is why it is especially important to clearly define and describe the intended outcomes for smart mobility. It is equally important to provide an effective explanation of how the technologies are applied to achieve the outcomes. The outcomes perspective can be thought of as a "shop window" view. Everything that smart mobility has to offer is put on display in an attractive way by explaining the value and benefits. One of the objectives of this book is to grab attention just like a shop window, but then entice to come inside for a more detailed look. In developing this more detailed look, the authors have quite deliberately created what they call the "under the hood" view.

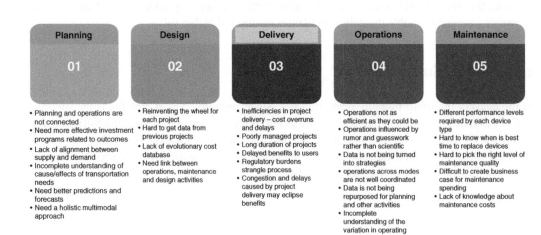

Figure 1.1 The transportation service delivery process.

This is based on real-world experience and lessons learned from smart mobility deployments. It has been created by working with practitioners and solution providers to find out what really happened, what works well, and what not so well.

The authors intend to take the reader behind the scenes to provide an understanding of how the technology works, why it is important, and the value that will be delivered. It is valuable that those involved in developing and delivering smart mobility understand the range of powerful technologies at work behind the scenes to deliver results to the end user. Constraints and limitations associated with the technologies must also be understood. There are also technological choices and business model options that need to be explored and understood. The authors have found that the best applications of technology are supported by the best ways to do things, always guided by an unwavering drive to achieve predefined outcomes. While technologies represent the building blocks of smart mobility, enabling products, services, and solutions, it is the service delivery enabled by the solutions that deliver value to the end users and to the system operators.

To develop and establish smart mobility services, it will be necessary to deliver a higher quality of user experience, reduce the cost of urban services, and support the seamless delivery of multimodal transportation in the city: all while managing the carbon footprint of smart mobility. Take electronic payment systems, for example. Electronic payments enable mobility service providers to offer users conditional discounts: "if you do that, then we will give you this." They might offer the best deal for the situation and support dynamically changing price. For example, if a person uses the transportation system in a smart city for a day, they might be charged one price. If, however, they use it again for a second day then the system can determine if a weekly pass might be more cost-effective for the traveler and automatically change the ticket to that type. Frequent traveler discounts and money-back guarantees regarding service conditions are also distinct possibilities. Such systems also offer considerable cost reductions compared to paper-based ticketing and a new flexibility in adapting ticketing and payment structures to exactly align with the demand for transportation and user needs.

In addition to an outcomes-based mindset, new levels of agility and flexibility are also the hallmarks of smart mobility. This agility enables a focus on the needs of the customer and a customer-friendly approach to payment strategies and operational models.

Another important aspect of smart mobility is the rich stream of data that will emanate from smart mobility systems. This data will be used to provide travelers with a customized experience and yet be handled using smart data management techniques to guarantee privacy and security of data. Moving from data to its conversion into information, and then development into insight and understanding, which in turn are the basis for new strategies and responses, the full power of smart mobility will become obvious.

It is worthwhile getting to know what smart mobility projects are currently in development and what is likely to be achieved in the future. Smart mobility already plays an extremely important role in today's smart cities, and the authors believe that this importance will grow in the smart cities of tomorrow. Successful growth will draw heavily on lessons and experiences from the past.

Some people have described data as the "new oil," meaning that data is an extremely valuable commodity. It is a very appropriate analogy, because when it comes out of the ground, crude oil is dirty, smelly, and nobody wants to touch it. However, when the crude oil is refined through fractional distillation, then a range of much sought-after products emerges with significant added value. Data is a building block for value, and it is important to understand how to handle data to harness that value.

It is worth noting at this point that the authors have some concerns with the use of the term "mobility" when referring to the quality of urban transportation services. In our minds, the term mobility means the ability to make more trips. The authors believe that the term "accessibility" may better describe the overall objectives of applying technology to transportation. Our definition of accessibility relates to the ease or difficulty of moving from one part of the city to another, measured by an accessibility index. An accessibility index can combine distance between origin and destination, expected travel time, expected travel time reliability, and the cost of travel as a proportion of household income. Since the community at large is comfortable with the term mobility, the authors will adopt it for this book, remembering that the overall objective may extend beyond the ability to make more trips. The authors will, however, avoid the exclusive use of the term "mobility" and balance its use with perspectives on "accessibility." The authors believe that accessibility optimization can provide a better approach to equitable smart mobility.

1.4 The Smart Mobility Revolution

While the term "revolution" can sometimes have negative connotations, our perception is that the smart mobility revolution is extremely positive. It will result in safer, cheaper, and more enjoyable travel. However, there will be rapid, radical change with new best practices, new business models, and new technologies to understand and apply. Our belief is that to get the best from the smart mobility revolution, it is necessary to reevaluate the current use of technology, assess the benefit of emerging technology, and reconsider best practices. Another important feature of the revolution is the speed at which change happens. This may be cause for concern when using published printed media as a channel for delivering our new information. However, having studied the ongoing revolution, the authors have concluded that there are lessons to be learned and best practices that should be described as "evergreen," since they remain valuable even as technology and capability change.

For example, a robust approach to risk management for projects will remain valuable even with changes in technology. There are also new lessons and new approaches that come along with new and emerging technologies. There will also be new skills and expertise required of those responsible for delivering smart mobility services and the consumer. For example, micromobility services will require that the user understands how to use the service, and this will involve the ability to operate smartphone and the information technology that controls access to the service. Consequently, the authors have decided to take a multichannel approach to for delivering our knowledge and information. This book will focus on the "evergreen" knowledge that stands the test of time. The authors will also collaborate with immersive publishing organizations to deliver a complimentary layer of dynamic content in line with changes in technology. This two-tiered approach, which the authors call the Smart mobility body of knowledge (SMBOK), will ensure the delivery of a solid core of practical advice accumulated over more than 50 years of experience while also keeping up to date with emerging technologies and best practices. The authors perceive this book as the cornerstone or foundation of this body of knowledge. It incorporates knowledge and experience that changes slowly and information from practical application that is timeless. The supplementary information on the immersive publishing channels will include topical, evolving subjects that build on the book content. This approach will also enable the authors to customize additional content for specific readership segments. While the content and structure of the book will effectively address a wide range of readers, it will also be valuable to provide additional content to address the evolving roles of the different readership segments. The Smart Mobility Body of Knowledge can be found at www.smartmobilitycenter.com.

1.5   Smart Mobility Versus Smart Cities

Smart cities have become an important topic, partly because more than half of the inhabitants of our planet now reside in urban areas [1] and partly because the application of advanced technologies can have a significant impact on the quality of life in such areas. This book explores a particular facet of smart cities – the use of advanced technology to support a higher level of mobility within cities.

One reason for the importance of smart mobility within a smart city lies in the potential reach of the technology to influence almost everybody's life. Another reason is rather more basic. An effective and successful smart mobility system will save money and generate revenue that can be applied to the implementation of other smart city technologies. Smart mobility plays an important role in the first wave of smart city applications because of the revenue-generating possibilities and because of the potential ubiquity of the technology.

Utilizing a combination of advanced technologies that support the full spectrum of mobility service delivery, it is possible for a smart city to feature an extremely efficient and effective mechanism for the transportation of people and goods within the urban environment. This will also increase the safety of the individual traveler by removing potential hazards and risk factors and improving the efficiency of the transportation process. This is achieved by making best use of available capacity, carefully aligning supply and demand, and balancing the use of the various modes of transportation available.

1.6   Informational Objectives of This Chapter

The book has been structured around informational objectives for each chapter. These are listed at the beginning of each chapter. The authors come from an engineering background that involves structure, discipline, and rigor. This made the adoption of an instructional system design approach the obvious choice for the book. It involves the application of structured system engineering techniques to the development of the materials that are used to communicate predefined information and knowledge. The statement of informational objectives at the beginning of each chapter also provides guidance as the chapter is developed in terms of the structure and format that will best communicate the objectives. It also supports a simple mechanism for checking that the content of each chapter completely addresses the intended objectives. After drafting each chapter, the authors make use of the informational objectives list as a quality management checklist, ensuring that the informational objectives are completely addressed before completing the chapter.

For this introductory chapter, the intention is to provide an overall background to the book and an overview of the contents that the reader can expect to find.

1.7   Background

When the authors initially discussed the concept of writing this book, the authors were working together in a consultant–client relationship. In the course of our work, the authors discovered our shared view of both the current situation and the future of transportation. Our working relationships deepened into friendships, and the authors came to a consensus that a revolution is underway and that it is especially important for the future of transportation to enable knowledge about the revolution to be shared widely and effectively. The authors understood that our different perspectives and backgrounds enabled the creation of something unique – a global view of smart mobility founded on a combination of experience in central and local government and the private sector. Our experience spans multiple continents including North America, Europe, the Middle East, and the Asian Pacific rim. This wide background equipped the authors to identify and understand the revolution that is already underway, providing the resources to turn the book concept into reality.

It should be noted that this book is not a how-to guide for the design and implementation of smart mobility in a smart city. Rather, the objective is to make our readers aware of the essential elements of a smart mobility system, the role it will play in a smart city, and the associated opportunities and challenges. A strong focus is placed on practical lessons learned based on experiences from prior implementations.

1.8   Why This Subject and Why Now?

This book addresses a knowledge gap, providing a bridge between advances in technology and advances in our strategic approach to smart mobility. It also offers a common mental model for smart mobility that all stakeholders in smart mobility can share. Many policymakers and planners have a perception that smart mobility is too difficult to manage, but too big to ignore. Many transportation operators and engineers may not fully appreciate that technology is, in fact, not the most important element of smart mobility success. It is crucial that policymakers understand that smart mobility can be managed and that engineers understand that it should be managed.

The book provides tools that will allow readers to answer the question, "What does smart mobility success look like?" By answering this big-picture question, it is a book for all smart mobility stakeholders, ranging from high-level policymakers to technically minded scientists and engineers. In addition to the focus on strategy and management, it also provides specific information for the more technically minded by addressing the smart mobility technical information needs for certain segments of the intended readership.

As the industry embarks on the next stage of smart mobility, a reference source on what has gone before, combined with guidance on how to move forward, is particularly valuable. The recent emergence of smart city programs around the world has placed a renewed focus on the application of advanced technologies within urban environments. Smart mobility has an important role to play in the successful and effective application of advanced technologies that will improve the lives of citizens and visitors in urban areas.

There is a tipping point at which the focus of such projects shifts from making the technology work to devising appropriate strategies to extract value from the technology. The industry is at that tipping point, and with this book, the authors aim to communicate the appropriate strategies to the community of smart mobility practitioners.

1.9   Intended Readership for the Book

The intended audience for the book can be split into public, private, and academic segments, as illustrated in Figure 1.2. Examples of specific roles within each audience segment are described in the following sections.

Public sector	Private sector	Academia
Practitioners	**Solution providers**	**Professors and teachers**
1. Best practices 2. Lessons learned 3. Procurement approaches 4. Business models 5. Technology capabilities and constraints	1. Problems to be addressed 2. Market characteristics 3. Customer characteristics 4. Value delivered 5. Outcomes expected	1. Practical lessons learned 2. Techniques applied 3. Matching objectives to technologies 4. Innovation and procurement 5. New business models
Non-technical decision-makers	**Investors**	**Students**
1. Optimize program development 2. Guidance for investment 3. How to address policy objectives 4. Clear communication of value	1. Problems to be addressed 2. Market characteristics 3. Technology trends 4. Promising approaches 5. New business models	1. Practical lessons learned 2. Techniques applied 3. Matching objectives to technologies 4. Innovation and procurement 5. New business models

Figure 1.2 Intended audience.

1.9.1 Public Sector

1.9.1.1 Smart City Planners, Designers, Operations Managers, Transportation Engineers, and Associated Technical Staff

This group will use the book to guide plans and decision-making for the incorporation of smart mobility technologies into smart city plans. Knowledge gained from reading the book will include an understanding of technological capabilities, constraints, and lessons learned from prior practical experiences.

1.9.1.2 Executive Directors of Public Sector Transportation Authorities

This reader group will use the book to identify uses for data generated from a smart mobility system, extract insight and information, and apply these to the creation of new strategies and responses. This group will also benefit from an understanding of how the capabilities of a smart mobility system can deliver additional value to their organizations and to the traveling public while preserving legacy investment in transportation management systems. Stewardship of existing investments can be achieved while embracing new technologies. The connections that the authors make between policy objectives, service delivery, and technology capabilities will also provide some valuable insight. This readership group can expect that the book will assist in improved decision-making.

1.9.1.3 Chief Technology Officers of Public Sector Transportation Authorities

This reader group will use the book as a technical reference source for understanding the application of smart mobility technologies within a smart city. It will also provide guidance on how to expand from an initial legacy deployment to a comprehensive smart mobility system. Furthermore, the book will provide information on how private-sector innovation and technologies can be deployed in smart cities via partnerships between private- and public-sector organizations.

This readership group can expect that the book will assist in improved decision-making.

1.9.1.4 Transportation Planners and Traffic Engineers

This group has always been at the forefront of the application of advanced technology to transportation. This includes the whole spectrum of transportation service delivery from initial planning to detailed design, to the delivery and implementation of projects, to the ongoing operation and maintenance of systems. Within the realm of intelligent transportation systems, this group has risen to the challenge of applying information and communication technologies to transportation, based on their intimate understanding of the traveler's needs. This group will benefit from the discussion in the book regarding the incorporation of legacy and sunk investments as new smart mobility technologies are applied within the smart city. This group will also gain knowledge regarding advanced technology capabilities and new possibilities for the use of data.

1.9.2 Private Sector

1.9.2.1 Smart Mobility Technology Startup Companies

This readership group will find it valuable to see how the technologies are applied within the context of a smart city and understand the overall project concepts and architectures contained in the book. Information on what works and what does not work based on practical experience will also be important. For this segment of our readership, perhaps the most valuable knowledge to be gained from the book will be an understanding of the needs, issues, problems, and objectives that must be addressed by smart mobility to be successful.

1.9.2.2 Executive Directors of Private Sector Operating Companies

This reader group will also use the book to identify uses for data generated from a smart mobility system, extract insight and information, and apply these to the creation of new strategies and responses. In addition, this group will benefit from the definition of challenges and opportunities associated with smart mobility and a discussion of the type of questions that can be addressed. Our explanation of the types of value and revenue that can be unlocked using our approach to smart mobility technologies will be of particular interest.

1.9.2.3 Private Sector Technology Solution Providers

Private companies wish to serve the smart mobility and smart cities market with emerging technologies, such as cryptocurrency in blockchains, mobility-as-a-service, and urban air mobility.

This readership group will gain an understanding of how emerging technologies can fit within a smart mobility framework. Insight will be provided on the best way to match emerging technology capabilities with the needs, issues, problems, and objectives of the public sector within a smart city. The authors also explain how technology-based solutions can be incorporated into infrastructure-based investment plans.

1.9.3 Academia

1.9.3.1 Students and Teachers in Planning and Technology Deployment for Smart Cities

This group will benefit from the broad overview that is provided on smart mobility and how it has evolved from earlier deployments of information and communication technologies applied to transportation. The book also defines some areas where additional research is required to deliver the tools and techniques required to support the future of smart mobility. The intent is to enable students and teachers to incorporate a robust and complete view of smart mobility into existing and planned courses relating to planning and engineering for smart cities and infrastructure.

1.10   Book Overview

The content of the book has been carefully selected and curated to provide a broad, yet complete overview of smart mobility technologies applied within a smart city. It addresses technology, organizational, and business perspectives. This requires suitable treatment of a wide range of subjects.

1.10.1   Chapter 1: Introduction (Current Chapter)

It introduces the subject of smart mobility and defines what is meant by the smart mobility revolution. This is where the background to the development of the book will be found and an explanation of why the authors believe that smart mobility is an important subject at this time. The chapter also defines the intended readership for the book and provides an overview of the contents, along with an explanation of the importance of the topics. It also provides a roadmap explaining where topics can be found in the book.

1.10.2   Chapter 2: Smart Cities Overview

It begins with an explanation of the context within which the smart mobility system will operate: the smart city. The scope of smart cities extends beyond smart mobility. It is important to understand this wider context. Considerable synergy exists between these two realms of city planning, particularly in the sharing of implementation, operations, and management costs.

1.10.3   Chapter 3: Smart Mobility: A Problem Statement

Having set the scene with an overview of smart cities, the book then moves on to describe the geographical, demographic, environmental, commercial, and political challenges that are inherent in any transportation service delivery activity. The authors call this our "problem statement" for smart mobility. Here again, the subject is wider than smart mobility, but it is essential to have an understanding of the context within which smart mobility will operate. These challenges will influence the strategic approach, desired feature set, and deployment style of any given smart mobility project.

1.10.4   Chapter 4: The Values and Benefits of Smart Mobility

Having defined the opportunities and challenges, the value and benefits of smart mobility are then addressed. Given the cost involved, identifying the expected benefits and determining return on investment are crucial elements of the overall value proposition of smart mobility. The development of cost and benefit estimates using a robust methodology is explained in terms of a working sample that relates the approach to practical application. This is an especially important subject in smart mobility, because defining, understanding, and communicating the values and benefits to be achieved is central to successful planning and implementation. This information is essential to the business justification for investment in smart mobility. Therefore, it has been given detailed treatment in the book. In the treatment of this core topic, the overall categories of value and benefits that can be delivered are defined, and a framework is presented that can be used to evaluate life-cycle cost analysis for various elements of a smart mobility system.

1.10.5   Chapter 5: Smart Mobility Progress Around the World

Building on this, the progress of smart mobility around the world is addressed, including the outcomes and results achieved. Anyone embarking on smart mobility system implementations within a smart city would be wise to conduct some due diligence to understand what others have learned before and to understand the capabilities and limitations of the technologies. To support this, lessons and practical experiences from prior implementations of smart mobility technologies have been collated. These are distilled and summarized to communicate knowledge and information to future implementers. These lessons and practical experiences are reinforced through the provision of detailed information on prior implementations that address both national and international applications of smart mobility. These case studies also provide valuable information on the application of smart mobility technologies in very specific cases. While these are focused on smart mobility, the deployments also hold valuable lessons for the implementation of integrated systems in general. The case studies are designed to provide an "under the hood" view that provides insight into what was good about the implementation and what was not so good. The intention is to provide a robust view of these prior deployments as foundational knowledge for future progress. The case studies have been selected to achieve a representative set of technology applications across a wide geographical area. The selection was also driven by the particular set of professional relationships that have been developed in the industry. This afforded an "under the hood" perspective on many of these projects. The selection of case studies is also designed to illustrate the concepts and technologies discussed elsewhere in the book and bring these to life through practical examples.

1.10.6   Chapter 6: Planning for Smart Mobility

The best results have been obtained from the application of smart mobility techniques and technologies when a robust planning approach has been adopted. The essential elements of planning for smart mobility are then addressed, illustrating how emerging technology can be managed and how smart mobility can be woven into a wider investment program including infrastructure.

1.10.7   Chapter 7: The Essential Elements of Smart Mobility

In this chapter, the authors describe the essential elements of smart mobility as required to fully support the operation of urban transportation as a single system. This requires smart mobility features with the following elements:

1) Clarity of purpose
2) Connectivity
3) Adaptability
4) Monitorability
5) Ability to sense threats
6) Intelligence to react and respond appropriately

Each of these essential elements is described in detail and examples are provided to illustrate the role of the elements within smart mobility. The planning and operation of urban transportation as a single system will also be explained, along with how these various elements fit together to support this.

1.10.8   Chapter 8: Smart Mobility Technologies

Attention is then turned to smart mobility technologies themselves. This includes a comprehensive catalog of smart mobility technologies that are currently in use or at an advanced stage of emergence. To separate commercial availability from research, however, the list of technologies has been restricted to those that have already been deployed, while a significant emphasis has been placed on smart mobility technologies, the authors are also at pains to point out that these are just the building blocks and that smart mobility services, based on and enabled by technologies, actually deliver the value and benefits.

1.10.9   Chapter 9: Smart Mobility Opportunities and Challenges

Building on the problem statement defined in Chapter 3, the discussion on opportunities and challenges is revisited. This time, the focus is placed on the opportunities and challenges that are specifically related to smart mobility and not the wider context.

1.10.10   Chapter 10: A Framework for Smart Mobility Success

The implementation of a smart mobility system goes beyond advanced technology and must feature an integration with existing transportation facilities and investment programs. Success requires a multilevel framework approach that considers not only the characteristics and constraints of technology but also the wider aspects of transportation investment, operations, and planning within a smart city. A proposed framework to raise the probability of success for smart mobility initiatives is described.

1.10.11   Chapter 11: Smart Mobility Performance Management

Building on the definition of outcomes or values and benefits discussed earlier by taking a deeper dive into the performance management aspects of smart mobility. Performance management is addressed from two perspectives: summative and formative. Summative performance management takes a retrospective look at previous performance and provides information required to guide future practice. Formative performance management provides rapid feedback to enable incremental course corrections to ongoing deployments. This also involves the definition of key performance indicators and an assessment of how to measure the data required for the indicators. The treatment of performance management will include a detailed assessment of how to measure performance and how to use the data to manage performance. As stated earlier, it is important to travel from data collection to information processing, to the development of insight and understanding, to the definition of action because of the new insight and understanding. This is central to our treatment of performance management and recognizes the new possibilities that are now available, thanks to modern information technologies and data science.

1.10.12   Chapter 12: Mobility for People

As mentioned in Section 1.3, smart mobility techniques enable the industry to take a complete view of mobility and transportation services, which defines success in terms of outcomes. This means that the complete trip is addressed from origin to ultimate destination, the management of multiple modes of transportation, and the enablement of flexibility and adaptability to respond to changes in transportation demand and operating conditions. It also enables a detailed understanding to be achieved regarding people characteristics and preferences. This is essential if mobility and accessibility are to be provided to all people on an equitable basis.

1.10.13 Chapter 13: Smart Mobility Policy and Strategy

An area that often lags the development and introduction of technology is the development, definition, and communication of mobility policy and strategy. The relationship between policy and technology is examined before setting out a proposed approach to coordinate policy and strategy development and to enable close alignment with current and emerging technology capabilities. The policies and strategies that can be supported by a smart mobility system are one of the most important aspects of the book and indeed of the implementation of a smart mobility system. Therefore, the subject of smart mobility strategy is addressed from the perspective of practical application. The use of smart mobility technologies to effectively operate strategies is then discussed.

1.10.14 Chapter 14: Smart Mobility Organization

This also requires a discussion on how to match technology capabilities with organizational concerns. The initial technology discussion is supplemented with an organizational perspective that considers how the public sector and the private sector can work together within a framework for the planning, design, delivery, and operation of smart mobility. Different types of business models that define roles and responsibilities for both public and private organizations are explored. Advice is also provided on how to benchmark and baseline organizational capability to ensure that the organization is adapted optimally to the proposed technology-based systems. Our treatment of this features the use of business process modeling and a process classification framework approach that sets the scene for benchmarking and baselining. This activity also provides a foundation for confirming where key performance indicators fit within the overall process and paves the way for the application of artificial intelligence and automation. The treatment of this subject will also include a discussion of the types of strategies, business models, and solutions that can be supported, thanks to the existence of smart mobility technologies. For public-sector organizations, it is explained how these strategies, business models, and solutions can enable the achievement of a set of defined and agreed policy objectives. For private-sector organizations, a similar set of objectives related to business and financial success is discussed.

1.10.15 Chapter 15: Smart Mobility Operational Management

The penultimate chapter in the book provides insight and advice on smart mobility from an operational management perspective. Operational management is crucial because it represents the primary interface between the users and the operators and is also the gateway through which much of the data must pass. Building on the discussion of business process modeling and process classification frameworks, it is explained how these tools can be used to improve operational management. This includes the use of analytics and how these can be applied to guide operations and the various stages of implementation of a smart mobility system. The discussion covers all stages of implementation from planning to design and project delivery, to operations and maintenance, with a particular focus on operational management. The valuable role that data plays as the raw material for analytics, along with the importance of applying smart data management techniques, is also discussed.

As the topics described above are addressed, the intention will be to address the following additional factors:

- Understanding the importance of smart mobility technologies within an overall smart city framework

The absolute best systems not only provide support for internal functions but also completely address the needs of the wider context within which they operate. For example, a smart mobility system can provide a rich stream of data regarding prevailing conditions in the smart city and the demand for transportation. This will provide a valuable input to transportation planning and operation systems beyond smart mobility system itself.

- Understanding the value of an analytical approach based on smart mobility data

Smart mobility data systems can not only accommodate faster, cheaper, and easier ways to pay for transportation services but also support an analytical approach to smart mobility operations, based on the data that emanates from them.

- The definition of suitable strategies to impact transportation demand and smart cities

In addition to understanding how the technologies work, it is vital to have a grasp of what they can do. This is a critical part of the outcome that is expected from a smart mobility system implementation. What strategies can now be applied because of the existence of a smart mobility system? How can the price of transportation be adjusted in real time to achieve policy objectives as prevailing transportation conditions change? How can the use of the different modes of transportation available within the smart city be optimized and balanced to deliver the best experience to citizens and visitors? How can the capacity of our various modes of transportation be recognized, ensuring that it is carefully matched to the prevailing demand for transportation? All these questions and more are addressed through the definition and development of a range of smart mobility use cases. Going beyond this, it is also important to have an effective way to communicate the strategies, explain the intended outcomes, and provide advice to the users on getting the best experience.

- Understanding the benefits and costs associated with a smart mobility system

This is a relatively simple issue that requires a sophisticated approach. In simple terms, if practitioners cannot communicate the benefits and costs associated with a smart mobility system, it will be exceedingly difficult to convince people to support the proposals. While the industry is at the early stages of smart mobility system deployment, it is relatively difficult to obtain observed benefit and cost data. However, this is no excuse for avoiding the subject. It is possible to make estimates based on prior experience and use these within a robust methodological approach.

- Defining technology procurement strategies that incorporate emerging technologies and approaches, such as cryptocurrencies and block chain, while also delivering robust, sustainable solutions

One of the biggest challenges in procurement is to adopt an approach that is thorough, fair, and agile. As technology changes rapidly and the pace of technology accelerates, it is necessary to define procurement strategies that ensure that what is required is not only delivered but also accommodates advances in technology.

1.10.16   Chapter 16: Summary and Conclusions

In conclusion, a summary chapter is provided that captures the essential points of the book and provides some suggested next steps for smart mobility knowledge gain for smart city practitioners. The important lessons learned from practical experience are also summarized in this chapter. The chapter also provides a summary of the consequences of the smart mobility revolution for both the public and private sectors.

Reference

1 United Nations, Department of Economic and Social Affairs. https://www.un.org/development/desa/en/news/population/2018-revision-of-world-urbanization-prospects.html#:~:text=Today%2C%2055%25%20of%20the%20world's,increase%20to%2068%25%20by%202050, retrieved December 30, 2020, at 5:13 PM Central European Time.

2

Smart Cities Overview

2.1 Learning Objectives

After reading this chapter you should be able to:

- Define the value of a smart city
- Understand the essential characteristics of a smart city
- Provide a plausible definition for a smart city
- Describe the service categories to be found in a smart city
- Define some examples of smart city services
- Understand the effects a smart city will have on its inhabitants and visitors
- Explain the importance of whole system and circular approaches to the management of smart city services

2.2 Introduction

Smart cities have become increasingly important as the proportion of the global population living in cities rises. According to the United Nations, 55% of the world's population lives in urban areas and this is expected to grow to 68% by 2050 [1]. Smart cities services represent ways to significantly improve the quality of life in urban areas, through the application of advanced technology. They have the ability to increase safety by reducing the number of fatalities generated as a byproduct of city processes. They have the potential to improve the quality of water, energy, and mobility in urban areas. This is achieved by using technology to have a better understanding of threats, opportunities, changes in operating conditions, and changes in the demand for services. The potential for advanced technology to enable us to harness the benefits of city services and processes, while avoiding the undesirable side effects is one of the major reasons to apply technology. Smart cities have also become a rallying call or banner, under which the application of advanced technology has been progressing around the globe. This book focuses on smart mobility and this in turn is a major component of smart cities. There is a close relationship between the operation of a smart mobility system and the operation of other services within a smart city. This makes it worthwhile to provide an overview of the smart city services portfolio as a way of illustrating the relationship and also setting smart mobility within the wider context.

Smart Mobility: Using Technology to Improve Transportation in Smart Cities, First Edition. Bob McQueen, Ammar Safi, and Shafia Alkheyaili.

A high-level view has been provided of smart city services, with the exception of the energy, mobility, and water categories category. As one of the important goals addressed by smart cities relates to sustainability, both environmental and economic, these categories have been selected for more detailed treatment. Water and energy are both good examples of cyclical processes in which multiple smart city services can be used to support improvement at each step. Mobility is the major focus of the book.

2.3 Smart City Services

Figure 2.1 provides an overview of the services to be found in a smart city. Some or all of these smart services can be found in combination within a smart city. Each of the categories described in the table is described in the following sections, followed by information on the service examples provided. For the energy, mobility and water areas, a deeper treatment has been applied that highlights the circular nature of the process in addition to the opportunities to apply smart city services.

2.4 Affordable Housing

One of the important issues in urban areas is the availability of reasonable quality housing that people can afford. This is an important equity issue within a city. It could be argued that not only should the city be smart, but it should also be compassionate and caring for all demographics within the population. This category relates to the initiative-taking management of housing to achieve this objective.

2.4.1 Modular Construction

Economies of scale can be achieved by taking a systematic approach to house a construction. This is especially appropriate when a large number of similar houses are to be constructed. In a similar manner to the software recycling discussed in this section and circular economy management, designs can be specially configured to reuse modules. This also facilitates the use of factory techniques to preassembled, or prefabricated modules that can then be transported to site. This enables the use of sophisticated tools to increase productivity and to maintain the accuracy that can be achieved with in a factory environment, compared to the on-site environment. Computer-aided drafting and manufacturing techniques can also be employed to make the overall process from planning through design to construction as efficient as possible [2].

2.4.2 3D Printing

The use of 3D printing techniques for home construction can reduce the cost of the building and can be used hand-in-hand with the modular construction techniques described in the previous section. The 3D printers used are like giant versions of the inkjet printers used in offices, except that instead of ink, the printer extrudes plastic or concrete. The material is laid down in successive layers to create a structure. Recent experience indicates that a $95\,m^2$ home can be constructed in three days using this technique, at a higher level of accuracy than conventional techniques. The technique can also cut the CO_2 emissions associated with the construction process by 75% [3]. Figure 2.2 shows an example.

	Service category	Service examples									
1	Affordable housing	Modular construction	3D printing	Housing management software							
2	Air quality	Air quality sensors on streetlights	Citizen & visitor information	Spatial planning							
3	Buildings	Smart buildings	Dynamic resource assignment	Advanced HVAC control							
4	Citizen and Visitor engagement	Smartphone applications	Virtual and augmented reality	Online meeting technology & multiple information delivery channels							
5	Construction	Smart construction sites	Automated construction vehicles	Delivery management							
6	Education	Digital services	Interactive learning	On-demand learning							
7	Employment	Opportunity network crowd sourcing	On-demand work	Local service counters							
8	Energy	Generation	Wholesale & trading	Transmission operations	Transmission ownership	Distribution	Large customer retail	Mass-market retail	Shared services	Waste heat capture	
9	Finance	Electronic payment	Blockchain	Road user charging							
10	Government	Smart kiosks	Open data	Smart city platforms							
11	Green spaces	Maintenance management	Park Wi-Fi	Energy and water conservation management							
12	Health	Wearable devices	Mobile in-home treatment and telemedicine	Surgical robots							
13	Manufacturing	Supply chain management	Delivery logistics for raw materials	Private cellular networks							
14	Mobility	Connected and automated vehicles	Connected citizens and visitors	Integrated payment systems	Smart grid and electric vehicles	Mobility management	Analytics	Enabling services			
15	Recycling	Smart garbage cans	Circular economy management	Energy recycling							
16	Retail and logistics	Location intelligence	Customer experience management	Real-time inventory management							
17	Safety	Early warning systems	Next-generation emergency management	Personal security systems							
18	Security	Real-time crime centres	Police officer wearables	Mobile video capture and recording							
19	Tourism and leisure	Event management	Tourist information services	Medical tourism							
20	Water	Collection	Screening and straining	Chemical addition	Coagulation & flocculation	Sedimentation & clarification	Filtration	Disinfection & fluoridation	Storage	Distribution	

Figure 2.1 Smart city service categories and examples.

Figure 2.2 3D printed building. *Source:* 3D printing to build a house: building innovation Bouygues.

2.4.3 Housing Management Software

In the same way that information technology and telecommunications networks are used to match supply and demand for products and services in smart cities, these can also be utilized to match supply and demand for housing. That are many examples where public agencies have developed specific systems for this purpose, but the most prominent example of this technique is in the private sector. A web application known as Zillow applies information technology and telecommunications techniques to real estate by providing a network the matches those who wish to sell houses, with those who wish to buy. The advantage of this technique is that additional information can be provided to users in addition to the availability of the home. The characteristics and location of the home can also be described along with analytics to help to smooth the transaction and the decision-making process. This can also include sophisticated search functions to make it easier for home seekers to find opportunities. Zillow is also an interesting example of a privately funded business model. The Zillow service is available for free to users, supported by advertising [4].

2.5 Air Quality

Another significant quality of life issue with direct relationships established between the quality of air in urban areas and the overall health of the population residing within the urban area and traveling through it. According to the United Nations more than 80% of people living in urban areas are exposed to air quality levels that exceed guidelines set by the World Health Organization [5]. Air quality can be measured, and guidelines established for parameters including the percentage of particulate matter in the air, and greenhouse gases (GhGs). These are considered to be ozone, nitrogen dioxide, sulfur dioxide, and carbon monoxide. Of course, air quality also has a relationship with smart mobility, smart manufacturing, and circular economy management [6].

2.5.1 Air Quality Sensors on Streetlights

Smart street lighting has become an important part of smart city initiatives. This consists of installing sensors on street lighting columns. While the bulbs and the lamps are being changed from older lighting technology to light emitting diodes (LEDs). There is a significant cost-saving advantage as LEDs consume much less electricity than previous technologies such as high-pressure sodium lamps. It is extremely cost-effective to add additional sensors to the lighting columns while the bulbs are being changed. The sensors can be powered from the electricity available at lighting column. Communications between sensors and a suitable back office can be achieved with wireless technology. Sensors that can be installed in street lighting can include motion sensors, seismic sensors, gunshot detection, and air quality sensors. Typically, an integrated sensor gateway that is installed on the lighting column can accept multiple sensors on a modular, plug-and-play basis. This can also be an Internet of Things (IoT) device with edge processing capability. Figure 2.3 shows an example of the type of sensor installation that can be fitted to lighting columns.

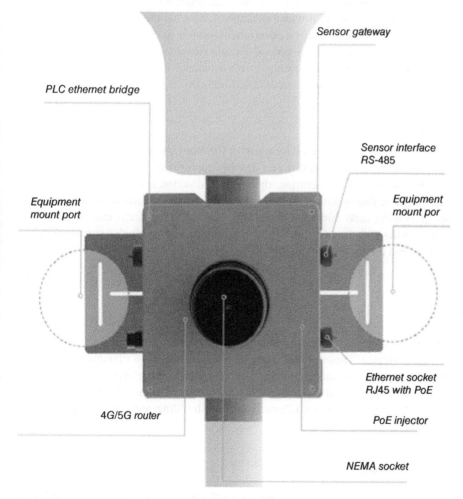

Figure 2.3   Smart street lighting IoT device.

The type of sensor installation that can be fitted to lighting columns. This features communication equipment as well as suitable sockets for installing various sensors, which can also include video cameras. The installation of multiple sensors on lighting columns provides the ability to create a network of low-cost air quality monitoring sensors that can be deployed to monitor air quality and weather parameters. This is a much more comprehensive approach than single-point special-purpose air quality sensors. Such sensors can typically measure dust particles, GhGs, noise, lighting levels, and weather conditions. These include rain, temperature, humidity, and pressure [7].

2.5.2 Citizen and Visitor Information

Air quality sensors and back-office analysis allow current conditions to be analyzed and suitable strategies identified. The strategies can be implemented through citizen and visitor information systems. These make use of wireline and wireless communications to deliver a variety of information services. These can include roadside dynamic message signs and smartphone apps. Such information can be provided as part of a wider package of information to citizens and visitors including traffic and transportation advice and government services information in addition to air quality advice. Advice can take the form of a general alert on current average conditions using an index, or can be specific suggestions for re-routing or area avoidance. In many ways, an air quality event can be considered as an incident and treated accordingly [8].

2.5.3 Spatial Planning

The data received from air quality sensors can form part of the input into a spatial planning system. These assess regional variations in air quality and develop appropriate strategies. They can also take account of emissions from manufacturing, the residential sector, and public transportation along with private vehicles. The spatial planning system can suggest actions such as limitations and prohibitions concerning fuels and vehicular traffic. Trends and patterns can be identified through the use of analytics and visualizations which can also be used as valuable communication tools for policymakers and the general public [9].

2.6 Buildings

There are a lot of possibilities to make buildings smarter and living easier through the application of advanced technologies. These include sensors to provide data to manage heating, air conditioning, building energy consumption, and the density of people within the building. Data from the sensors are combined into a building management system that can provide a complete picture of the variation in conditions inside the building and suggest appropriate strategies.

2.6.1 Occupancy Monitoring

This is a process of collecting occupancy data using sensors and telecommunications in order to understand how the building or facility is being utilized. Making use of an advanced data management and analysis system, informed changes and optimizations can be made to improve the efficiency of the building and the conditions in the building. Sensors can be placed on the desks and doorways and in meeting rooms to detect and count the number of people in each area. In COVID times, this data can also be used to ensure that social distancing compliance is achieved, improving health and safety [10].

2.6.2 Dynamic Resource Assignment

Building on occupancy monitoring, dynamic resource assignment can be achieved as a result of the insight and understanding gained from occupancy monitoring. This includes the selective switch off of lights, heating, and air conditioning in unoccupied areas. Predictive algorithms can also be used to enable common space and room conditions to be adjusted in advance of planned use. This is particularly effective when combined with meeting calendars and other systems that record planned meetings and appointments.

2.6.3 Advanced Heating Ventilation and Air Conditioning (HVAC) Control

Figure 2.4 depicts a typical advanced heating ventilation and air conditioning control system set within the context of an overall building management system [11].

The equipment or devices are controlled by unitary controllers. These do not talk to each other, and there is a need to coordinate the control actions. This is achieved in the building management system. A series of Proportionate Integral Derivative (PID) loops are the main processes used to control building automation systems and implement the control strategies. The control strategies can be based on manual observation, rule-based, or through the application of artificial intelligence and machine learning techniques.

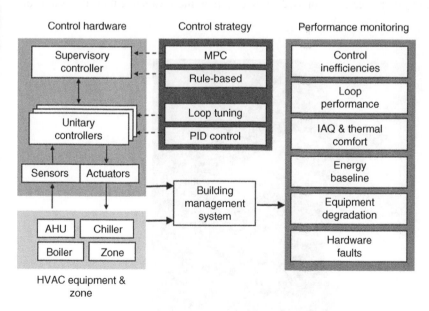

Figure 2.4 Typical HVAC control system set within a building management system.

2.7 Citizen and Visitor Engagement

A crucial aspect of smart city development is to ensure that the needs of the citizen and visitor are completely understood and considered in the implementation of smart city services. This might seem like an obvious point, but in many cases, technology aspects and characteristics have eclipsed proper consideration of the people's needs. Effective citizen engagement will require the development and delivery of appropriate information through a number of channels that address the needs of awareness, interest, desire, and action.

2.7.1 Smartphone Applications

Multiple information delivery channels can be used to support citizen engagement. These can include the web, TV, radio, and of course smartphone applications. As smartphones are in wide use and are personal devices, this is the most effective means of delivering information and encouraging citizen engagement. According to a recent study, the number of smartphone users in the world in 2022 was more than 6.6 billion which translates to almost 84% of the world population. When this is widened to include mobile phones there are more than 7.2 billion mobile and smart phones used in the world today by about 92% of the world population [12].

For the purposes of equity, it should not be assumed that all citizens have a smartphone or a mobile phone requiring that other delivery channels be used in combination [13] (Figure 2.5).

2.7.2 Virtual and Augmented Reality

Virtual and augmented reality are also services that are enabled by advanced technologies. Virtual reality involves the simulation of reality that users can view through specially developed goggles. This is particular useful when trying to explain a proposed project or implementation to the public. The ability to see and experience the proposed changes can be more valuable than thousands of

Figure 2.5 Smart mobile phone user. *Source:* Jonas Leupe/Unsplash.

words. Augmented reality adds items that will not physically exist to the view seen through a smartphone camera. For example, the smartphone camera view of a bus stop could have schedule and route performance information overlaid onto the image. The following figure shows an application of virtual reality [14] (Figure 2.6).

The following figure shows an example of the application of augmented reality to enable the citizen to visualize a proposed project [15] (Figure 2.7).

Figure 2.6 Example of virtual reality. *Source:* XR Expo/Unsplash.

Figure 2.7 Example of augmented reality. *Source:* Patrick Schneider/Unsplash.

2.7.3 Online Meeting Technology and Multiple Information Delivery Channels

Solutions such as Microsoft TEAMS, WebEx and Zoom can be used to support remote meetings. These, and other solutions, can also enable interaction between the remote audience and the speaker. This enables polls to be taken and responses to be elicited from the audience to make the experience more interactive. The information to be shared with the audience and to the public can also be delivered over multiple channels including web-based, smartphone, TV, and radio [16].

2.8 Construction

Cities almost behave like living organisms that continually evolve. Land use patterns change, buildings are constructed, and buildings are demolished. The infrastructure that supports transportation in urban areas also evolves over time including new roads, new transit facilities, and multimodal hubs. These all require construction activity that has to be planned, designed, implemented, and monitored. Here again there are a range of sensors available to provide the data for effective management of all aspects of construction activity.

2.8.1 Smart Construction Sites

Building on construction sites requires close coordination of a range of activities, including the delivery of the material required for construction. Labor resources, tools and machinery, and materials must all be coordinated to deliver an optimum project. This provides a large potential for the application of the kind of technology used in smart city services. One of the most powerful approaches is known as building information modeling (BIM), this involves the application of a virtual 3D modeling techniques to depict a construction project. BIM enables a collaborative cross-disciplinary approach to be taken to design, and construction, serving as an important tool to support the kind of coordination required. BIM makes use of technology to support the work processes involved in construction. It spans design, construction, and later operation of the building. During design, architects and engineers can use the 3D model to fine-tune the design and detect any potential issues. During construction, it becomes a reference to ensure that the building is constructed according to the design. After construction, it served as a platform for facilities management. BIM has evolved from the adoption of sophisticated computer-aided design and manufacturing software adopted by leading architects. In addition to serving as a communication tool, this type of software can also significantly reduce costs by minimizing defects and automating the process from design, through procurement, to building, to operations. This type of approach also enables more sophisticated designs to be realized, sometimes ones that look impossible. See the figure below of the Walt Disney Concert Hall in Los Angeles [17–19] (Figure 2.8).

2.8.2 Automated Construction Vehicles

The automated vehicle is still emerging as a smart city technology. However, it is likely that automated delivery vehicles will play a significant role in future construction sites. This includes fully automated vehicles to move construction materials around inside the construction site and

Figure 2.8 The Disney concert Hall, Los Angeles. *Source:* WikiImages/Pixabay.

facilitate delivery of the materials from beyond the construction site. This technology can also be applied to deliver automated dozers, excavators, as well as trucks. It is likely this technology will be introduced on a gradual basis, initially providing human assistance and decision support before becoming fully automated. This can even include remote operation of construction machinery from a central location [20, 21].

2.8.3 Delivery Management

Delivery management is important for construction but is also important for wider commerce in urban logistics.

With respect to construction, the efficient delivery of materials to construction site can have a significant impact on the overall progress of the project. This is particularly important with respect to perishable materials such as concrete and asphalt, which are to be delivered and utilized within a specific time window. Information technology can be used to develop a platform, which combined with vehicle and infrastructure sensors can manage the delivery process. The wider application of this to urban freight and logistics including FedEx and UPS also presents an opportunity.

2.9 Education

The traditional model where students travel to attend classes in specific facilities, has evolved over time to include a component of distance or remote learning. The impact of the COVID 19 virus has been to intensify the use of distance learning. Smart elements of education can include this but can also include smart application of mobility techniques to the transportation associated with educational activity. Education is a significant generator of transportation demand so should be considered carefully within any smart mobility approach.

2.9.1 Digital Services

This is use of multiple digital technology applications to support more effective learning and better teaching. It is sometimes referred to as Technology Enhanced Learning (TEL) or e-learning. Techniques can include the use of texts and chat facilities to interact with students during lectures. Lectures and learning activities need not be entirely online or physical, with use of so-called blended learning techniques. These enable some parts of the learning process to be face-to-face, for example, class discussions, while other parts, such as assignments, are delivered online.

2.9.2 Interactive Learning

This falls under the umbrella of digital services and includes tools and techniques to support interaction between the teacher and the student. This can involve the use of special tools that enable students to provide feedback and questions to the teacher. It can also include the use of visualization techniques to explain what-if scenarios and support kinetic learning.

2.9.3 On-demand Learning

An interesting feature of the application of advanced technology to education is the possibility that the formal structure can be adapted to allow landing on an on-demand basis. This opens up learning opportunities to a wider range of students by not imposing a rigorous timetable. Students don't need to be in a certain place at a set time to receive the learning. This also falls under the category of digital services, and it is apparent that a combination of all three approaches would provide a powerful new tool for the ability to transfer knowledge effectively.

2.10 Employment

In addition to the generation of new jobs, a smart city should provide opportunities for economic growth and should have a focus on attracting industry to the city because of the quality of life and efficiency of the smart city services. The delivery of smart city services can also generate job opportunities across a wide spectrum of capability levels.

2.10.1 Opportunity Network Crowdsourcing

Establishment and support of networks is a key component in smart city services. Often, networks are considered to be the hardware and software required to connect people together, are in the case of IOT connect devices together. A more complete view of a network would include multiple perspectives including hardware, software, data, and people. The latter is a particularly important component as without the people, the rest can have little effect. The examples of

AirBnB, Uber, and to some extent, Amazon have demonstrated that that is power and value in a network that matches supply and demand. Air B&B could be considered to be the largest hotel operator in the world but in fact have no hotel rooms and do not own any physical infrastructure. Uber could also be described as the largest taxi company in the world, but does not own any vehicles and doesn't employ any full-time drivers. Uber does not describe itself as a taxi company, but as a big data company. This reflects the value of the data that is collected about the network and the demand for transportation services. Amazon, in the early stages, also worked on the basis of establishing and supporting a network that matches the supply of products to the demand for the same. More recently Amazon has extended the model to include ownership of physical infrastructure through the acquisition of retail operators such as Whole Foods Amazon is now demonstrating the use ofa a blended approach that includes a cyber network and physical assets working in combination.

These examples can also be applied to more effective matching of those who have work to be done to those who have the capability to do the work. Here again, Amazon has pioneered the concept through the establishment and operation of the Mechanical Turk. This service enables a crowdsourcing approach to be taken in which businesses who have work assignments can identify and hire remotely located corroded workers to perform discrete on-demand tasks, that computers are currently unable to perform. The mechanical Turk was a fake chess-playing machine created in the late 18th century. It did things that computers were not able to do at the time, that is play expert level chess. It was in fact a large box concealing a human chess expert who was responsible for moving the chess pieces. The revelation that it was effective reinforces the point that that are still some things that computers cannot do even today, hence the need for a network that matches such jobs to human labor capital [22, 23].

2.10.2   On-demand Work

Establishment and operation of a network that connects job opportunities to those seeking employment on a crowdsourcing basis can also support the concept of on-demand work. In fact, Uber could also be described as an example of this. In the course of various Uber trips, the authors have conducted brief interviews with Uber drivers to determine the reasons they drive for Uber and what they like about it. In most cases, the opportunity for on-demand work is the number one reason. It enables a kind of flexibility that can better fit with an individual's lifestyle. But example, single parents can mold the amount of time they drive and the time of their employment, to suit the other needs of a new life. Full-time students can supplement their income by coordinating on-demand work opportunities such as Uber, with the demands of their learning activities. Digital education services also play a role here to enable a more flexible pattern that integrates working and learning [24].

2.10.3   Local Service Counters

The crowdsourcing network that matches employment opportunities with jobseekers can also be complemented with physical infrastructure. Local service counters can be incorporated into the network to allow face-to-face interaction and enable identification and confirmation of job opportunities. These local service counters do not need to be specifically designed for employment opportunities but can be one function within a shared multifunction service counter arrangement. Some transportation agencies in the United States are looking at the possibility of smart mobility hubs that are designed to support interchange from one mode of transportation to another. This infrastructure could also be used to support local service counters for employment networks.

2.11 Energy

Another important facet of smart cities involves the balance of energy supply across traditional energy sources such as oil, gas, and electricity and new sustainable energy sources such as wind power and solar. Smart city technologies have the potential to be applied across the full spectrum of energy activities from generation, through distribution to industrial and residential use. To illustrate this, one variation of the energy cycle: the electricity cycle is considered as illustrated in Figure 2.9.

2.11.1 Generation

The process of generating electricity can benefit from optimization of generating plants and portfolios of energy sources. This can be achieved with a combination of sensors, telecommunications, and command and control.

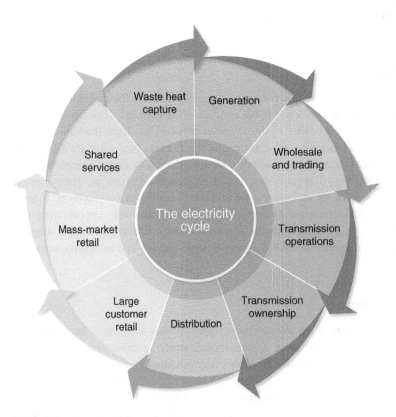

Figure 2.9 The electricity cycle.

2.11.2   Wholesale and Trading

Trading activities associated with electricity can be managed using advanced analytics, data exchange platforms, and trading platforms.

2.11.3   Transmission Operations

Transmission operations can be monitored to ensure maximum efficiency. This includes energy losses over the length of cables and the generation of heat from the electricity transmission process.

2.11.4   Transmission Ownership

Transmission ownership is a possibility to have Blockchain and other transactional management systems applied in order to maximize the efficiency of transmission ownership and ensure that fees collected are matched to those that own and provide service.

2.11.5   Distribution

This part of the cycle lends itself to the creation of networks that match supply and demand. These will be similar to Uber and mobility as a service approach.

2.11.6   Large Customer Retail

The provision of support services for large electricity users can be accomplished using services remarkably similar to smart city services. These include customer information, understanding the customers through advanced marketing, and ensuring a high-level customer service using monitoring techniques.

2.11.7   Mass-market Retail

In the same way as large customer retail, customer service can be optimized for the mass market, on a larger scale. This can be done by providing customer information, enabling a detailed understanding of customer needs. This underpins customer service using a combination of automated data collection, manual data collection, information processing, and response generation capability.

2.11.8   Shared Services

The wide range of services is shared across the other elements of the cycle, and these can all benefit from the application of advanced data management and information processing capability.

2.11.9   Waste Heat Capture

Most depictions of the electricity generation and supply process do not include this final step. However, there is the possibility of recycling energy by capturing waste and reconverting it back into electricity for input into the generation cycle.

2.12   Finance

The use of smart city services to gain a detailed understanding of the operation of the city can provide significant input into financial planning. Smart city services can also reduce the cost of transaction processing and make payment for services easier and safer. The use of information technology can also provide new data and information regarding transaction flows and the demand for services. There is a growing awareness that smart city services designed for specific functions can also generate data that can be used to inform other functions within the city.

2.12.1   Electronic Payment

Global electronic payment infrastructure already exists in the form of the credit card companies and bank-to-bank transfer systems such as the society for worldwide Interbank financial telecommunications (SWIFT). The latter works in the background and is only visible to consumers when the system is used to transfer money. The former has high visibility to consumers and could be considered to be ubiquitous. Transportation service providers such as transit companies have also developed and implemented their own electronic payment system using smart card and host card emulation techniques.

2.12.2   Blockchain

Blockchain is still a relatively new technology and has considerable hype associated with it. In many cases, there is a misunderstanding about the difference between Blockchain and cryptocurrencies such as Bitcoin. In fact, Blockchain is the underlying enabling technology behind Bitcoin, but it is not the same. IBM has recently been emphasizing the difference by referring to Blockchain as "Blockchain for Business." Blockchain is an incorruptible digital ledger of economic transactions that can be programmed to record not just financial transactions but virtually everything of value.

While the technology is still emerging, it does hold significant potential for transportation, especially with regard to applications that support an integrated payment system for smart cities.

Although Blockchain is available as an open-source software platform, an investigation of the technology reveals that there will be costs associated with the implementation of Blockchain. More detail on the Blockchain technology is contained in Chapter 8 – technologies.

2.12.3   Road User Charging

Road user charging is based on the principle that road users should be charged according to the congestion and inconvenience that they cause other drivers, the wear-and-tear caused to infrastructure, and the environmental impact. Using dedicated short-range communication technologies to support a two-way communication between roadside transceivers and a transponder installed in the vehicle, it is possible to implement a variable fee according to where the vehicle is driven, the distance traveled, and the time of day. There is renewed interest in road user charging techniques due to the difficulties now being encountered with taxation on fuel. The introduction of substantial numbers of electric vehicles and the continuing improvement in internal combustion engine efficiency means that less tax revenue is available to fund transportation. Road user charging or distance-based taxation could be a viable alternative.

2.13   Government

Government services play a significant role in the quality of life within a city. It is important that the services are delivered in an effective and efficient manner with the right level of customer support. The use of smart city services can assist in maintaining the appropriate service levels. This can include enabling citizens and consumers to pull information from government data sources, reducing the cost to government in delivering the information. The services can also be applied to evaluate the effectiveness of service delivery and outcomes achieved.

2.13.1   Smart Kiosks

While smartphone apps have been one of the more dominant methods of information delivery, smart kiosks have the advantage of being accessible to all including those who do not possess a smartphone or mobile phone. Smart kiosks can be placed in existing service points within government offices, in public locations, and in places where significant numbers of people can have access to them. An interesting approach to smart kiosks is to incorporate service features into existing automated service points such as automated teller machines. The figure shows an illustration of the typical interactive kiosk or smart kiosk. It is a ruggedized computer terminal that provides information and supports application for communication, commerce, entertainment, or education. It represents a significant interface between smart city services and the end user. The ability to display a large map and provide additional information about the city means that an interactive kiosk also supports the delivery of tourist information services. Other functions can also be built into the kiosk such as sensors for weather data collection and communication to enable the user to summon help if required [25] (Figure 2.10).

Figure 2.10   Example of a smart kiosk. *Source:* OMS Intelligence Solutions.

2.13.2 Open Data

Open data is more of a concept than a technology application; however, it has an important role in enabling multiple smart city services. It enables data to be brought from multiple sources that are accessible to the city, to a data management system, and made available in an open basis to the public and to the private sector. This can function as an economic stimulus by providing the raw data that private sector companies require in order to launch and support smart city apps. This makes good use of public and private sector resources by providing public support for private enterprise while leaving the private sector to develop consumer apps, something the public sector agencies have little experience in. An interesting private sector model to consider with respect to open data is the concept of two-tier data access. The first tier is openly available to all comers, whereas the second tier, which has added detail and therefore added value, could be offered in return for a transaction fee. This recognizes the public sector's role in funding the cost of data collection and converting it to information. This is the subject of some debate as many public sector agencies believe it is their job to collect data and make it openly available to all comers. With the public bearing the cost of collecting and processing the data. However, that is a growing belief that a public-private partnership can be struck which enables the public sector to recover some of the cost of data collection and processing and hence provide revenue to improve both. Open data is universally accessible and stored in a standard format to facilitate use by anybody. While it is open data, it is usually licensed under an open license that provides a legal framework for the provision of the data and specifies terms and conditions. The UK government data.gov.uk project is a good example of how government might establish and operate open data initiatives. It contains more than 30,000 data sets from a range of UK government departments. Data has been depersonalized and provided in a standard format [26].

2.13.3 Smart City Platforms

Smart city platforms can be purpose-specific such as electronic payment systems, mobility as a service, water management, energy management, or building management. There is also the possibility of a super platform that acts as an umbrella across multiple services. These have the advantage of using data from a larger number of sources and opening the possibility for cross-domain insights derived from analytics. For example, if it is expected that there is a relationship between the peak demand for transportation in the city and peak energy consumption, then such a super platform would be able to source data from energy companies and transportation service providers, enabling advanced analytics to determine trends and patterns. The results could improve the supply of transportation and energy by more closely matching supply and demand. The figure depicts the characteristics of a super platform (Figure 2.11).

This illustrates how the platform can ingest data from multiple sources, in this case, transportation and mobility data from connected vehicles, smartphone location intelligence, and other city and regional data. The data is brought together into a single unified data set with an analytics engine and accessibility arrangements to make the data widely accessible for analytics while providing the appropriate operational and security arrangements. Note that the ingestion process also includes a data quality management system to assess and maintain appropriate data quality levels. The resulting data is then fed upwards into a series of analytics activities. These are results-focused. In this particular case, eight specific analytics focus areas have been identified to be supported

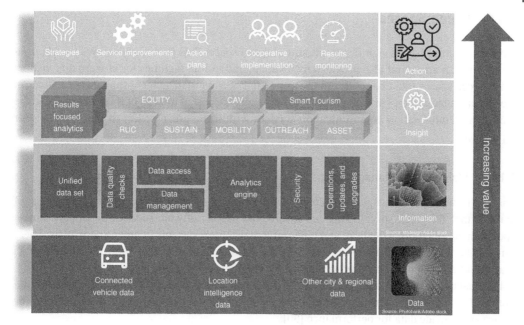

Figure 2.11 Example of a super platform.

through a combination of the analytics engine and suitably experienced analytical staff. For this particular project, the following analytics focus areas were selected:

Equity: analytics to uncover the current situation with respect to equity across the city, uncover trends and patterns, and support the development of strategies for equity improvement.

CAV: the use of connected and automated vehicle data to understand variations in the demand for transportation and in operating conditions. The vehicles function as probes enabling wide-area data collection to support the understanding of citywide trends and variations.

Smart tourism: the use of process data and analytics to support the provision of decision-quality information to tourists in order to maintain the quality of tourism and visitor user experience. Data can also be gathered to understand the variation in tourist activity and their trip-making behavior.

RUC: Road user charging. The use of analytics to understand the relationship between the cost of using roads and driver behavior. This enables implementation strategies to be assessed and different business models to be evaluated. For example, different fee levels for different roads could be analyzed to understand the effects on demand. Different road user charging strategies such as a fee for crossing a cordon around the city, or a zonal presence system can be evaluated. A good example of the latter is the congestion pricing system in London [27].

The former has been implemented in Singapore as an alternative to ownership-based vehicle constraints [28].

2.14   Green Spaces

Most major cities have a significant amount of green space. This not only improves the environment in urban areas, but also provides recreational and leisure facilities for citizens and visitors alike. Green spaces must be subject to the same kind of asset management the other resources in the city are subjected to. This can range from managing the cost of upkeep of the green space to the provision of information communication services within the green space, to improve the user experience.

2.14.1   Maintenance Management

The application of technology-based services can have a significant impact on the overall cost of maintenance for green spaces. This can involve the use of software services to support remote operation, automation, irrigation, and environmental control systems in green space areas. These can also be used to schedule maintenance in an optimized manner. The figure below shows an example of an automated lawnmower design for commercial use [29] (Figure 2.12).

This makes use of lidar and radar sensors, combined with robotic operation and wireless communications. It can be controlled from a command center with onboard GPS used to locate the device and provide data for performance [30].

2.14.2   Park Wi-Fi

Public Wi-Fi is widely available in most cities. In many cases, it is available on a free-access basis. This can to extend the utility and value of green spaces by providing visitors with Wi-Fi capability. This has had some interesting effects in some areas where high school and college students visit

Figure 2.12   Autonomous mowing machine. *Source:* Graze, Inc.

Figure 2.13 Public Wi-Fi in green spaces. *Source:* TarikVision/Adobe Stock.

the park in order to have access to an Internet connection to do assignments and homework. This is a good example of the use of technology to improve equity by making Internet access available on a wider basis [31, 32] (Figure 2.13).

Note that this can be combined with the smart lighting described earlier to provide Wi-Fi routers on street lighting poles.

2.14.3 Energy and Water Conservation Management

The conservation of energy and water can also be achieved through a combination of sensors, telecommunications, and climate control. Specially developed software can be used to analyze trends and patterns, take account of weather conditions, and provide optimized strategies for conserving energy and water.

2.15 Health

There is exceptionally large potential to use smart city services to improve the health category. This can range from better management of medical appointments, improving accessibility to medical treatment across all demographics, and managing the quality of service delivered. Since hospitals and other medical facilities are significant generators of transportation demand, smart mobility can also play a role in smart city health services.

2.15.1 Wearable Devices

There has been considerable growth in this area with a lot of companies entering the market [33]. Wearable devices can take many forms including smart watches, smart rings, and sensors to

Figure 2.14 Example of a wearable device. *Source:* Simon Daoudi/Unsplash.

measure health parameters. Market-leading devises include the Apple Watch and Fitbit. Ear buds for smart phones are aslo being introduced with integrtaed sensors for measuring health parameters. Many devices can work in combination with the smartphone or computers to enable a more sophisticated analysis to be conducted on the data. Beyond wide-area networks and local-area networks, the concept is now emerging of a body area network. There is also the possibility that such wearables would interact with other smart city sensors and infrastructure, connecting humans to the wider network. Wearables can also extend to glucose monitoring systems and even tiny sensors that can be implanted to measure the pressure inside the eye. Instead of the occasional visit to the doctor for a health checkup, there is now the possibility that health monitoring can be conducted on a continual basis. Imagine combining wearable devices with smart refrigerators. The wearable device monitors your vital signs, the smart refrigerator monitors the food you that you have and if necessary automatically orders healthier food from online grocery delivery services. Wearables, in combination with computers and web technology, can also offer services that complement the wearable product. For example, Fitbit as a product can be supported by a range of services providing advice on diet and exercise. This may sound a bit like a big brother or a nanny state, but at the end of the day, it will be consumer choice whether to use these or not [34–35] (Figure 2.14).

2.15.2 Mobile In-home Treatment and Telemedicine

Mobile in-home treatment reverses the traditional pattern of patients having to go to a medical facility or hospital. Instead, the medical facility comes to them. In future, it could even be an automated vehicle that comes to the home medical facilities such as X-ray, MRI, or even dialysis [36] (Figure 2.15).

Telemedicine involves the distribution of health-related services via electronic information and medication technologies. It enables remote patient and doctor contact, for the delivery of care, advice, education, and intervention. It can also include health monitoring and automatic reminders to take medication and other activities [37] (Figure 2.16).

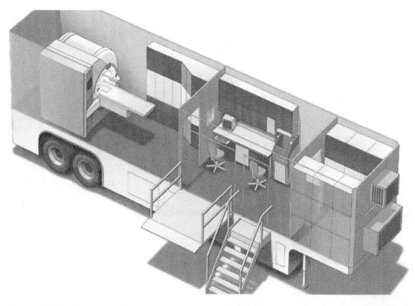

Figure 2.15 Example of an autonomous mobile medical facility. *Source:* macrovector/Adobe Stock.

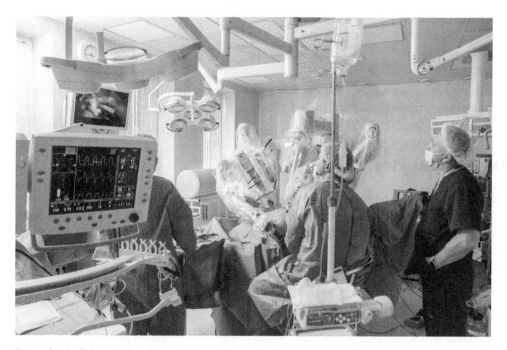

Figure 2.16 Telemedicine. *Source:* National Cancer Institute/Unsplash.

2.15.3 Surgical Robots

Combination of robotics and remote control can create new possibilities for surgery. The robotic control can be used to enhance the accuracy of the surgery or could even allow the surgery to be conducted at a distance. A specialist heart surgeon in Boston, for example, could operate on a

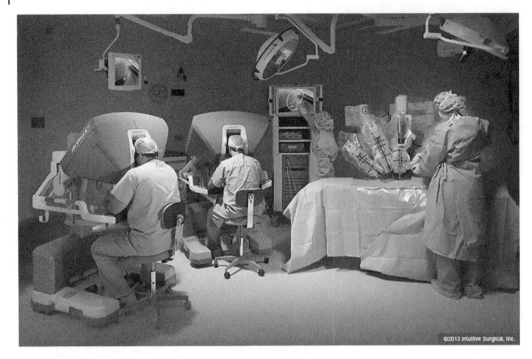

Figure 2.17 Robotic surgery. *Source:* romaset/Adobe Stock.

patient in Paris, by using robotics and remote control. This is a possibility to make surgery more efficient and less expensive [38, 39] (Figure 2.17).

2.16 Manufacturing

A city with a vibrant economy will feature a significant element of manufacturing, as well as service industries. Smart city services can be applied to improve the efficiency of manufacturing including the supply chain that provides the raw materials, and the delivery chain that distributes the finished products. Smart city services can be applied within the manufacturing facility to improve productivity and efficiency through the use of robotics.

2.16.1 Supply Chain Management

In the same way that the energy and water cycles can be managed through the application of technology-enabled services, the supply chain for manufacturing can be approached in the same way. A smart factory requires efficient input of raw materials and efficient output of finished products. However, this is only one part of the overall chain, the figure shows an example of a supply chain (Figure 2.18).

This depicts a supply chain as a linear process from supplier to customer. Note the possibility of a feedback loop from the customer through recycling back to the supplier, in the same manner as the energy and water cycles discussed earlier [40].

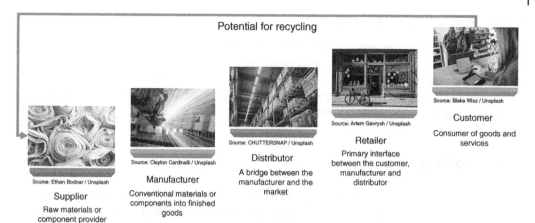

Figure 2.18 Supply chain example.

2.16.2 Industrial Robotics

Industrial robotics apply automation to the manufacturing process, automating routine and repetitive tasks on the production line. They can also be found as part of a wider cyberphysical system that integrates sensing, computing, and control networking into physical objects and infrastructure. In this way, industrial robots can be connected to each other and to a wider management system. The application of industrial robotics and cyberphysical solutions is often referred to as Industry 4.0, or the Fourth Industrial Revolution [41].

The first Industrial Revolution involved the application of steam engines. The second Industrial Revolution was powered by electricity. The third Industrial Revolution involved preliminary automation and machinery. The fourth Industrial Revolution is expected to be manifested by a combination of technologies including robotics, artificial intelligence, and 3D printing, among others [42] (Figure 2.19).

Figure 2.19 Example of industrial robots. *Source:* Khaligo/Adobe stock.

2.16.3 Private Cellular Networks

Cellular wireless networks are covered in Chapter 8 and are technologies that are worthy of mention here because of their ability to provide the connectivity required for industry 4.0 and cyberphysical systems. Unlike public or cellular wireless, these systems are dedicated to specific functions in offices or factories. They can provide high-speed connectivity between devices, between computers, and even between factories [43].

2.17 Mobility

This category tends to be inordinately important within the overall smart city context. For example, it is estimated that at least 20% of the energy generated in the United States is used to power transportation systems. There is also no point in having smart hospitals, smart offices, smart factories, and smart education facilities if they are not easily accessible. Smart mobility services while delivering value in their own right, also reinforce the value of the other services as an enabling mechanism. The figure below illustrates eight services that could be deployed to support smart mobility within a smart city (Figure 2.20).

2.17.1 Connected and Automated Vehicles

Connected vehicles have the capability to support a two-way wireless conversation between roadside and cloud infrastructure and vehicle, this could also include vehicle-to-vehicle communications. On the other hand, automated vehicles are capable of driving without a driver. Automated vehicles are being introduced on a transitional basis achieving different levels of automation until full automation without a driver is reached. In the interim, the level of decision support for drivers is increasing.

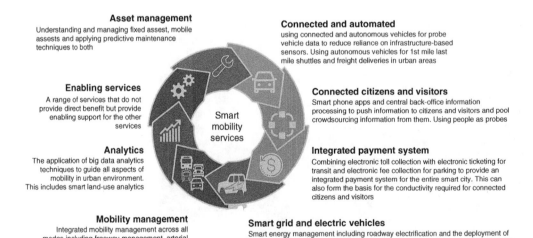

Asset management
Understanding and managing fixed assest, mobile assests and applying predictive maintenance techniques to both

Connected and automated
using connected and autonomous vehicles for probe vehicle data to reduce reliance on infrastructure-based sensors. Using autonomous vehicles for 1st mile last mile shuttles and freight deliveries in urban areas

Enabling services
A range of services that do not provide direct benefit but provide enabling support for the other services

Connected citizens and visitors
Smart phone apps and central back-office information processing to push information to citizens and visitors and pool crowdsourcing information from them. Using people as probes

Analytics
The application of big data analytics techniques to guide all aspects of mobility in urban environment. This includes smart land-use analytics

Integrated payment system
Combining electronic toll collection with electronic ticketing for transit and electronic fee collection for parking to provide an integrated payment system for the entire smart city. This can also form the basis for the conductivity required for connected citizens and visitors

Smart mobility services

Mobility management
Integrated mobility management across all modes including freeway management, arterial management: urban surface streets control, transit management, smart metro

Smart grid and electric vehicles
Smart energy management including roadway electrification and the deployment of electric vehicles. Requires analysis on the best placement of electric vehicle charging points and analysis of the art impact of electric vehicles on both mobility and energy consumption

Figure 2.20 Smart mobility services.

2.17.2   Connected Citizens and Visitors

The use of smartphone apps to enable communication to and from citizens and visitors. The information can include travel information such as mobility as a service or other general information regarding government services and the smart city locale.

2.17.3   Integrated Payment Systems

These use a single back office, account-based system, with multiple payment instruments including smart cards and smartphones. These have the advantage of being people-specific, enablng their use across different modes of transportation. Electronic toll collection is typically achieved using a dedicated short-range communication link between a transponder fitted to the vehicle and a transceiver at the roadside. Electronic toll collection systems are vehicle-specific.

2.17.4   Smart Grid and Electric Vehicles

The emergence of electric vehicles requires matching charging infrastructure to enable the vehicles to be refueled. This charging infrastructure is often connected to a smart grid technique to ensure that the electric vehicles have the appropriate charging characteristics available. Different electric vehicle charging infrastructure is required depending on the location of the infrastructure. For example, an in-home electric vehicle charging system may have up to eight hours to charge the vehicle, while those systems that are fitted at the roadside could be required to charge the vehicle in a much shorter period of time. Charging infrastructure at offices, factories or other buildings could also have three to four hours to charge the vehicle. The point is that the characteristics, as well as the location of the vehicle charging infrastructure, need to be carefully designed to meet the needs. To understand the scale of the task to provide a suitable number of electric vehicles charging stations, and replicate the ubiquity's availability of petrol and gas, there were over 136,000 petrol stations in the EU, Norway, Switzerland, and Turkey in 2020. While in the United States of America, there were approximately 115,000 gas stations. This is a growing challenge for drivers and fleet operators to switch from internal combustion engines to electricity to power their vehicles.

Electric vehicle charging stations also create a demand for electricity in places where electricity was previously not required (Figure 2.21). A recent study suggests that as demand for electric vehicle charging stations continues to rise and market penetration rises, a situation could be reached in which cumulative electricity grid investment needed to support electric vehicle charging exceeds several hundred euros per vehicle. This raises the question of using another technology application – electronic payment systems to adjust demand for electricity based on time of use, location, and energy consumed.

So, the introduction of electric vehicles is more than about electric vehicle technology and the availability of electric vehicles to the consumer. Like many things about smart cities, there is a wider system effect that must be addressed for maximum efficiency. Consider how valuable a DVD would be unless you also had a DVD player [44–47].

2.17.5   Mobility Management

This covers the entire spectrum of transportation services including private cars, traffic network companies, transit, freight, and nonmotorized transportation, shared micro-mobility services are also included in this. Ideally these will be managed using a single mobility management system. Currently most modes of transportation are still operated on a fragmented mode across the city.

Figure 2.21 Electric vehicle charging. *Source:* Ernest Ojeh/Unsplash.

2.17.6 Enabling Services

This category includes things that have no directly measurable benefit to mobility but are essential in order to underpin mobility operations. These include things like data governance, communications infrastructure, and performance management systems.

2.18 Recycling

This is a category of smart city services that benefit from the adoption of a system approach to smart city planning, design, and operation. With renewed focus on resource efficiency and material flows, the concept of a circular economy has emerged over the past few years. This approach decouples resource consumption from economic growth through a comprehensive approach to resource management. This involves the application of innovation and investments to migrate away from the current linear model to a circular economy, which considers the complete lifecycle of resources and possibilities for reuse and recycling. According to the World Bank, the world's cities generated more than two billion tons of solid waste in 2016. This equates to a footprint of 0.74 kg per person per day. Managing waste properly is essential for sustainability but remains a challenge for many developing countries and cities which often spend between 20 and 50% of budgets on effective waste management [48] (Figure 2.22).

2.18.1 Smart Garbage Cans

That are several aspects of the smart garbage can. The first is connectivity. The smart garbage can is connected to a central management system using a combination of wireless and/or wireline telecommunications technologies. This connection is used to let the central management system know when the garbage can needs to be emptied. This enables the possibility of dynamic itineraries for garbage crews rather than following a less efficient fixed route and schedule. A second aspect is the ability of the smart garbage can to compact waste, hence increasing the capacity of the

Figure 2.22 Waste recycling. *Source:* Nick Fewings/Unsplash.

garbage can, requiring it to be emptied by the garbage crews less often. The third aspect is support for effective recycling through the use of the camera and video image processing techniques to enable the user to identify the type of waste inputted into the appropriate compartment in the smart garbage can. There are numerous implementations of smart garbage cans around the world that feature at least one of these aspects, but not all smart garbage cans feature all three.

2.18.2 Circular Economy Management

The circular economy is a systematic solution framework for managing waste to achieve sustainability. The figure illustrates the solution framework (Figure 2.23).

It depicts a cyclical approach from production and distribution of products to consumption and storage through waste management to materials recycling and the use of eco-design principles to create new products with better recycling possibilities. It also depicts the objectives of minimizing extraction and importation of natural resources and management of the emissions from incineration and landfills.

The framework helps to address global challenges such as climate change, biodiversity loss, waste, and pollution. It involves the application of the following principles:

- Elimination of waste and pollution
- Continuous recycling of products and materials
- Increasing resilience through diversity
- Use of renewable energy

The name for the approach emphasizes the contrast between the current linear process in which natural resources are used to create products, which are then used then discarded (waste). Applying this process, the burden of waste management is placed at the end of the process. The circular management approach, as the name suggests, takes a lifecycle or circular approach is illustrated in the figure above.

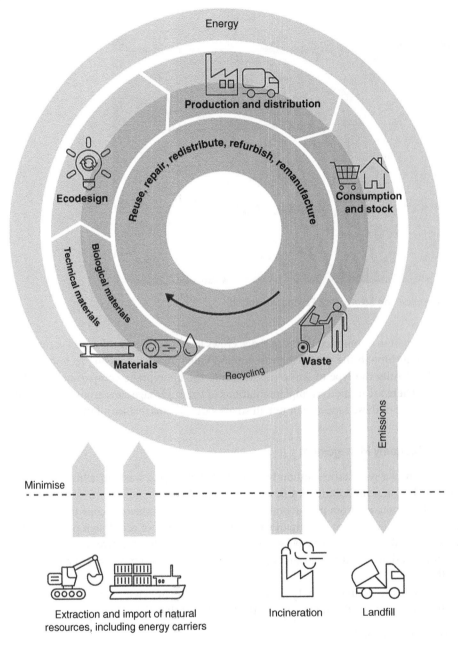

Figure 2.23 Circular economy management solution framework.

Adopting this process integrates waste management in multiple places in the cycle and supports coordination between a wide range of participants.

One interesting feature of circular economy management is the potential to make use of multiple smart city services within a systematic and structured approach. Another interesting feature is that the overall principle of the circular economy – adopting a systems approach to minimizing waste and maximizing the value for resources – is a core principle of smart cities. It is interesting to note that the software industry has also adopted the term software recycling to capture and preserve the value of the resources expended in creating software, by reuse of software modules for other purposes. It can also mean something as simple as reselling software packages that are no longer required by the original owner, in the same way as used cars and used books, for example. The most effective approach to software recycling would also include the original design of the software in a modular fashion to enable software recycling and reuse [49].

2.18.3 Energy Recycling

There are numerous ways in which energy can be recycled by applying an energy recovery process to use energy that would normally be wasted. It usually involves converting energy into electricity or capturing thermal energy. A good example can be found in electric vehicles. Electric vehicles are a smart city element that play a role in energy recycling. In many such vehicles, the braking process is achieved through the use of an electric motor on each wheel hub, or just the drive wheel hubs, front, or rear. The electric motor is used to convert electricity into kinetic energy to move the vehicle forward and convert kinetic energy into electricity in order to slow the vehicle down. This takes advantage of the fact that electric motors, when used in reverse, function as generators. The technique is also used for rail vehicles. Note that the regenerative braking process is also supplemented by conventional mechanical braking in order to provide braking at lower speeds and to immobilize stationary vehicles. The figure depicts the overall architecture of the Toyota approach to regenerative braking showing the separate circuits for regenerative braking and hydraulic braking and how they are configured to work together [50].

Energy recycling can also be achieved by enabling electric vehicles to connect to the smart grid and return stored energy that has not been used to the grid. This also opens the possibility that the electric vehicle could actually power the home in the event of a power disruption such as those occurring during hurricanes (Figure 2.24).

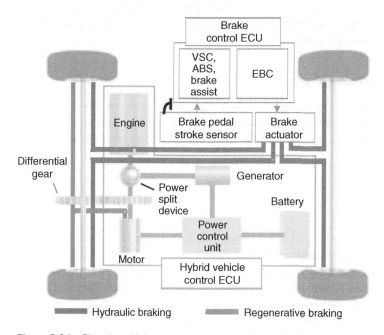

Figure 2.24 Electric vehicle regenerative and hydraulic braking systems.

2.19   Retail and Logistics

The interface between the customer and the retailer can benefit from smart city services. These enable a greater understanding of customer behavior and preferences by providing data that maps customer movement in stores and matches purchases to demographics. This can also influence the logistics chain to supply goods to the store and deliver goods to buyers.

2.19.1   Location Intelligence

The systems provide information on the location of the consumer and therefore information regarding the consumer's behavior. They can detect how long a consumer dwells at a particular exhibit item inside the store and can detect the pattern of visits that the consumer exhibits. Wi-Fi and smartphone and GPS technology can be used to provide this picture. Location intelligence can also be supplemented by sentiment analysis by interrogating the social media feeds. This can all be combined to give an understanding of consumer behavior and consumer sentiment regarding particular items inside the store.

2.19.2   Customer Experience Management

This goes beyond location intelligence to provide a total customer experience management approach. In addition to the location intelligence application described above, the customer's experience in using various services is assessed and appropriate action is taken. For example, if a customer is delayed in checking into a hotel, the system may recognize this and offer vouchers or a discount in compensation. The overall philosophy here is that a consumer or customer will have an experience whether it is managed or not. The customer experience management approach is a way to ensure that the customer has an optimum experience and a high confidence and perception of the quality of the services provided. This raises an interesting point that will be discussed later, in that one aspect of customer experience management may be to offer a moneyback guarantee if the level of service promised is not delivered.

2.19.3   Real-Time Inventory Management

Retail stores are exceptionally good at assessing how much they have in stock and how much they have on shelves. This goes one step further by assessing the type of products and services being purchased by consumers in different stores. Stocking levels at different stores can then be adjusted to suit the local demographics [51].

2.20   Safety

Another important indicator of the quality of life in the city relates to safety levels. It is important to note that perceived levels of safety are just as important as actual safety levels. Smart city services can improve communication abilities for citizens and visitors enabling information and alerts to be provided and help to be summoned more efficiently. Information technology can detect trends and patterns enabling longer-term safety improvements to be designed. In smart mobility applications, the focus on safety of vehicles and drivers is extended to include nonmotorized and pedestrian activity.

2.20.1 Early Warning Systems

These provide early warning of impending threats and can use a variety of technologies to deliver the message. Smartphone apps would be one channel, along with TV, radio, and dynamic message signs along the highways. The type of events that could generate warnings could be weather events, special events, or terrorist activity. Wrong way driving would also fit into this category where sensors detect that a vehicle is being driven in the wrong direction along the highway and alert drivers appropriately.

2.20.2 Next-Generation Emergency Systems

Emergency systems match those that need emergency support (requester) to the assets and services that provide it (provider). Current approaches involve the designation of specified three-digit telephone numbers that requesters can use to summon assistance. Requested are relayed to a public safety answering point (PSAP) where an operator deals with the request. For example, 911 in the USA and 999 in the United Kingdom are reserved for access to PSAPs. While these systems can provide wide access to emergency services they have some drawbacks.These include the provision of assistance to deaf and hard of hearing citizens, support for multiple languages, establishing accurate location of the service requester, and situations where the requester is incapacitated and unable to make a call. Such systems also have difficulty in supporting the use of voice-over IP telephony techniques. Next-generation emergency systems will fully harness the capability of internet-enabled technologies including fully digital communications. The combination of audio, real-time text, images, and video will support a complete picture of the situation. This will also include automated integration of data from operators, devices, and GPS. Interoperability and connectivity from different emergency services andresource providers will also be supported, creating enahnced connectivity between the requester and the provider [52].

2.20.3 Personal Security Systems

These systems are designed to automatically summon help. They can trigger a local alarm, or through a wireless communication link, alert a central dispatching facility. They can be standalone purpose-specific systems that are incorporated into a home security system, which covers the security of the entire home. They can also be combined with a key safe that allows responders to access the home using a pre-agreed code. The central dispatching approach has the advantage of communicating the individual's location using GPS, and enabling responders to have access to the user's medical records in advance.

2.21 Security

System engineers often regard security and safety as a closely related pair. The perspective is that safety protects the user from the system and security protects the system from the user. A growing reliance on data and information makes it more important than ever that security is protected and that appropriate data governance arrangements are in place to protect the data and to protect individual privacy.

2.21.1 Real-time Crime Centers

These are remarkably similar in function to traffic or transportation management centers where field personnel and devices are coordinated through a central location. This enables instant access to information and data to help identify patterns and stop emerging crime. Cloud-based technologies are significant enablers of real-time crime centers. Real-time crime centers can contain centralized technology platforms to extract data from multiple sources including automated and anecdotal data providers. They can also bring live video feeds together from any source creating a central hub that raises situational awareness and makes investigation more effective. Many existing transportation management centers feature the co-location of transportation personnel with police officers to provide assistance in the incident response process. The future of urban mobility could include close integration between real-time crime centers and transportation management centers [53].

2.21.2 Police Officer Wearables

This extends the concept of people based probe data to include police officers and enforcement officials. The use of body cam techniques is already prevalent in police forces around the world to enable police officer activity to be monitored and to record an actual account of proceedings for evidential purposes. This can be extended to include sensors to measure biometrics and automatically detect and verify the location of a gunshot [54].

2.21.3 Mobile Video Capture and Recording

In addition to police officer wearables video cameras, the concept can be extended to include cameras on patrol vehicles. Video from these sources could be integrated with video from fixed infrastructure CCTV to provide a complete picture of the environment and activities.

2.21.4 Event Management

Event management could be considered to be a variation of overall incident management supported by a transportation management Center. Incidents can be caused by unexpected weather events or unexpected human-related events such as terrorist attacks. Events can also be considered to be an incident but of a preplanned, scheduled nature. For example, a sports game or a concert. The overall management needs of any kind of incident are remarkably similar, with the following process supported (Figure 2.25).

2.21.5 Step One Detection

Involves the identification of a situation that requires some kind of response. This can be done using automated sensors or anecdotal data.

2.21.6 Step Two Verification

Depending on the source, this step verifies that an incident is in progress and that a response is required.

2.21.7 Step Three Plan Selection and Response

This involves the interrogation of a library a predefined responses and the application of that response plan to the situation.

Figure 2.25 Incident management process.

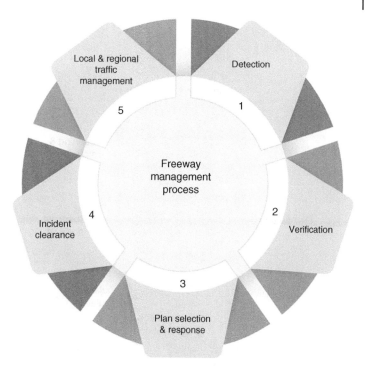

2.21.8   Step Four Incident Clearance

This involves activities and resources required to clear the incident and return the transportation network to a normal situation. In the case of a preplanned event, this would involve the management of the unusual traffic patterns in the course of the event, especially the influx of people ahead of the event and their floor people after the event.

2.21.9   Step Five – Local and Regional Transportation Management

The activities required to fully support coordination of resources according to the transportation conditions generated by the event.

For both events and incidents, sensors and data sources are required, data is communicated to a central location such as a transportation management or event management center. At the event management center, there would be a capability to blend data from different sources and conduct advanced analytics, revealing insight and understanding to be applied to response strategies. In some situations, there will be a need for ad hoc responses in addition to preplanned ones. Responses can also include the redeployment of event and incident management resources in response to the change in the transportation network conditions and demand.

2.22   Water

Water offers a splendid example of how smart city services can have a broad impact across the board. It is also a microcosm that can be used to illustrate how the most effective approach to sustainability lies in treating urban services as a complete cycle (Figure 2.26).

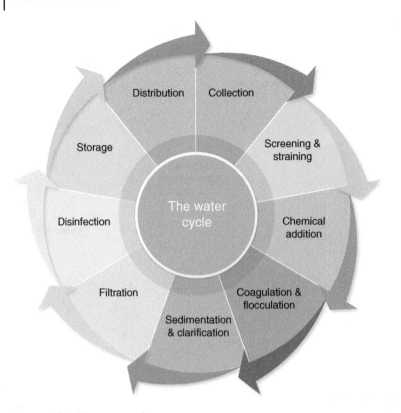

Figure 2.26 The water cycle.

Efficient and effective distribution of water in the city is also a valuable quality of life factor. Commercial consumers can use the water for industrial purposes. The overall process of managing catchment areas, water treatment, water distribution, and customer service can benefit from the application of smart city services. These range from performance management systems to payment systems that enable efficient payment in return for water services [55–57]. Here are examples of how smart city services and technologies can be applied to the different steps in the water cycle.

2.22.1 Collection

Water quality and flow sensors can be used to maintain an accurate picture of the condition of natural water bodies. This can also include biological testing for water quality.

2.22.2 Screening and Straining

The incoming water is screened to remove large objects and debris. Sensors, telecommunications, and control technologies can be used to ensure that this process is running efficiently and also signal the need for human intervention if required. Video and image processing can be used to automate inspection to ensure that large objects and debris are successfully removed at this step in the process.

2.22.3   Chemical Addition

Various chemicals are introduced to the water to promote coagulation and flocculation at the later stages. The exact volume of these chemicals can be optimized using sensors, telecommunications, and command and control.

2.22.4   Coagulation and Flocculation

The chemicals added to the water in the previous step promote coagulation of smaller particles, which then group together into larger heavier particles that are easier to remove by sedimentation or filtration. Here again, sensors and actuators can be used in combination with command control to ensure that the optimum amount of chemical has been added to achieve the desired result.

2.22.5   Sedimentation and Clarification

The flocculated particles are allowed to settle and are then removed from the water supply. This is achieved in sedimentation tanks where the water moves slowly causing the heavier particles to settle to the bottom, for removal as sludge. This is collected and removed to drying lagoons. Process control can be applied to ensure that the correct amount of chemicals is added, and the chemicals are mixed appropriately. A range of sensors can also be applied to check the quality of the water. Here again, process control can be applied to ensure that the results being achieved are those that were intended. Video and electronic sensors can be used to maintain a continuous quality assessment and effectiveness measurement of the process. Sensors can collect data regarding the sedimentation process, transmit it to a central control center, where analytics can be used to detect trends and patterns and deliver optimum strategies.

2.22.6   Filtration

Water is passed through gravel and sand filters in order to provide an extra level of clarification. This step of the process can again be subjected to process control of an appropriate quality level. Note that the filters are cleaned regularly by backwashing or reversing the floor. Process control can indicate the optimum time to do this and the length of time over which the filters should be backwashed. Sensors can also evaluate the condition of the filters which would have an influence on the backwashing time.

2.22.7   Disinfection and Fluoridation

This involve the addition of chemicals to the water to achieve a pre-defined level of quality. Sensors and actuators can be used to adjust the volume of chemicals to ensure that the desired effect is being achieved.

2.22.8   Water Storage

There are multiple opportunities to use sensors to monitor water storage and also adjust storage facilities in light of weather conditions and predicted demand. It is also possible to use advanced analytics to assess water storage levels and adjust the storage of water in each facility on a network basis.

2.22.9 Distribution

One of the ways in which water use can be monitored is through the use of smart meters that provide feedback to the water supplier and also to the consumer. Sensors and information platforms can be also used to monitor water pressure in pipes on a continuous basis. For dramatic leakage, sudden pressure drops can indicate that a leak has occurred and requires action. On a routine basis, water pressure can be monitored and the difference between the volume of water entering the pipe and the volume of water leaving the pipe can be assessed. The pressure in the pipe can be reduced by partially closing valves and hence reducing the total volume of water lost through chronic leakage. This approach can also indicate when it is time to replace the pipe due to unacceptably elevated levels of leakage.

2.23 Smart City Service Integration

It is apparent that there is no single city around the globe right now that is completely smart, although many cities have made substantial progress in specific areas. This is not surprising as the range of smart city services required to make a city completely smart is large. This will be addressed in detail in Chapter 5 on smart mobility progress around the world.

Analogies can be helpful, especially when attempting to explain a complex subject or an innovative subject when trying to find the way forward. An analogy has the advantage of building a bridge between our current and previous experience and the nature of the new thing under investigation. Smart cities are one of those new things where it is helpful to have a good analogy. To be able to relate past or current experience in a particular area and apply it to smart cities. Recent experiences of one of the authors brought to light an interesting analogy for smart cities comparing all the elements of a smart city with a human body. Now before this is examined in more detail, the point must be made that there is no way that were going to replicate the complexity and intricacy of the human body in a smart city system. It is not a recipe for a smart city; however, it can be a great model that can be used to optimize the chances of success. The amazing design and the designer of the human body is a subject of continual amazement to us, but also the subject for another publication at another time.

The authors are not experts in medicine or anatomy; however, our basic understanding of the human body and detailed understanding of smart mobility allows us to make some useful comparisons.

From an engineering perspective, the human body has the following characteristics:

- Deployment and use of sensors
- Deployment and use of high-speed communications
- A significant capability in distributed processing
- A strong capability to collect data, turn into information, gain insight and understanding, then develop actions and responses
- The human body is also capable of monitoring the effectiveness of those responses and taking appropriate actions.

Every human body, to varying degrees, makes use of the following sensors, or senses:

Sight: the ability to collect visual data and make sense of it
Hearing: the ability to collect audible data and make sense of it
Touch: the ability to collect tactile data and make sense of it

Smell: the ability to collect olfactory data and make sense of it

Taste: the ability to collect taste data and make sense of it

Each of the sensor types is not used in isolation; however, they are used in combination and take account of the data from each sensor type or channel, when determining insight and action. There is an amazing data fusion process going on that makes use of big data and data processing. It is estimated that the human retina, the part of the eye that receives the light as part of the conversion process to neurons, can transmit ten million bits per second [58]. If a gigabit ethernet connection will operate at around one thousand Mb per second, or one million bits per second, you can understand how fast the human retina communicates. Bits per second. This is an estimate, but it does indicate the use of big data. Human sensors can collect big data delivering the challenge of how to turn incredibly large volumes of data into information insight and action.

2.23.1 Edge Processing

The human body takes significant advantage of what is known today as "edge processing." For example, light is converted into neurons or electrical impulses in the eye. This conversion process is not conducted in the brain but in the peripheral device, the eye.

Some incredibly well-designed features of the eye, cones, and rods, translate light into neurons. This includes light intensity and light color. Of course, as it is a total design, the peripheral device delivers information to the back office, or the brain in a format and content that is expected. It is also important to note that the brain makes use of distributed processing techniques to analyze the incoming data in different areas of the brain, enabling parallel processing and a strong ability to deal with a large volume of data. It is also beginning to be understood that as well as the conscious mind, use is made of the subconscious mind which has an extremely fast ability to process information, based on preprogrammed responses. If a truck is approaching your high-speed your conscious mind does not have the time to determine the appropriate response, your subconscious simply says "jump out of the way" and your body obeys. If your sense of touch detects that you put your finger into a flame, then at extremely high speed your subconscious instructs your arm to remove your finger from the flame. Malcolm Gladwell authored a whole book on the subject called Blink [59] in which he explores the role of the subconscious mind through a series of examples from tennis coaches to firefighters.

2.23.2 Stereoscopic Vision

The eye gets even more amazing when it is understood how both eyes work together. At a basic level, the need for two eyes would of course relate to having a backup. It is significant risk management. People use both eyes along with other capabilities of the body, in combination enabling a three dimensional approach to determining distance and size of objects. This typically involves a combination of motor skills and sensors. The eye is capable of a degree of motion, the human head is also capable of motion and the human body can also move. Using our motor skills to control these in combination with the sensory capabilities of the eye gives us considerable and powerful abilities to see and sense. This gets even more interesting when eyes are used along with the other four senses in combination to get the total picture of the external environment. The use of sensors in a multisensory coordination approach can also support multiple actions.

As discussed earlier, the human body can be used as a model, not a recipe to get the best out of a smart city systems. There are a number of lessons from the human body analogy that can be particularly valuable including the following:

2.23.3 Data Processing

The body uses edge processing techniques and distributed parallel processing. The lesson here is not to use these techniques to avoid processing load but to focus it on the appropriate place. By carefully managing the processing capability of a smart mobility system, it can be ensured that the best use is made of automated techniques and that the appropriate level of resources is applied to each stage in the process. This also suggests that an ability for smart mobility system to dynamically reassign resources would be extremely valuable in the light of changing demand and operating conditions associated with mobility.

It would be optimal for the data fusion to be conducted across agencies and all the stakeholders involved in smart mobility management. Transportation practitioners tend to specialize in individual modes to achieve efficiency through focus. However, many of the smart mobility applications and technologies work best when they are shared across the city or region and applied in a coordinated, harmonious fashion. The body makes use of data from all five senses. Here again, the lesson can be learned that data should be combined from all available sources to achieve a complete picture of the demand for transportation and variations in the operating conditions.

2.23.4 Stratified Processing and Response

A smart mobility system would benefit from a subconscious as well as a conscious ability. What is meant by this? Well, there could be a number of preprogrammed preplanned responses that can be implemented in an extremely fast way. There can also be, using today's data processing techniques, real-time processing of data streams to calculate analytics on the fly. Using both techniques in combination, it should be possible to develop extremely fast responses for appropriate situations, while also dealing with those situations where a longer timeframe and more considered approach can be taken. This is a valuable area for the application of artificial intelligence to smart cities to enable decision-making to be automated and to use advanced machine learning techniques to detect trends, patterns, and develop preprogrammed responses.

2.23.5 Multisensory Coordination

There is a growing awareness of the value of using multiple sensors and bringing the data from the sensors back to a central point for processing and fusion. The IoT is starting to be used to support edge processing to manage the bandwidth that is required to bring such data back to a central location. Learning from the human body, the use of multiple sensors can be augmented by the use of smart mobility "motor skills" to take advantage of mobile data sources. This will include probe vehicle data from connected and automated vehicles. Motion may also come from refocusing static roadside sensors onto various locations and the use of portable and temporary sensors to assist us in matching the sensor distribution pattern to the data collection needs. This could also involve the application of a "swarm" technique where multiple agencies work together in a coordinated fashion to make use of sensors in an optimized way, whether these be static roadside sensors or mobile probe vehicle sensors.

2.23.6 Effects Management

It is also important to remember that the wide range of sensor techniques and technologies available must be used to collect data initially but also be used to monitor the effects of the actions and strategies that are developed. Continuous use of the sensors to provide feedback to ensure that the appropriate effect has been achieved, or otherwise, is vital to the operation of smart mobility as a single system.

One of the favorite keynote speeches that the authors refer to often is the one that Sam Palmisano, then Chief Executive Officer of IBM gave at the ITS America annual meeting in Houston in 2013. The speech is paraphrased as follows.

IBM have been in the business of designing and developing systems for over one hundred years. This includes implementation and operations. Based on that experience essential characteristics of a system that prove that an application is operating as a single system can be summarized in the following four statements:

- Clarity of purpose: the range of objectives to be achieved by the system should be clear and agreed by all stakeholders
- Adaptability: the system should be capable of adapting to changes in the demand for services and the operating environment
- Connectivity: the major essential components of the system should be connected
- Monitorability: key system performance measures should be able to be determined at appropriate times to enable timely intervention

This is a wonderful way to summarize the critical aspects of the system from hardware and software perspective. Building on the discussion on the human body analogy, it can be added that monitorability should involve the use of multiple sensors in an agile manner. Connectivity should include the balance of data supply with the ability to turn the data into information. Adaptability should take full advantage of all the organizational capabilities for smart mobility stakeholders across the region or city. Adaptability should also include the application of preplanned responses just like the subconscious mind. This will require that such responses be reevaluated, defined, and agreed in advance.

2.24 The Smart Mobility Body of Knowledge

This book represents one element of a body of knowledge for smart mobility. When researching and developing our thoughts on smart cities in the human body analogy, it occurs that the smart mobility body of knowledge is more than a curated set of information. It also captures thinking about smart mobility, which could be known as mobility philosophy. This sets the scene for holistic mobility management through connection, adaptation sensors, and objectives-driven action. It reinforces the point that a convenient set of well curated information brought to a central and accessible point is not the only thing that is required. To be successful, it is also important to capture thinking about smart mobility, successful approaches, and effective strategies. When the smart mobility revolution is discussed one of the focus areas is how progress can be accelerated to achieve benefits as quickly as possible. Another area of focus is how to optimize the use of resources to get as large a positive effect as possible with the investment available. This also includes the ability to apply experience after all wisdom can be defined as the successful application of knowledge. The authors believe that information, knowledge, philosophy, and best practices are all great tools to accelerate and optimize smart mobility.

2.25 Teamwork and Multiagency Coordination

The study of the human body analogy also brought us to the conclusion that the standalone capabilities of a human can be dramatically enhanced by working in close cooperation and coordination with other humans. This is known as teamwork which is a vital element in determining a successful approach to smart mobility in urban areas. There is a need for individuals and organizations to work together toward a common set of goals. The ability to do this represents a force multiplier that will deliver results that the community and industry should strive to achieve. Appropriately coordinated and managed teams of people equipped with the appropriate resources and tools provide an even more powerful force multiplier.

In addition to providing an overview of smart cities, this chapter also considered what can be learned from the human body as an analogy to smart mobility in smart cities. The lessons derived from the analogy can be used to accelerate progress on a global basis to complement accomplishments and to determine that is still required to get to the point where the world has completely smart cities.

References

1 United Nations, Department of Economic and Social Affairs. The 68% of the world population projected to live in urban areas is by 2050 says United Nations. *Newsletter*, https://www.un.org/development/desa/en/news/population/2018-revision-of-world-urbanization-prospects.html, retrieved April 3, 2022, at 11:42 AM Central European Time.

2 McKinsey and Company. Modular construction: from projects to products. https://www.mckinsey.com/business-functions/operations/our-insights/modular-construction-from-projects-to-products, retrieved April 3, 2022, at 11:43 AM Central European Time.

3 Bouygues Construction. 3D printing to build a house: building innovation website. https://www.bouygues-construction.com/en/innovation/all-innovations/3d-printing-build-house-0, retrieved April 3, 2022, at 11:48 AM Central European Time.

4 Investopedia Article. Why Zillow is free and how it makes money. https://www.investopedia.com/articles/personal-finance/110615/why-zillow-free-and-how-it-makes-money.asp, retrieved April 3, 2022, at 11: 50 AM Central European Time.

5 World Health Organization. Global air quality guidelines. https://www.who.int/publications/i/item/9789240034228?Ua=1, retrieved April 3, 2022, at 11:51 AM Central European Time.

6 United Nations. Sustainable development goals. https://www.un.org/sustainabledevelopment/blog/2016/05/un-health-agency-warns-of-rise-in-urban-air-pollution-with-poorest-cities-most-at-risk/, retrieved April 3, 2022, at 11:53 AM Central European Time.

7 Citysys. Smart IoT hub. https://citysys.oms-is.eu/catalogue/communication-bus/smart-iot-hub/, retrieved April 3, 2022, at 12:55 PM Central European Time.

8 Oizom Cloud. Air quality monitoring for smart city infrastructure. https://oizom.com/usecase/smart-city-air-quality-monitoring-system/#:~:text=For%20Smart%20cities%2C%20it%20is,air%20quality%20and%20meteorological%20parameters, Retrieved April 3, 2022, at 12:01 PM Central European Time.

9 Stöglehner, G., Sciendo. Conceptualizing quality in spatial planning. January 2021, https://sciendo.com/journal/RARA, retrieved April 3, 2022, at 12:06 PM Central European Time.

10 Irisys True Occupancy Website. https://www.trueoccupancy.com/, retrieved, April 3, 2022, at 12:15 PM Central European Time.

11 ScienceDirect. Advanced HVAC control: theory versus reality. Proceedings of the 18th World Congress, the International Federation of Automatic Control. https://www.sciencedirect.com/science/article/pii/S1474667016440887, retrieved April 3, 2022, at 1:57 PM Central European Time.

12 Bankmycell Website. https://www.bankmycell.com/blog/how-many-phones-are-in-the-world, retrieved April 3, 2022, at 12:20 PM Central European Time.

13 Unsplash. https://unsplash.com/photos/wK-elt11pF0, retrieved April 3, 2022, at 12:23 PM Central European Time.

14 Unsplash. https://unsplash.com/photos/wK-elt11pF0, retrieved April 3, 2022, at 12:26 PM Central European Time.

15 Unsplash. https://unsplash.com/photos/87oz2SoV9Ug, retrieved April 3, 2022, at 12:29 PM Central European Time.

16 Bee Smart City Website. https://hub.beesmart.city/en/strategy/how-smart-cities-boost-citizen-engagement#:~:text=Active%20citizen%20participation%20is%20an,in%20the%20public's%20best%20interests, Retrieved April 3, 2022, at 12:30 PM Central European Time.

17 Mecanica Website. Frank Gehry design. https://www.mecanicasolutions.com/en/project/frank-gehry-design/, retrieved April 3, 2022, at 12:32 PM Central European Time.

18 Illumin Magazine. https://illumin.usc.edu/curves-of-steel-catia-and-the-walt-disney-concert-hall/, retrieved April 3, 2022, at 12:33 PM Central European Time.

19 Technical University of Denmark. Smart construction technology cards, smart construction site. https://www.technologycards.net/the-technologies/smart-construction-site, retrieved April 3, 2022, at 12:35 PM Central European Time.

20 Caterpillar, Cat Command. https://www.cat.com/en_US/by-industry/construction-industry-resources/technology/command.html, retrieved April 3, 2022, at 12:38 PM Central European Time.

21 Technical University of Denmark. Smart construction technology cards, autonomous construction vehicles. https://www.technologycards.net/the-technologies/autonomous-construction-vehicles, retrieved April 3, 2022, at 12:40 PM Central European Time.

22 Pew Research Center. How mechanical Turk works. https://www.pewresearch.org/internet/2016/07/11/research-in-the-crowdsourcing-age-a-case-study/pi_2016-07-11_mechanical-turk_1-02/, retrieved April 3, 2022, at 12:45 PM Central European Time.

23 Amazon Mechanical Turk. https://www.mturk.com/, retrieved April 23, 2022, at 12:51 PM Central European Time.

24 Open Edition Journals. Field action science reports, smart cities, and new forms of employment. https://journals.openedition.org/factsreports/4290, retrieved April 3, 2022, at 12:52 PM Central European Time.

25 Citysys. Smart Info kiosk. https://citysys.oms-is.eu/catalogue/infrastructure/smart-info-kiosk/#:~:text=The%20interactive%20Kiosk%20is%20a,and%20significantly%20supports%20their%20engagement, retrieved April 3, 2022, at 12:55 PM Central European Time.

26 Wikipedia. https://data.gov.uk, https://en.wikipedia.org/wiki/Data.gov.uk, Graze mowing website, retrieved April 3, 2022, at 12:56 PM Central European Time.

27 Transport for London. Congestion charge website. https://tfl.gov.uk/modes/driving/congestion-charge#:~:text=The%20Congestion%20Charge%20is%20a,by%20setting%20up%20Auto%20Pay, retrieved April 3, 2022, at 12:57 PM Central European Time.

28 Development in Asia. The case for electronic road pricing. https://development.asia/case-study/case-electronic-road-pricing, retrieved April 3, 2022, at 12:58 PM Central European Time.

29 Graze Mowing Website. https://www.grazemowing.com/#technology, retrieved April 3, 2022, at 1:01 PM. Central European Time.

30 Compta Emerging Business. https://www.ceb-solutions.com/products/bee2green/, retrieved April 3, 2022, at 1:03 PM Central European Time.

31 Citizen Matters. Public Wi-Fi hotspots in Chennai: whom is it helping? https://chennai.citizenmatters. in/public-wifi-hotspots-of-chennai-awareness-and-utility-24092, retrieved April 3, 2022, at 1:04 PM Central European Time.

32 Unsplash. https://unsplash.com/photos/2wFoa040m8g, retrieved April 3, 2022, at 1:09 PM Central European Time.

33 TechCrunch Wearables. https://techcrunch.com/wearables/, retrieved April 3, 2022, at 1:05 PM Central European Time.

34 Fitbit Website. https://www.fitbit.com/global/us/home, retrieved April 3, 2022, at 1:07 PM Central European Time.

35 Apple Shop. Apple watch series 7. https://www.apple.com/shop/buy-watch/apple-watch, retrieved April 3, 2022, at 1:08 PM Central European Time.

36 Health Facilities Magazine. FGII provides updated guidance in designing mobile medical units. https://www.hfmmagazine.com/articles/3547-fgi-provides-updated-guidance-on-designing-mobile-medical-units, retrieved April 3, 2022, at 1:11 PM Central European Time.

37 Unsplash. https://unsplash.com/photos/L8tWZT4CcVQ, retrieved April 3, 2022, at 1:13 PM Central European Time.

38 Science Direct. Smart healthcare: making medical care more intelligent. *Global Health. J.* 2019, 3, (3), 62–65. https://www.sciencedirect.com/science/article/pii/S2414644719300508, retrieved April 3, 2022, at 1:14 PM Central European Time.

39 AARP. Is robotic surgery right for you?https://www.aarp.org/health/conditions-treatments/info-12-2013/robotic-surgery-risks-benefits.html, retrieved April 3, 2022, at 1:15 PM Central European Time.

40 Procurious Website. Supply chain process: where does your supply chain begin and end?https://www.procurious.com/procurement-news/where-does-your-supply-chain-begin-and-end, retrieved April 3, 2022, at 1:16 PM Central European Time.

41 World Economic Forum. Global Agenda, "what is the fourth Industrial Revolution?". https://www.weforum.org/agenda/2016/01/what-is-the-fourth-industrial-revolution/, retrieved April 3, 2022, at 1:17 PM Central European Time.

42 Unsplash. https://unsplash.com/photos/8gr6bObQLOI, retrieved April 3, 2022, at 1:19 PM Central European Time.

43 Nokia.Implementprivatewireless.https://www.nokia.com/networks/private-wireless/why-consider-private-wireless/?did=d000000006s7&gclid=Cj0KCQjw6J-SBhCrARIsAH0yMZgbSkXIWmt7WVLSu-j73P0_C4mCq_N9-KbWn2QM42WkVcgSsx7QwiIaAvCCEALw_wcB, retrieved April 3, 2022, at 4 PM Central European Time.

44 Unsplash. https://unsplash.com/photos/aEytUoE1Tkc, retrieved April 3, 2022, 1:23 PM Central European Time.

45 Fuels Europe. 2021 Statistical report. https://www.fuelseurope.eu/publications/publications/fuelseurope-statistical-report-2021, retrieved April 3, 2022, at 1:26 PM Central European Time.

46 MarketWatch. How many gas stations out in the USA? How many will there be in 10 years?https://www.marketwatch.com/story/how-many-gas-stations-are-in-us-how-many-will-there-be-in-10-years-2020-02-16, retrieved April 3, 2022, at 1:29 PM Central European Time.

47 McKinsey and Company. The potential impact of electric vehicles and global energy systems. https://www.mckinsey.com/industries/automotive-and-assembly/our-insights/the-potential-impact-of-electric-vehicles-on-global-energy-systems, retrieved April 3, 2022, at 1:30 PM Central European Time.

48 World Bank. Solid waste management. https://www.worldbank.org/en/topic/urbandevelopment/brief/solid-waste-management#:~:text=In%202016%2C%20the%20worlds'%20cities,3.40%20billion%20tons%20in%202050, retrieved April 3, 2022, at 1:31 PM Central European Time.

49 INC. Magazine. How to recycle your software. https://www.inc.com/software/articles/200808/recycling.html, retrieved April 3, 2022, at 1:33 PM Central European Time.

50 Sarip, S., Technical University of Malaysia. Lightweight friction brakes for the road vehicle with regenerative braking, thesis. https://www.researchgate.net/publication/282656165_Lightweight_friction_brakes_for_a_road_vehicle_with_regenerative_braking, retrieved April 3, 2022, at 1:35 PM Central European Time.

51 Mulesoft Retail. https://www.mulesoft.com/integration-solutions/saas/retail, retrieved April 3, 2022, at 1:37 PM Central European Time.

52 EMYNOS. Next-generation emergency communications. https://www.emynos.eu/documents/whitepaper, retrieved April 3, 2022, at 1:38 PM Central European Time.

53 Wikipedia. Real-time crime Center. https://en.wikipedia.org/wiki/Real_Time_Crime_Center, retrieved April 3, 2022, at 1:39 PM Central European Time.

54 Police1 by Lexipol. Body worn tech trends improving police operations. https://www.police1.com/police-products/police-technology/articles/body-worn-tech-trends-improving-police-operations-vqtxffjy1qtsgkgn/, retrieved April 3, 2022, at 1:40 PM Central European Time.

55 Hunter Water. Water treatment processes. https://www.hunterwater.com.au/our-water/water-supply/water-quality/how-we-protect-our-water-supply/water-treatment-processes, retrieved April 3, 2022, at 1:42 PM Central European Time.

56 Researchgate. Real-time control of urban water cycle and the cyber physical systems framework. https://www.researchgate.net/publication/339000263_Real-Time_Control_of_Urban_Water_Cycle_under_Cyber-_Physical_Systems_Framework/figures?Lo=1, retrieved April 3, 2022, at 1:46 PM Central European Time.

57 Allerin. Smart wastewater management systems in smart cities. https://www.allerin.com/blog/smart-wastewater-management-systems-in-smart-cities, retrieved April 3, 2022, at 1:47 PM Central European Time.

58 Edward Tufte, homepage, retina communicates to bring it 10 million bits per second: implications for evidence displays?, https://www.edwardtufte.com/bboard/q-and-a-fetch-msg?msg_id=0002NC, retrieved June 6, 2024 at 9:07 AM Central European Time.

59 Malcolm Gladwell, Blink: The Power of Thinking Without Thinking, Hachette Book Group.

3

Smart Mobility: A Problem Statement

3.1 Informational Objectives

After reading this chapter, you should be able to:

- Define what is meant by a problem statement
- Understand the value of a problem statement
- Define smart mobility
- Describe the benefits of transportation
- Explain how to avoid undesirable side effects
- Describe smart mobility challenges
- Describe some advice on developing a specific problem statement for a city

3.2 What Is a Problem Statement?

As the name suggests, a problem statement is a concise summary of the problem to be addressed by the smart mobility implementation. Typically, this contains a range of topics that is wider than just objectives. In our experience, the problem statement should address needs, issues, problems, and objectives. The need is defined in the problem statement as transportation needs for the city or region under consideration. Needs will vary considerably from one city to the other depending on geographical context, local political motivation, the nature of previous investments in infrastructure and technology, and other local factors. Issues describe the specific local factors that must be considered when defining an effective and successful solution. These can include lack of available budget, specific weather conditions, specific local economic development factors, and an appreciation of the success or failure of previous initiatives. This latter set of factors is particularly important in learning lessons from the past and applying them to future initiatives. Problems lie at the heart of the problem statement and define specific things to be addressed such as congestion, unreliable transit, inflexible operations, insufficient capacity, or an inability to effectively connect job opportunities with residential areas. This is intended to be an illustrative list and is not a comprehensive catalog of all possible needs, issues, or problems. The last category, "objectives" builds on the first three by defining what the target, or successful outcome will be. For example, if congestion is identified as a problem, then congestion reduction measures would form part of the

Smart Mobility: Using Technology to Improve Transportation in Smart Cities, First Edition. Bob McQueen, Ammar Safi, and Shafia Alkheyaili.
© 2024 The Institute of Electrical and Electronics Engineers, Inc. Published 2024 by John Wiley & Sons, Inc.

objectives. These will include a reduction in vehicle-miles traveled, an increase in travel time reliability, and a reduction in the number of stops, for example. It could be argued that all these categories could be rolled up under one simple needs heading, but experience indicates that it is better to keep them separate and allow a multiple-perspective approach to the whole subject of defining what it is worth trying to do and what it is to be achieved.

3.3 Why a Problem Statement Is Important

The development of a problem statement might seem like a trivial exercise; however, the authors have found that it represents a fundamental, foundational element in any successful smart mobility program. The problem statement is effectively a mission statement as characterized in aerospace and defense. Of course, the most famous mission statement is the one from the "Star Trek TV series" "to boldly go where no one has gone before." It is also one of the most famous examples of a split infinitive grammatical error. But that is another story.

The problem statement is designed for sharing. There have been situations in the past where technologists and other technical people get together and define the project concept, with little or no communication with the end user. Of course, the end user can be the public, the traveler or the people responsible for the delivery of transportation services. To be successful, the problem statement is a key communication tool to make sure that what is to be developed and designed is what people want. It also requires that information is accumulated from across the full spectrum of intended smart mobility users on their perspective and their views on the needs, issues, problems, and objectives. The problem statement is also used to help to keep a focus on the result. Technology projects can be tricky and new technologies can be fascinating to the point where the original objectives are obscured. It is important to establish regular review points at which the problem statement will be reviewed and compared against the progress and direction of the actual implementation.

A problem statement is also the hallmark of a disciplined approach. Sometimes people are reluctant to state what they intend to do if it does not work out. The willingness and ability to create a problem statement demonstrates confidence in achieving the result and in our opinion, is a major signal of capability and expertise. The overall objective is to minimize surprises and maximize the probability of getting what is wanted and avoiding what is not. In fact, a complete problem statement would also include the definition of undesirable outcomes and things to be avoided to make sure that the approach taken either avoids undesirable outcomes or manages them. This is an element of risk management that will be described and discussed later in the book.

3.4 What Is Smart Mobility?

To effectively discuss the problem statement being addressed by smart mobility it is first necessary to define what is meant by smart mobility. This is a challenge. It is apparent that smart mobility means different things to different people, and the risk would be run of losing some of the meaning if one simple definition was agreed. Drawing on a very useful research paper [1] it is possible to present a multi-perspective definition of smart mobility as illustrated in Figure 3.1. The table also indicates the perspective from which the definition was developed.

Not that some of these definitions are derived from a perspective of the intended outcomes such as accessibility, sustainability, or energy conservation. Others are derived based on the applications

Perspective	Definition	References
Smart city	It is the area of a smart city representing broadly defined mobility, the components that comprise the traditional understanding of the transport of people and goods, and the dissemination of information by digital means	[2]
Sustainability	Intelligent mobility is a comprehensive concept that makes the transport network's sustainability more achievable due to the search for improvements in transport services, balancing the application of technology with social, economic, and environmental aspects	[3, 4]
Connected vehicles	It is the ability to access transportation services from integrated platforms that aggregate the community and present intense processing of data from users to match the demand forecast. The infrastructure must be smart with connected and sustainable vehicles	[5]
Data	It is the integration of sustainable, intelligent, and cooperative vehicle technologies with a cloud server and vehicle networks based on big data	[6]
Citizens	A sustainable and safe environment that meets the mobility needs of citizens and integration with intelligent systems to provide traffic information	[7]
Accessibility	Smart mobility is related to transport and use of communication and information technologies to promote accessibility and increase the quality of life	[8]
Energy	It is the application of solutions that combine behavioral economics, e-participation, and crowdsourcing to obtain better energy consumption when moving around the city	[9]
Digital technologies	It is the adoption of digital technologies that make mobility services in a city or territory more accessible and easier for citizens	[10]

Figure 3.1 Multi-perspective definition of smart mobility.

of technology that are envisioned such as connected vehicles, digital technology, and data. One definition looks at it from the perspective of the end user, namely the citizen, and one definition places smart mobility within the wider context of a smart city.

3.5 Why Bother with Smart Mobility?

What exactly justifies interest and investment in smart mobility? Since the very beginning of the transportation era, probably with the commercial introduction of the automobile, the world has had a love–hate relationship with transportation. On the one hand, people are enamored by the benefits that can be delivered by a good transportation system. On the other hand, they are also concerned about what could be called the "undesirable side effects," in terms of the impact on the environment and the effects of too many people trying to use the system at the same time. Smart mobility offers the potential to take advantage of the benefits while managing the side effects much better than has ever been possible before. Smart mobility effectively adds an intelligent dimension to transportation. Smart mobility involves the application of advanced technology to enable the management of transportation like a system. This was very clearly articulated by the then chairman of IBM, Sam Palmisano, during a keynote speech at the Intelligent Transportation Society of America's annual meeting in Houston in 2013. To paraphrase, he stated that IBM has been in the business of developing systems for more than 100 years. Consequently, they have a good understanding of the essential characteristics, or abilities of a system, namely:

1) Clarity of purpose
2) Connectivity
3) Adaptability
4) Monitorability

Clarity of purpose means that the objectives of the system are clearly stated and agreed. In other words, an agreed problem statement. Connectivity is achieved when all the major components and subsystems are connected appropriately. The subsystem may refer to a smart mobility subsystem, a mode of transportation, or a particular agency responsible for the delivery of a particular transportation service. Adaptability is the capability of the system to adapt to changes in operating conditions or changes in demand. This could also be referred to as agility and this can be an issue as transportation systems involve a considerable amount of fixed infrastructure that is relatively inflexible.

Monitorability is the ability of the system to provide data regarding its status at any given time. This may not be an English word, but it has been used to enable all the system characteristics labels to be as concise as possible.

The subject will be revisited in Chapter 7, the essential elements of smart mobility.

The seminal work conducted in the US National ITS Architecture Program defined the essential characteristics of an Intelligent Transportation System [2] including the major interfaces required and the data flows between subsystems. It illustrated for the first time that larger systems are comprised of subsystems and transportation can be considered as an assembly of subsystems. This is an important point as the responsibility for transportation delivery within a typical city is divided between different specialist organizations. Each organization typically has its own system, defining the challenge in terms of connectivity, adaptability, and monitorability across the subsystems. Of course, clarity of purpose across the subsystems is also an important aspect and something that requires significant effort to make sure that the individual organizational objectives for each entity involved in transportation are aligned for maximum effect. This will be discussed in Chapter 14 where organizational aspects of smart mobility and how to organize for success are discussed.

One way to view the enablement of different entities responsible for transportation within a city installation is "swimming like a school of fish." Just like the fish, each entity has its own autonomous operation, objectives, and choices. However, the fish swim together for mutual benefit. An essential element of smart mobility in a city will be to encourage and support entities to move together for mutual benefit. It is also likely that the individual departments within each transportation entity will require to be coordinated as well.

To illustrate this point, consider the following example, drawn from the experience of one of the authors. While working with the major transportation agency, he was asked to conduct life-cycle cost analysis under various intelligent transportation systems within the agency. This involved information gathering through a series of interviews with the different departments within the organization. Embarking on this data collection exercise, interviews were arranged with various departments including the department responsible for the planning and operation of express lanes, and the Department responsible for the planning and operation of transit. One of the questions to be asked during the interview related to the overall objectives, in transportation terms, for the Department. When talking with the express lanes department, it was stated that the primary objective of the express lane's operation was to adjust the cost of using the express lanes to stimulate a 5% modal shift from the use of private cars to the use of transit. In other words, the Department objective was to shift 5% of all the trips made regionally, from the private car to transit. During the second interview with the transit department, it was mentioned that the express lanes department held this objective. The transit department was surprised, in fact, horrified at the prospect of a 5% shift from private cars to transit. This was because the transit system currently only handles 5% of the regional trips and a shift of an additional 5% would effectively double the demand for their transit services. After this initial reaction, the transit department staff then asked the obvious question. "Does this mean we can get more money for transit vehicles and services?" This is an example of misaligned objectives between departments. This really happened and the authors suspect that it is the tip of the iceberg of misaligned objectives in city and regional transportation. This will be revisited in Chapter 15 where smart mobility operational management will be addressed. The way to address this misalignment is to ensure a tiered set of performance measures based on mutually agreed objectives. The higher tier of performance objectives is easy to communicate and can be shared across the whole region and all users, the second-tier performance objectives are highly specific to the various job categories within transportation service delivery organizations.

To summarize, smart mobility is important, in fact, essential for a smart city because of its abilities to add value to existing infrastructure. It has been said that information technology will reduce our dependence on asphalt, concrete, and steel infrastructure. While this may be true, the fact remains that such infrastructure is a vital component of transportation. Smart mobility has the flexibility, adaptability, and monitorability that can make infrastructure smarter and enable to get better value for money.

3.6 Modes of Transportation in a City

It is noted that one of the major benefits of smart mobility is the ability to coordinate transportation service delivery across modes and to enable modal use optimization. So, what exactly is a mode of transportation and what are the modes of transportation in a city?

What transportation could be viewed as a subsystem. A collection of hardware, software, data, and people converging on an agreed set of objectives. A mode of transportation comprises the following elements:

3.6.1 Infrastructure

The asphalt, concrete, and steel required to support the delivery of transportation service using the mode. This also includes the mechanical and electrical systems required.

3.6.2 Network

The infrastructure is connected in a certain way to allow connectivity of service and support travel from multiple origins to multiple destinations. The network can also include the computing and electrical connections required.

3.6.3 Hubs

Major interchange locations where travelers can one service to another within mods or from one mode to another.

Terminals the name suggests, the endpoint, or termination of the service. Typically bus stations, railway stations, or ferry terminals.

3.6.4 Vehicles

The most obvious example of a vehicle is a private car which makes use of highway infrastructure to support the total mode of transportation. Other vehicles could include buses, trains, trams, subways, and ferries. Vehicles can also include micromobility, bicycles, and of course human powered pedestrians.

3.6.5 Containers

Containers are used to bring together a collection of items for movement. The authors suppose that a bus or a train could be regarded as a container for people, but this is not the traditional use of the term.

3.6.6 Workforce

The staff required to operate and manage the mode.

3.6.7 Propulsion Systems

The means by which the vehicles are propelled along the infrastructure. This can include electrical power, gasoline, and other forms of energy. Note that of the total energy generated in the United States approximately 28% is used to propel transportation systems.

3.6.8 Power Supplies

This can include batteries for electric vehicles or can be generating plants. The balance between centralized power supplies and distributed power supplies is something that is of considerable interest when addressing the efficiency of transportation systems. Of course, human–computer power supply for cycling and walking is also included in this category.

3.6.9 Operations

The set of business processes and activities involved in maintaining the operation of the mode.

To the authors knowledge, the following major modes of transportation, summarized in Figure 3.2, can be found in a typical city. Note that not all modes of transportation are in use in all cities and that the dominant modes of transportation vary from city to city.

3.7 The Benefits of Smart Mobility

In addition to the positive effects of smart mobility discussed previously in this chapter, there is a range of specific outcomes that can be supported and enabled by smart mobility. Deeper insight into objectives and outcomes will be provided in Chapter 15 on smart mobility operational

Modes of transportation used in cities	Subcategories			
Aerial tramways	Cable cars	Monorails		
Animal powered	Horses	Carts		
Automated vehicles	Automated cars	Automated shuttles	Automated buses	Automated trucks
Bicycle				
Boats				
Buses				
Cable transport	Cable cars			
Cars				
Ferries				
Flying	Urban air mobility	Helicopters		
Freight	Urban logistics	Trucks		
Funicular railways				
Inland waterways				
Micro-mobility	Electric scooters	Electric bicycles		
Multimodal trip chains	Can encompass all the other modes			
Pipelines				
Ride sharing				
Subways				
Trains				
Walking				

Figure 3.2   Transportation modes in cities.

management. The major effects on benefits of smart mobility can be summarized in four major categories: safety, efficiency, enhanced user experience, and environment. Figures 3.3–3.6 summarize the expected benefits of smart mobility under each of these categories.

Note that this figure represents an overview of the values and benefits that can be delivered by smart mobility. This subject will be addressed in Chapter 15 on smart mobility operational management, where detailed key performance indicators will be defined based on these categories.

Safety
Lives
Injuries
Cost of damage
Insurance costs
Faster response

Figure 3.3 Smart mobility safety benefits.

Efficiency
Improve the planning and efficiency of public transport means.
Reducing congestion
Getting more from infrastructure investments
Optimizing transportation as a single system'
Complete trip management
Modal coordination
Alignment of objectives
Connectivity
Adaptability
Monitorability
Aligning supply and demand
Effective Public Private Partnerships (PPP)
Optimize parking spaces and their management.
Optimize transit services
Optimize freight
Managing performance
Making great use of data
Systems are wonderful data generators

Figure 3.4 Smart mobility efficiency benefits.

Figure 3.5 Smart mobility enhanced
user experience benefits.

Enhanced user experience
Reduce congestion and citizen frustration
Improve the quality of life of the citizen
Prioritize the citizen in the field of mobility
Travel time reliability
Stops
Complete trip support
Original origin
Ultimate destination
First mile
Last mile
Middle mile
Equity
Moving to a market segment of 1

Figure 3.6 Smart mobility environmental
benefits.

Environmental
Reduce the environmental impact
Managing carbon footprint
Reduce emissions by smoother traffic
Reduce emissions by better use of all modes
Empty return buses

3.8 Avoiding any Undesirable Side Effects

Perhaps one of the most important qualities of smart mobility is the potential to balance the achievement and delivery of value and benefits, with the management of undesirable side effects. The application of advanced technology can provide increased flexibility and enable a higher level of capability in managing the side effects. While it is particularly important to obtain the full benefits of smart mobility, there are several potential side effects that must be managed. They should either be mitigated or avoided if possible. A few potential side effects have been identified and summarized in Figure 3.7. These are intended to be examples of side effects and not an exhaustive catalog. The following section discusses each of these potential side effects in turn.

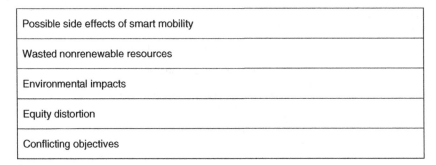

Possible side effects of smart mobility
Wasted nonrenewable resources
Environmental impacts
Equity distortion
Conflicting objectives

Figure 3.7 Possible side effects of smart mobility.

3.8.1 Wasted Nonrenewable Resources

A nonrenewable resource can be defined as one for which there are finite quantities available and include things like oil, gas, coal, and fossil fuels. It can also include metals and minerals. Of course, many nonrenewable resources can be recycled, extended the period over which they are available; however, the quantities are still finite and not unlimited. An example of wasted nonrenewable resources would be additional road construction, instead of making better use of existing capacity. This would cause additional consumption of oil in the manufacturing of the asphalt and energy in the production of the cement and concrete. The use of inductive loops for traffic signals involves burying a nonrenewable resource, copper, in the pavement. Alternatives such as LIDAR and video image processing would help to minimize the use of copper. At a more general level, the additional fuel consumption caused by congestion and unreliable traffic conditions makes a major contribution to the consumption of nonrenewable resources.

3.8.2 Environmental Impacts

The transportation process can have significant impacts on the environment. These can range from visual intrusion, noise, energy, and emissions effects. Visual intrusion can be the result of large-scale transportation infrastructure projects such as major highways and transit facilities. These can also generate significant noise. The expansion or creation of a major airport can also have significant implications with respect to noise. Major transit and highway facilities also have the potential to divide existing communities by making it difficult to cross the new corridor. This can also apply to animals when new facilities cut through existing habitat and make migration paths and day-to-day animal movements difficult or impossible to use. As discussed in the section on nonrenewable resources, transportation side effects can include energy consumption and emissions. Emissions have the potential to have detrimental effects on air quality as CO_2 and other greenhouse gases are emitted as a side effect of the transportation process. Diesel emissions can also release carbon particles into the atmosphere.

3.8.3 Equity Distortion

The concept of "Lexus lanes" sums up this side effect. The term refers to the provision of transportation facilities intended for use by segments of the population that have enough money to pay. It was originally coined for express and high occupancy toll lanes in the United States.

It was observed that existing high occupancy vehicle lanes, which were dedicated for use only by vehicles with a minimum number of passengers, were underutilized. By introducing a strategy that enabled those vehicles that did not meet the minimum vehicle occupancy requirements to use the facility in return for a fee or toll, excess capacity could be utilized. These facilities are based on the principle that a higher level of service should cost more money. While these facilities provide an overall reduction in congestion, they can discriminate against those not able to pay the toll. The term has since been more widely used to refer to any transportation facility that does not support fair and equitable use. Of course, it can be argued that the provision of transportation facilities designed for a higher demographic will in fact provide benefits for all segments of the population, through better utilization of capacity and congestion reduction effects. Equity issues can also extend to accessibility to the transportation facility for disabled people, including accommodation for wheelchairs This is a particularly important issue when establishing public–private partnerships, in which the public sector seeks to attain transportation policy objectives in partnership with the private sector, who are trying to establish a revenue-generating business. This will be discussed in Chapter 10, a framework for smart mobility success, where public–private partnership challenges and opportunities will be explored.

3.8.4 Conflicting Objectives

Because of the distributed nature of transportation management within cities and regions, it is possible that different transportation service delivery organizations will have different objectives. The problem arises when these objectives are in conflict. A specific example of this was provided earlier in this chapter when "clarity of purpose" was discussed as one of the important attributes of transportation operating as a single system. It is the authors opinion that distributed and decentralized transportation service delivery planning and operations is a critical element in overall efficiency as it supports specialist resources and approaches. However, the possibility of conflicting objectives can be a very real side effect of distributed and decentralized approaches.

3.9 Challenges

Based on experience from previous implementations and discussions with practitioners, a long list of potential challenges related to planning, operations, and service delivery in transportation has been identified. This is an important part of the problem statement as it frames the primary challenges to be addressed by smart mobility.

3.9.1 Geography

Each city is unique. The land use pattern of the city has probably been influenced by topographical features such as rivers, mountains, and other natural features. Very few cities are purpose-built from the ground up with the specific objective of optimizing transportation.

Examples of cities that were planned in this way include Canberra Australia, Brasília Brazil, and Washington, DC. Although the latter contains elements of organic growth as well as planned growth.

3.9.2 Politics

Just as decentralized planning and operations for transportation service delivery in the city can pose challenges, so too can city and local politics. Most smart mobility implementation cuts across jurisdictional boundaries creating a need for political coordination and cooperation. In addition to the need to support effective cooperation, there is also a need to inform politicians on the nature of smart mobility. Many of the technologies used in smart mobility are either relatively new or emerging areas, making it essential to brief local politicians on what the technologies are, what they can do, what the expected outcomes are, and any undesirable side effects. This is essential, clear communication of the value and benefits of smart mobility to political leadership is the key to obtaining funding and support required for the implementation.

3.9.3 Demographical, Environmental, Commercial

Each city will have slightly different demographics and different mixes of land use. Environmental issues vary considerably from one city to another depending on local priorities and the current condition of the environment, for example, air quality. Transportation has also come to be seen as a critical element in economic development. This involves making a city or a region attractive for inward investment and for companies to relocate headquarters from another city. Economic development can also involve initiatives to increase the number of job opportunities in the city and transportation can help this by improving accessibility between residential areas and areas where job opportunities exist. Similarly, commercial challenges can involve appropriate access arrangements for business, commercial, and manufacturing. For example, smart factories will require a stable supply of raw materials going into the manufacturing facility and appropriate support for finished products leaving the factory.

3.9.4 Accelerating Technology Development

Since the dawn of the PC era, the challenge has always been to ensure the attainment of a reasonable return on investment in information technology before it is time to refresh. Information technology seems to be replaced on a cycle related to the emergence of new technology options, rather than the state of repair or operational capability of the prior investment. Technology is emerging at an accelerated pace which continues to accelerate. Causing some concern about the viability and stability of investments in technology. Take this book as an example, by the time the manuscript has been completed and published, some new technologies will have emerged that could not be covered in the book. This is one of the reasons to develop a complementary smart mobility body of knowledge that provides additional information to that found in the book. This includes topical information on new and emerging technologies and insight from practitioners and lessons learned and practical experience gained. The state of flux of technology is one of the reasons why is important to do a good job of benefit-cost analysis as part of the plan to invest in information technology. The benefit-cost analysis should reveal that it is possible to achieve a reasonable return on investment before it is necessary to upgrade the technology. A good benefit-cost analysis will also include arrangements for upgrading the technology on a reasonable cycle. Benefit-cost evaluation work conducted by the authors for several clients has typically included a technology refresh at the 7- or 10-year mark. This means that after 7 or 10 years of operation, there is money reserved in the financial plan to reinvest and upgrade the technology. The pace of technology change is also a good reason why it is important to be up to date on what technologies exist, the state of their maturity, what they can do, and what they can do. This will be addressed in Chapter 8 on smart mobility technologies.

A feature of the accelerating pace of technology is a shortening of the time from initial invention to commercialization. When Alexander Graham Bell invented the telephone, there was a period of about 25 years from that very first telephone call to full-scale availability of the technology on a commercial basis. These days it can take as little as 25 months as our skill and experience have improved in establishing and operating start-up companies and commercializing technologies at a rapid pace. The number of technology categories involved in smart mobility has also increased as new ways have been found to adapt and adopt technologies designed for other industries, or some technologies designed specifically for the sector. This wider range of technological options, while providing new opportunities, also presents a challenge in terms of keeping up to date on what is available. The challenge involves the need to get accurate information about technological capabilities and constraints including information and practical experiences from prior implementers.

3.9.5 Appropriate Use of Data

There has been an explosion in the availability of data. In addition to providing outcomes in terms of safety, efficiency, user experience, and environmental effects, most smart mobility implementations are also significant generators of data. Such data can be used in a feedback loop to improve the quality of planning and operations for smart mobility, or in some cases can form the basis for new enterprises that monetize the value of the data. The challenge is to understand what data is available, the appropriate use of the data, and how to effectively convert the data to information, gain insight and understanding from the information, and then define an action plan based on the new insight.

3.9.6 Harnessing the Resources and Motivation of the Private Sector

About 20 years ago, the smart mobility revolution would have been led by the public sector. The public sector invested significantly in the early days of intelligent transportation systems because of the obvious benefits to the public. While those public benefits still exist today, a wide range of private-sector enterprises have become involved. If the average smart mobility program around the globe were to be characterized, is likely to be led by a private sector company. The private sector has become acutely aware of the revenue and return opportunities that are presented by the application of technology to the transportation market. This revolution of the private sector role also represents a challenge to the public sector. Traditional procurement with the public sector providing overall direction of the project, in some cases may no longer be appropriate. The public sector may need to acquire new skills and capabilities as advocates and influencers, rather than owners and directors.

3.9.7 Finding the Right Starting Point

This may seem trivial, but the selection of the appropriate starting point for a smart mobility initiative is crucial to its success. There are many factors that should be considered when selecting the best starting point including prior investment, local priorities, the availability of technology, and the capability of local and regional service and solution providers. This subject will be addressed in detail in the chapter on planning for smart mobility.

3.9.8 Defining Agree in the Right Endpoint

This is complimentary to the prior challenge as a suitable starting point, requires a suitable endpoint. The endpoint should be defined as the achievement of all the objectives set out for the initiative and be presented as a future vision.

3.9.9 Developing a Roadmap

The journey from starting point to endpoint should also be illustrated by developing a roadmap. Challenges associated with the creation of a roadmap include the definition of appropriate milestones, ensuring that each milestone has independent stand-alone benefits, and that the roadmap is achievable with foreseeable resources from both public and private sector. It is important to show incremental progress and results along the way, as a survival strategy until full implementation is achieved. A Western analogy comes to mind here. In many Western states in the United States it is popular to have annual rodeos. An especially popular event in rodeos involves bull riding. Specially trained riders attempt to stay on a wild bull for as long as possible. Inevitably, even highly trained riders can be thrown off by the bull. Additional participants known as "rodeo clowns" are brought into play when a rider is thrown off and is endangered by the bull. The primary function of the rodeo clown is to distract the bull, to enable the rider to escape safely. This may be rather cynical, but the smart mobility program, especially if it is a long one, may need some rodeo clowns. In this case, the rodeo clowns are early winner and milestone projects that deliver value and photo opportunities along the way to final completion of the whole initiative.

3.9.10 Keeping it Going Once you Start

This might seem strange, but there seems to be more of a challenge to identify funding for operations and maintenance related to smart mobility, than finding the large sums of money required for initial implementation. Most investment curves for smart mobility initiatives have what could be called a "whale shape." The body of the whale takes the form of the capital investment expended over the implementation. The tale of the whale takes the form of the operations and maintenance funding required to keep the smart mobility program operational and deliver the services.

3.9.11 Keeping a Focus on the Objectives

In a changing world, the environment in which the smart mobility initiative is implemented can change dramatically. As discussed earlier, new technologies are emerging at an accelerated pace and there are many factors that can distract from the original objectives. While it is necessary to keep the objectives under review to ensure that they are still relevant to the situation, there is a challenge in making sure that the original objectives are front and center when taking decisions and actions to implement the program. There have been many occasions when the objectives, for one reason or another, have been obscured, or forgotten, leading to a suboptimal outcome. The challenge also extends to clear definition and agreement of objectives among all regional and city stakeholders. It may also be necessary to align objectives across different agencies responsible for transportation service delivery, or even within departments in the same agency.

3.9.12 A Mindset for Success

At first glance, this could seem like a trivial challenge; however, the authors have experienced what they refer to as the "spot the dog" syndrome. Consider the following poem:

> *This is the fable of a dog named Spot*
> *He was fast but didn't plan a lot*
> *He liked to chase things on the road*
> *Especially buses, that's his mode*
>
> *He went out chasing them every day*
> *And usually, they got away*
> *Buses are faster than our dog Spot*
> *He's very fit but he doesn't catch a lot*
>
> *But then one day when it was foggy*
> *Spot became a lucky Doggy*
> *He managed to finally catch the bus*
> *And oh, did that cause a lot of fuss*
>
> *He 'd put all his effort into the chase*
> *Didn't realize what he was to face*
> *He had no plan for after the fact*
> *He just didn't know how to react*
>
> *This has been the fable of a dog named Spot*
> *Obsessed by the chase without a thought*
> *He didn't plan for his success*
> *His achievement left him in a mess*

This is a light-hearted way to treat a serious subject. Dogs have a propensity to chase buses but are unlikely to have a plan for catching one. A focus on activities required for a project or an initiative can often cause distraction regarding the outcome. It is also important to approach large initiatives with a success-focused mindset. There may be challenges, and there may be risks, but the planning work should assume that success will be achieved.

3.9.13 Assessing Capability

In simple terms, can the project initiative being planned be delivered? Does the organization have the appropriate skills and experience to underwrite success? If there are skill gaps in the organization then is there a plan in place to fill the gaps by using external resources such as service providers and consultants? A careful assessment should be made of organizational capability as it is likely that the organization will be challenged by the different technologies and new ways of doing business associated with smart mobility. The authors have had considerable success in adapting the Capability Maturity Model to structure an organization-wide capability assessment for smart mobility.

3.10   Developing a Custom Problem Statement

An obvious starting point that is often overlooked in the development of a problem statement for smart mobility is the identification of the problem owner. The simple question "who's problem, is it?" must be asked as the owner of the problem must be involved in the development of the solution and may be a leading candidate to provide the funding or at least contribute funding to the cost of the project. It is likely that the owner of the problem will be a major beneficiary of the smart mobility initiative. Sometimes it is also necessary to take a wider view of who might benefit from the smart mobility initiative. For example, a major European city installed variable message signs at bus stops. The signs indicated how many minutes before the next bus on a particular route would arrive at a bus stop. The signs were deployed in an urban area, and in many cases, the bus stops were near retail outlets. Observed behavior after the implementation of the system indicated that if travelers were aware of the estimated arrival time for the bus, they might decide to go shopping in the local retail outlets.

Therefore, an unexpected beneficiary of the system was local retailers. This is an important insight as there could be a possibility that operations and maintenance costs will be shifted from the public agency to the retailer. By installing message signs in shop windows, the retailers can take responsibility for operations and maintenance, while also finding a new way to attract customers. This example shows that there is value in taking the time to conduct an in-depth benefits analysis not just by quantifying the benefits, but also identifying the beneficiaries.

Remembering the "spot the dog" fable, it is also important to understand the proposed use of the problem statement once it has been created. The identification of the intended use for the problem statement and the intended readership are especially important factors in developing an effective problem statement. The intended readership would also be a candidate to be involved in the development of the problem statement in terms of suggesting input and reviewing drafts. To ensure that the problem owner's and other beneficiaries of the smart mobility initiative are aware of what is being developed on their behalf and agree that the problem defined, is worthy of a solution.

To conclude this chapter, it is appropriate to discuss the application of technology to assist in the definition and communication of problem statements. In the COVID-19 era, there has been a dramatic move to the use of social media and other touchless technology approaches for public consultation and receiving input on proposals and designs. A new degree of interactivity is possible between those planning and designing smart mobility and those on the receiving end. The challenge in a higher level of interactivity is an increased burden on the subject matter experts who are planning, designing, and advocating the smart mobility solution. High-frequency interaction requires spontaneous responses to unpredictable questions, this in turn requires a high level of expertise and experience. It may also be necessary to employ special-purpose tools to enable the proposals to be visualized and communicated effectively to the stakeholders. As discussed, when talking about political challenges, it is particularly important to be clear on the nature of the technologies and what they can do. It is often difficult to solicit meaningful input from stakeholders if they do not understand the context. If a fence is not established around the sandbox, then inputs can go wildly astray.

3.11   Summary

In this chapter, several aspects of the problems that can be addressed by smart mobility initiatives, have been explored. Beginning with the definition of a problem statement, a description of the value of a good problem statement has been explored. The value of smart mobility in offering the capability of balancing the attainment of benefits associated with transportation, while avoiding, or managing, undesirable side effects, has been explained. A discussion was offered on the benefits of transportation and potential undesirable side effects. Smart mobility challenges and opportunities were addressed, and a list of challenges associated with smart mobility was provided. The chapter concluded with some advice and information on how to create specific problem statements for a particular city or region.

References

1 Maldonado Silveira Alonso Munhoz, P.A.; da Costa Dias, F.; Kowal Chinelli, C.; Azevedo Guedes, A.L.; dos Santos, J.A.N.; da Silveira e Silva, W.; Soares, C.A.P. Smart mobility: the main drivers for increasing the intelligence of urban mobility. *Sustainability* 2020, 12(24), 10675. Published: 21 December 2020.

2 Orlowski, A.; Romanowska, P. Smart cities concept: smart mobility indicator. *Cybern. Syst.* 2019, 50, 118–131.

3 Zawieska, J.; Pieriegud, J. Smart city as a tool for sustainable mobility and transport decarbonisation. *Transp. Policy* 2018, 63, 39–50.

4 Badii, C.; Bellini, P.; Cenni, D.; DiFino, A.; Nesi, P.; Paolucci, M. Analysis and assessment of a knowledge based smart city architecture providing service APIs. *Futur. Gener. Comput. Syst.* 2017, 75, 14–29.

5 Docherty, I.; Marsden, G.; Anable, J. The governance of smart mobility. *Transp. Res. Part A Policy Pr.* 2018, 115, 114–125.

6 Kim, J.; Moon, Y.-J.; Suh, I.-S. Smart mobility strategy in Korea on sustainability, safety and efficiency toward 2025. *IEEE Intell. Transp. Syst. Mag.* 2015, 7, 58–67.

7 Salvia, M.; Cornacchia, C.; Di Renzo, G.C.; Braccio, G.; Annunziato, M.; Colangelo, A.; Orifici, L.; Lapenna, V. Promoting smartness among local areas in a southern Italian region: the smart Basilicata project. *Indoor Built Environ.* 2016, 25, 1024–1038.

8 Stolfi, D.H.; Alba, E. Red swarm: reducing travel times in smart cities by using bio-inspired algorithms. *Appl. Soft Comput.* 2014, 24, 181–195

9 Maier, E. Smart mobility – encouraging sustainable mobility behaviour by designing and implementing policies with citizen involvement. *JeDEM* 2012, 4, 115–141.

10 Dell'Era, C.; Altuna, N.; Verganti, R. Designing radical innovations of meanings for society: envisioning new scenarios for smart mobility. *Creat. Innov. Manag.* 2018, 27, 387–400.

4

The Value and Benefits of Smart Mobility

4.1 Informational Objectives

After reading this chapter, you should be able to:

- Explain the difference between features, benefits, and values
- Define the categories of benefits that can be delivered by smart mobility
- Explain the challenges associated with the estimation of benefits and cost for smart mobility applications
- Describe the opportunities associated with the estimation of benefits and costs for smart mobility applications
- Explain a suitable approach philosophy that addresses the challenges and embraces the opportunities
- Describe some examples of the estimation of benefits and costs for smart mobility applications
- Describe the value and limitations of the approach
- Understand how to apply the estimation framework to specific smart mobility applications

4.2 Introduction

Estimation of benefits and costs for smart mobility is a difficult area. It is fraught with difficulties and challenges regarding availability of data and robust approaches to "before" and "after" effects assessment. Despite this, it is well worthwhile to address this complex subject in a straightforward and robust manner that supports a pragmatic approach to this difficult subject. The approach that has been defined is simple to understand and apply, requires little resources, and provides a sound basis for high-level decision-making when prioritizing smart mobility applications and investments. The suggested approach is not rigorous from an economics point of view and has some obvious deficiencies and weaknesses. However, it does serve to address a gap that the authors have perceived in the ability to develop practical estimates of benefits and costs for smart mobility applications. It should be noted that there have been, and continue to be, considerable efforts to study and develop tools for benefit-cost analysis for traditional transportation needs [1]. The gap that is addressed is in the availability of strategic-level sketch planning tools to support the prioritization of deployment sequences for smart mobility. Many of the current and previous efforts focus on detailed estimation of specific initiatives.

Smart Mobility: Using Technology to Improve Transportation in Smart Cities, First Edition. Bob McQueen, Ammar Safi, and Shafia Alkheyaili.
© 2024 The Institute of Electrical and Electronics Engineers, Inc. Published 2024 by John Wiley & Sons, Inc.

4.3 Features, Benefits, and Value

Brian Felker summed this up beautifully in his 2017 article [2] in which he differentiated between the features of a solution, the benefits delivered by the feature, and the value perceived by the solution user. Note that the user can be the end user or traveler or could be the user in the middle of the transportation service delivery process such as an operator or service provider.

4.3.1 Solution Features

Solution features are aspects of the solution, whether it be a product or a service, which are designed to deliver benefits. For example, one feature of a modern automobile could be parking assistance. The features are the ability of the vehicle to park itself and the value provided to the user is an enhanced user experience, improved safety, and increased efficiency. The enhanced user experience is generated by a reduction in stress by removing the need for the driver to park the vehicle. Improved safety results from a reduction in minor collisions caused by driver error and increased efficiency through a reduction in the time required to park the vehicle.

4.3.2 Benefits

The benefits that are delivered by features can span a wide range. The full range of benefits will be explained and defined in Chapter 11 on performance management for smart mobility. The complete range of performance measures or key performance indicators (KPIs) can be rolled up or summarized under the major benefits categories summarized in Figure 4.1.

4.3.2.1 Improved Safety
This involves smart mobility features that reduce crashes and the consequential loss of life, injuries, and property damage.

4.3.2.2 Increased Efficiency
Efficiency means saving time and saving money. This can take the form of reduced travel times, or just as important, more reliable travel times.

Major benefit category	
1	Improved safety
2	Increased efficiency
3	Enhanced user experience
4	Equity effects
5	Climate effects
6	Economic development effects

Figure 4.1 Smart mobility benefit categories.

4.3.2.3 Enhanced User Experience
This could be described as customer service; however, the term used in the information technology industry is user experience. This seems more appropriate as some of the services delivered as part of smart mobility will be used by paying customers, while others will be used on a free basis by users. The term user experience is also a reminder that users will have an experience whether it is managed or not and that one of the primary factors in success is to ensure a positive user experience.

4.3.2.4 Equity Effects
Recently there has been a growing concern that the delivery of smart mobility services should be conducted in a fair and equitable manner. This requires that the demographics of

the target city or region be fully understood and fully accounted for in the design of smart mobility and the features provided. For example, consider traffic signal timings. These include an allocation of green time to enable pedestrians to safely cross the road, as well as ensure the efficient movement of traffic from the different arms of the intersection. Classical approaches to this involve an assumption regarding the average speed at which pedestrians are able to cross the road. Like many aspects of smart mobility, averages do not serve the purpose adequately and do a poor job of capturing the variation crossing speed capabilities across the demographic range. It is highly unlikely that a senior citizen will be able to walk across the road at the same speed as an Olympic athlete. Of course, this is a ridiculous comparison, but it makes a point about the variability of walking speed capability across a typical demographic in the city. The use of advanced sensors and signal timing algorithms will enable the measurement of actual road crossing speed, detect pedestrian characteristics, and set the scene for close customization of pedestrian green time to the characteristics of the crossing users.

4.3.2.5 Climate Effects
Smart mobility solutions have the ability to enable the delivery of the desired effects of increased mobility and accessibility while minimizing and managing the undesirable side effects. Climate effects fall into this category. Nonrenewable energy consumption and emissions from the transportation process can have a significant effect on the climate, requiring that these be carefully managed. Smart mobility can deliver significant benefits in terms of managing climate effects.

4.3.2.6 Economic Development Effects
Mobility and accessibility in the city can have a direct influence on economic activity. The ability to connect people with job opportunities and education can have significant economic development effects. These affect the attractiveness of the city for businesses and industries to locate their activities within the city. Ease of mobility and accessibility across the city also have an impact on the abilities for business and industry to attract the right staff.

Note that the various benefits categories are not mutually exclusive, and it is typically the case that the full range of objectives will be addressed by a single smart mobility deployment. In fact, one of the big areas of promise for smart mobility is the flexibility and customization to enable a range of sometimes conflicting benefits to be delivered simultaneously. Smart mobility is not so much a "trade-off" situation as a "trade-in."

4.3.3 Value

It is important to provide a clear definition of this term. There is often confusion between the two terms – value and benefits. Benefits are measurable, although there are challenges associated with the measurement. Benefits involve the quantification of the effect of outcomes of the smart mobility solution. However, there are many times when the quantified benefits do not come into alignment with user perception. This is the key to the difference between value and benefits. Benefits are measured, either directly or indirectly, while value is perceived. One of the big challenges in determining the value the user will place on a specific feature is the ability to gather data. Typically, data collection has been undertaken with small sample sizes and inherent bias. This is one of the big potentially promising areas for smart mobility – the use of advanced sensors and communication techniques to measure perception of value at a larger scale and at a more sophisticated level. An extremely interesting facet of smart mobility is that most of the applications are capable of achieving their intended purposes, while also providing a rich stream of data that can improve the quality of service delivery and raise the standard of customization. This also defines an enigma related to smart mobility. If the sophisticated data capture facilities inherent in the smart mobility application were available before it was implemented, then the implementation be better targeted to the end users and the intended demographic. This indicates that when applying smart mobility solutions, flexibility should be available not just to scale to a larger number of end users, but also to take full account of the new data provided by the implementation. A smart mobility solution should be considered as a long-term operational management challenge and not a one-off project.

4.4 The Challenges

There are many challenges associated with determining costs and benefits associated with smart mobility. Seven of the more significant challenges have been identified and summarized in Figure 4.2 and are described in the following sections.

4.4.1 Lack of Consistent Before and After Data

The traditional approach would involve the definition of "before" and "after" conditions, enabling the difference between them to be used as a measure of benefit accrued. This would require a clear definition of the "before" and "after" conditions, the baseline from which benefits can be determined. In many cases, this baseline does not exist as a before study was not conducted. In other cases, the standards to which the data is collected are not consistent making it difficult to compare data from different projects.

4.4.2 Resource Intensive Data Collection Needs

The traditional before-and-after approach can also consume a great deal of resources and is prone to inaccuracy as result of small sample sizes and the limited duration of service. It is necessary to achieve a balance between the cost of evaluation and the cost of planning, design, and implementation of the smart mobility solution.

Major challenges
1 Lack of consistent before and after data
2 Resource intensive data collection needs
3 Evaluation can take a long time
4 Isolation of effects
5 Accounting for residual value
6 Defining a complete evaluation
7 The need for detailed design as the basis for accurate cost estimates

Figure 4.2 Smart mobility benefit-cost evaluation challenges.

4.4.3 Evaluation Can Take a Long Time

The data collection and analysis work required before and after the implementation of a smart mobility project can be extensive. To maintain momentum and move as quickly as possible to the implementation, it is necessary to carefully manage the time resources required for evaluation.

4.4.4 Isolation of Effects

Another challenge with respect to the definition of benefits from smart mobility is the isolation of effects to be attributed to the specific smart mobility application being studied. It is unusual for a smart mobility application to be implemented as a single standalone initiative. There are also many external factors that affect value and benefits including the economy and other local factors. For example, the introduction of travel information or a Mobility as a Service (MaaS) system could have an effect on other applications such as advanced traffic management.

4.4.5 Accounting for Residual Value

One of the elements of value associated for smart mobility and any other major project for that matter, lies in the residual value of the implementation. It can be exceedingly difficult to put an accurate estimate on residual value for infrastructure, and even more difficult for services. If the services are developed and delivered on a pay-as-you-go basis, then there would be no residual value.

4.4.6 Defining a Complete Evaluation

A thorough evaluation that addresses all aspects of costs and benefits associated with smart mobility requires extensive knowledge and understanding of the solutions to be delivered. It also requires an awareness of the interaction between the implementation and other investments. Interaction between implementations must also be considered as well as external factors that influence benefits and costs such as economic conditions.

4.4.7 The Need for Detailed Design as the Basis for Accurate Cost Estimates

Accurate cost estimates for smart mobility implementations can only be derived as a result of a detailed design. At the planning stages in implementation, a detailed design may not be available, influencing cost estimate accuracy. While high-level designs should be available, it should be borne in mind that these may not support the level of accuracy desired for the cost estimates.

Many of these challenges will be addressed as the use of big data and analytics techniques emerges. The use of big data and analytics can be significantly enabled through the construction of a central data repository. One of the big advantages of constructing a central and accessible database of traffic and transportation data is that this represents a historical record that can be used to determine before and after. Another approach to values and benefits is the creation of a mathematical simulation model that provides a representational digital twin of the physical situation and enables what-if scenarios to be investigated. These are typically derived from small sample sizes of the actual conditions to be simulated. That are also inherent inaccuracies in mathematical simulation as algorithms and expressions are defined and developed to represent real-life conditions. In many cases, these are approximations. Here again, the availability of big data and analytics techniques will improve the accuracy of simulation techniques over time. However, the current situation is that mathematical techniques can provide an estimate of current conditions and can be used to predict future conditions. It is also apparent that the current situation with respect to before and after studies is a little unstructured.

There are few common standards or guidelines for the collection of before and after data, making it difficult to compare one project to another. The European Union addressed this in the EasyWay project [3]. This project concluded that there is a need to provide formal structure and standards for performance evaluation on smart mobility implementations, particularly post-project where there seems to be less activity and less coherent data available. The project resulted in the development of a handbook of best practices that provides a working example of evaluation approach advocated for projects within the program, particularly those receiving financial support from the European Commission.

It is difficult to define and agree a range of key performance indicators (KPI) to measure performance. Here again, that is no common agreement on the key performance measures to be used for smart mobility implementations and typically the performance indicators are project or implementation specific. There is an interesting chicken and egg situation here. The implementation of smart mobility applications will in fact provide sources of data that can be used for future evaluation. It becomes easier to evaluate smart mobility after it has been implemented. This is an important factor in the design of smart mobility implementations and full advantage should be taken of this capability, to ensure that effective evaluations can be conducted. Considering the challenges described above, it is essential to attempt to define the values and benefits of smart mobility, even if this requires a higher level of abstraction. The EasyWay handbook notes that there are three important criteria for an evaluation approach:

1) Evaluation results should be transparent
2) The approach results should be easily understandable
3) The evaluation results can be compared easily with other results

4.5 The Opportunities

There are a wide range of opportunities associated with smart mobility implementation. A sample of five of the more significant opportunities has been defined and summarized in Figure 4.3, then described in the later sections.

Opportunities
1 Scientific approach to investment programs
2 Rational prioritization
3 Informing political input
4 Building the value proposition for smart mobility
5 Effective communications to stakeholders and the public

Figure 4.3 Smart mobility benefit-cost evaluation opportunities.

4.5.1 Scientific Approach to Investment Programs

In spite of all the challenges, there is an opportunity to take a scientific approach to investment programs for smart mobility. This involves the use of data and not opinion and takes advantage of data availability and data science, data management capabilities to conduct sophisticated analyses.

4.5.2 Rational Prioritization

A scientific approach can also feature a rational prioritization of resources. Even a high-level evaluation of the relative benefits and costs of different smart mobility applications can support rational prioritization that addresses a range of objectives, including informing political input, sharing the value proposition for smart mobility, and effective communications to stakeholders and the public.

4.5.3 Informing Political Decision-making

Benefit-cost analysis can provide scientifically documented facts that can be of considerable value in informing political decision-making. In fact, in the absence of effective benefit-cost analysis, the opportunity arises for bias and subjective opinion to creep into the decision-making process.

4.5.4 Building the Value Proposition for Smart Mobility

Mobility has to compete with many other priorities within cities including education, healthcare, water, electricity, and many other infrastructures that require investment. A well-funded benefit-cost analysis will help mobility to compete against each other priorities.

4.5.5 Effective Communications to Stakeholders and the Public

In addition to providing input to political decision-making, benefit-cost analysis can provide clear and concise information to the full spectrum of mobility stakeholders and the end users.

4.6 Approach Philosophy

Given the opportunities and challenges described above, a particular approach philosophy has been identified. This involves the use of published data and building relationships between this data and the parameters required to conduct a benefit-cost analysis. The use of published data rather than data collected from specific sources or specifically for the analysis makes it straightforward to justify the data which forms the foundation for the analysis. This also addresses the challenges of resources and time required to conduct data collection as specific data collection for the analysis is avoided. This does however introduce a degree of innacurracy into the analysis, to address this some major parameters are treated as variables and again be modified by the person conducting the analysis, or by an end client applying judgment and expertise that accounts for local situations. The approach philosophy also does not rely on the availability of a detailed design for the systems.

The cost estimates required for benefit-cost analysis are derived from higher level conceptual designs, or from specific pricing available from vendors. Here again, the cost estimates are treated as variables that can be adjusted by the person conducting the analysis, or by an end client being advised by a consultant. The overall approach can be described as a sketch planning approach to benefit-cost analysis as it is inherently high-level and may not provide the detailed benefit-cost analysis required for fine-tuning investment programs or prioritization of investments. However, the approaches have been defined to be extremely valuable in enabling clients to make broad choices between different investment priorities and also to communicate the benefits and relative cost of the proposed implementations. The technique has been successfully applied with a number of clients on different smart mobility applications. Such case studies are described in the following sections.

4.7 Smart Mobility Benefit-cost Analysis Case Studies

To illustrate the approach philosophy defined and described above, four case studies or example applications of the approach philosophy have been identified and described in the following sections. These applications are based on work for actual clients over the past 12 months. Each example focuses on a specific smart mobility application, therefore there is no intention of providing a comprehensive assessment of a portfolio of smart mobility benefits across the board. Instead, there is a narrow focus on the benefits that can be achieved by each application. The four applications are:

1) Roadside infrastructure sensors
2) Transit data analysis
3) Truck tire defects monitoring
4) Advanced traffic management

For each application, the following information is provided – the background to the work, a description of the specific smart mobility application, a summary of the expected benefits, and an overview of the parameters used in the benefit-cost calculations.

4.7.1 Case Study 1 – Roadside Infrastructure-based Sensors, Lifecycle Cost Comparison

4.7.1.1 Background
The client for this project was an innovative sensor solution provider with an approach that involves the use of existing fiber optics along highways as traffic speed sensors. A specially developed interrogator is attached to an existing dark fiber and specialized acoustic techniques are used to convert the fiber into a traffic speed sensor which collects traffic data every 50 m along the highway. The client was interested in developing a tool that would enable simple lifecycle cost comparison of this distributed fiber-optic sensing approach with current alternatives for traffic speed monitoring including video image processing, microwave vehicle detection, and inductive loops. Note that this first analysis does not address the benefits as these are assumed to be the same for each sensor type. The parameters used for the project are summarized in Figure 4.4.

The workflow for the calculations is shown in Figure 4.5.

The dashboard that was developed to communicate the relative cost analysis is illustrated in Figures 4.6–4.11, a total of six screens for the dashboard, one for each type of speed sensor, and a summary of lifecycle costs per year per mile.

This first screen for video image processing sensors allows data to be input about annual maintenance and operating cost per device, the number of devices per mile, the design life for the device, and the capital cost per device. With input from the client, the parameters were set at

No.	Parameter
1.1	Video image processing sensor annual maintenance and operating cost per device (dollars)
1.2	Video image processing sensor number of devices per mile
1.3	Video image processing sensor cost per mile
1.4	Video image processing sensor design life
1.5	Video image processing sensor capital cost per device
1.6	Video image processing annual cost
1.7	Video image processing annual cost per mile
2.1	Microwave vehicle detection maintenance and operating costs
2.2	Microwave vehicle detection sensor number of devices per mile
2.3	Microwave vehicle detection cost per mile
2.4	Microwave vehicle detection sensor design life
2.5	Microwave vehicle detection sensor capital cost per device
2.6	Microwave vehicle detection annual cost
2.7	Microwave vehicle detection annual cost per mile
3.1	Inductive loops sensor maintenance and operating costs
3.2	Inductive loops sensor number of devices per mile
3.3	Inductive loops sensor cost per mile
3.4	Inductive loops sensor design life
3.5	Inductive loops sensor capital cost per device
3.6	Inductive loops sensor annual costs
3.7	Inductive loops sensor annual cost per mile
4.1	Distributed fiber-optic sensor annual maintenance and operating cost per device
4.2	Distributed fiber-optic sensor number of devices per mile
4.3	Distributed fiber-optic sensor cost per mile
4.4	Distributed fiber-optic sensor design life
4.5	Distributed fiber-optic sensor capital cost per device
4.6	Distributed fiber-optic sensor annual costs
4.7	Distributed fiber-optic sensor annual cost per mile

Figure 4.4   Case study 1 – Roadside infrastructure-based sensors, lifecycle cost comparison parameters.

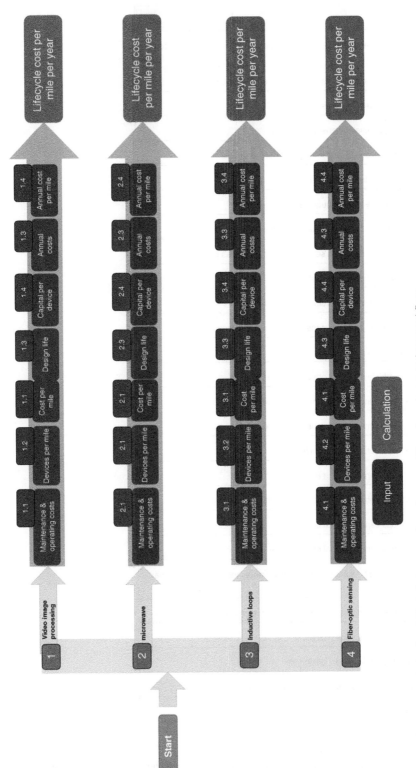

Figure 4.5 Case study 1 – Roadside infrastructure-based sensors, lifecycle cost comparison workflow.

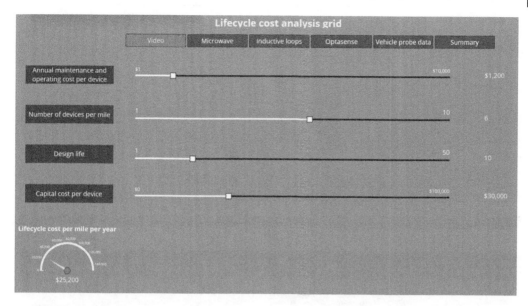

Figure 4.6 Case study 1 – Roadside infrastructure-based sensors, lifecycle cost comparison dashboard: video sensors.

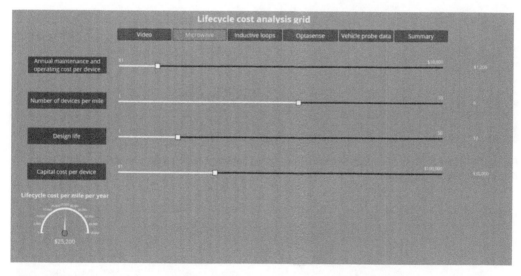

Figure 4.7 Case study 1 – Roadside infrastructure-based sensors, lifecycle cost comparison dashboard: microwave sensors.

$1200 per device, six devices per mile, design life of 10 years, and capital cost per device of $30,000, respectively, yielding a lifecycle cost per mile per year of $25,200.

This second screen for microwave sensors shows the client agreed values for each parameter at $1200 per device, six devices per mile, design life of five years, and capital cost per device of $30,000, respectively, yielding a lifecycle cost per mile per year of $25,200.

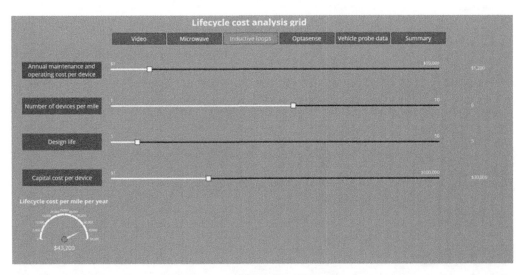

Figure 4.8 Case study 1 – Roadside infrastructure-based sensors, lifecycle cost comparison inductive loops.

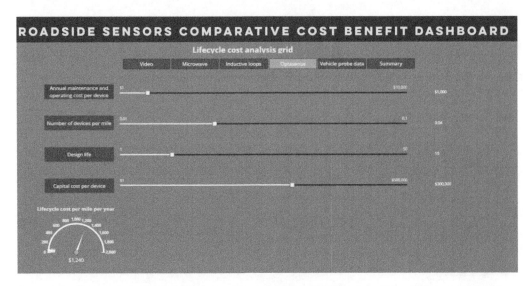

Figure 4.9 Case study 1 – Roadside infrastructure-based sensors, lifecycle cost comparison dashboard: distributed fiber optic sensors.

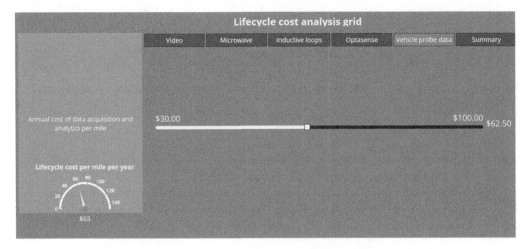

Figure 4.10 Case study 1 – Roadside infrastructure-based sensors, lifecycle cost comparison dashboard: probe vehicle data.

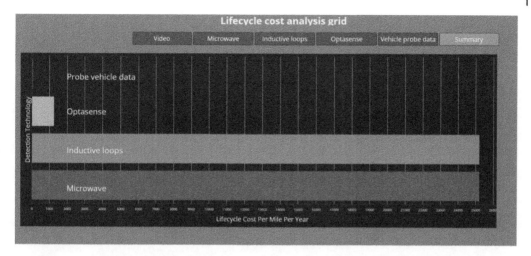

Figure 4.11 Case study 1 – Roadside infrastructure-based sensors, lifecycle cost comparison summary dashboard.

This third screen for inductive loop sensors shows the client-agreed values for each parameter at $1200 per device, six devices per mile, design life of five years, and capital cost per device of $30,000, respectively, yielding a lifecycle cost per mile per year of $43,200.

This fourth screen for the OptaSense distributed fiber-optic sensing technique shows the client agreed values for each parameter at $1000 per device, 0.04 devices per mile, design life of 20 years, and capital cost per device of $300,000, respectively, yielding a lifecycle cost per mile per year of $640. This assumes that a single OptaSense interrogator unit can address 25 miles of highway.

For vehicle probe data, the approach taken to determine the average lifecycle cost per mile per year, is different from the other sensor options. Probe vehicle data is acquired on an annual contract from a private source and does not require roadside devices. Therefore, this fifth screen for vehicle probe data shows the client agreed lifecycle cost per mile per year of $62. This was determined by dividing the annual vehicle probe data procurement total by the number of miles covered.

The sixth, and final screen, shows a summary of lifecycle costs per mile for each of the sensor technologies. It indicates that probe vehicle data is the most cost-efficient approach approximately $63 per mile per year, followed by OptaSense at $640 per mile per year, then microwave at $25,200 per mile per year, and finally inductive loops at $43,200 per mile per year.

4.7.2 Case Study 2 – Truck Tire Defects Monitoring Benefit and Cost Analysis

The development of a benefit-cost analysis for truck tire defects monitoring was conducted for a major international supplier of weigh-in motion and truck tire condition evaluation solutions.

The purpose of the work was the creation of a tool for staff within the company to use to gain a greater understanding of the benefits of their truck enforcement and tire condition monitoring system. Once internal staff were comfortable with the assumptions and characteristics of the lifecycle cost analysis model, it would then be used as a tool to support a dialogue with selected clients. The purpose of the dialogue would be to shift the client conversation from a cost-focused one, to a wider dialogue that also included the potential benefits of the application. The company recognized that the ability to have an informed discussion with the client on the benefits of the solution

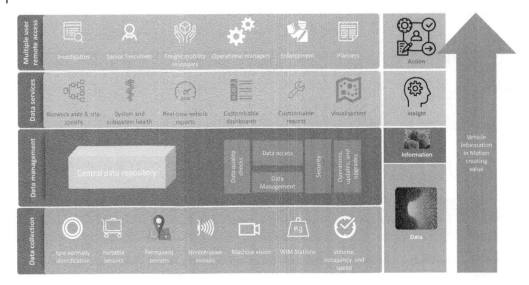

Figure 4.12 Case study 2 – Truck tire defects monitoring benefit and cost analysis conceptual design.

would significantly reinforce the ability to sell it. It was agreed with the client that published data should form the basis for the calculations and that the resulting sketch planning tool could also form the basis for further development work involving external consultants and academia. There was also a strong inclination to engage clients in the development of plans for further enhancement to the model, enabling client ownership in the work as it progressed.

The conceptual design for the application is summarized in Figure 4.12.

Data is received from roadside, infrastructure-based sensors that conduct multiple measurements of heavy trucks. An array of different sensors is used to collect the following data:

- Number of vehicles passing
- Vehicle speed
- Vehicle classification by weight and number of axles
- Vehicle weight
- Tire anomaly data

The data collected is combined with commercial vehicle operation vehicle-based data and operational data from individual weigh stations, in a central repository. Data management processes are then applied. This involves the resulting data set being subject to analytics and reporting that enable positive action to be taken. The data is collected in a nonintrusive manner that does not require the vehicle to be stopped. This enables a large sample size of vehicles to be subject to measurement, enabling the identification of vehicles of interest for further detailed manual inspections. The primary objective of the application is to preserve the safety of the road network by ensuring that heavy trucks are operated in a safe manner and are well maintained.

The outcomes have been categorized under three major headings – safety, efficiency, and user experience. Safety relates to providing insights into trucks exceeding their weight limits, as well as to the ability to reduce the number of defective truck tires on the road using advanced technology to automatically screen for tire anomalies. This would have a direct effect on the number of lives lost and injuries caused by truck tire defects. The second category, efficiency, relates to the fact that badly maintained tires result in an increase in truck operating cost. This includes driving the truck with defective flat tires or severely underinflated tires.

The third category, user experience, focuses on improving the day-to-day experience of both the truck driver and fleet operator, in other words, the end users of the application. One of the main benefits of this category is the avoidance of out-of-service events because of roadside inspection. The advanced technology approach to truck enforcement means that many more vehicles will be screened and subsequently subject to roadside inspections in case of violations. Providing information through reports, dashboards, and analytics on the results of screening for a multitude of safety and compliance criteria increases the transparency of the operations and shows the benefits of the investment in the technology to increase safety. Reducing the number and duration of out-of-service events resulting from these inspections significantly improves the user experience. The user experience also includes the data analyst who has the job of converting the data into information, conducting analytics, and generating reports.

Having defined the outcomes, the next question is, how "how do we communicate the value of these outcomes in a clear and simple manner." A model was created using published data, making a few conservative assumptions, the model provides estimates of the value of safety, efficiency, and user experience that can be delivered using an end-to-end solution including roadside infrastructure and back office.

4.7.2.1 Parameters Used

The lifecycle cost analysis model compares lifecycle costs including capital and operating, with several benefits categories including safety, efficiency, user experience, and analysis benefits as illustrated in Figure 4.13.

Capital and operating cost estimated adopted are summarized in Figure 4.14.

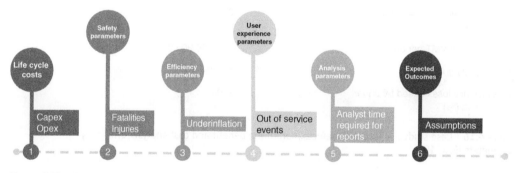

Figure 4.13 Case study 2 – Truck tire defects monitoring benefit and cost analysis cost and benefit categories.

	Individual site			
		Weight in motion	Tire condition	Total
1	CAPEX	$3,00000	$1,000,000	$4,000,000
2	Design life	10	10	
3	Capital life cycle	$300,000	$100,000	$400,000
4	OPEX	$30,000	$100,000	$130,000
5	Total life-cycle cost per station	$330,000	$200,000	$500,000

Figure 4.14 Case study 2 – Truck tire defects monitoring benefit and cost analysis, capital, and operating cost estimates.

The lifecycle cost analysis model uses a range of parameters from publicly available data under each category summarized in Figures 4.15–4.18.

Data	Value
1 Annual volume of trucks passing the location	100,000
2 Truck fatalities per year	3903
3 Truck injuries per year	111,000
4 Total US large truck fleet	12,229,216
5 Fatalities per year involving tire defects	738
6 Cost per fatality ($)	$1,400,000
7 Cost per injury ($)	$1,000,000

Figure 4.15 Case study 2 – Truck tire defects monitoring benefit and cost analysis safety parameters [4–7].

Data	Value
1 Annual volume of trucks passing the location	100,000
2 Truck operating cost per mile	$1.65
3 Percentage with tire safety defects	39%
4 Total truck VMT per annum	305,000,000,000
5 Total US truck fleet	12,229,216
6 Efficiency loss caused by under inflation (%)	20%

Figure 4.16 Case study 2 – Truck tire defects monitoring benefit and cost analysis, efficiency parameters [5, 8–10].

Data	Value
1 Annual volume of trucks passing the location	100,000
2 Annual roadside inspections	2,700,000
3 Out of service rate	20%
4 Average duration of out of service event (hours)	4
5 Total US truck fleet	12,229,216

Figure 4.17 Case study 2 – Truck tire defects monitoring benefit and cost analysis, user experience parameters [5, 9, 11–13].

	Data	
1	Trucks passing	100,000
2	Minutes per truck for analysis	5
3	Analyst hourly cost ($)	$25
	Calculations	
4	Hours per year	8333.33
5	Cost per of analysis per year	$200,321

Figure 4.18 Case study 2 – Truck tire defects monitoring benefit and cost analysis, data analysis parameters [14, 15].

4.7.3 Expected Outcomes

In addition to the parameter data for safety, efficiency, user experience, and analysis time, some assumptions about expected outcomes are required. It is an early day in the development of the application, therefore, before and after data is limited. However, assumptions can be made regarding the value of the expected outcomes. As a starting point, some extremely conservative outcome assumptions were determined and are summarized in Figure 4.19.

The approach was to make conservative assumptions and then engage the client and subject matter experts in a dialogue to ratify and validate these assumptions. Therefore, the assumptions should be treated as a starting point in an ongoing dialogue.

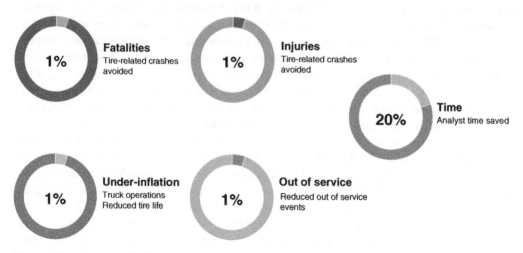

Figure 4.19 Case study 2 – Truck tire defects monitoring benefit and cost analysis, expected outcomes.

4.7.4 Benefit Cost Dashboard

The lifecycle cost analysis model then takes the data parameters and expected outcomes summarized above and creates a benefit-cost evaluation dashboard as illustrated in Figure 4.20.

Based on the parameters and assumptions described above, with an average truck volume of 100,000 vehicles per year passing the installation, the lifecycle cost analysis model calculates a benefit-to-cost ratio of 7.6. Of course, this value and the assumptions on which it is based will be discussed with the client as part of the strategy to engage the client in a positive dialogue regarding benefits as well as costs. The intended purpose of the tool is to support effective dialogue and conversation on how to realistically estimate the value that can be achieved from an end-to-end solution tire condition monitoring system.

4.8 Case Study 3 – Advanced Traffic Management

This is an example of benefit and cost evaluation for one of the most important applications of smart mobility technology. Advanced traffic management involves the application of sensors, telecommunications, and control technologies to manage the entire hierarchy of road networks within an urban area. Advanced traffic management is also used for rural areas and for intercity traffic management. This book will focus on its application for urban areas. Advanced traffic management can address the following types of roads.

1) Urban surface streets
2) Urban arterial roads
3) Limited access highways including freeways, motorways, and express lanes

For urban surface streets traffic signal control technology is employed to manage traffic flows on a corridor, or network-wide coordinated basis. Traffic management involves the coordination of traffic flows to minimize conflict between traffic streams. This typically occurs at intersections, although vehicle-to-vehicle and vehicle-to-pedestrian conflicts can manifest themselves in the area of roadway between intersections. This type of interference is often referred to as "side friction" and results in the need to calibrate any mathematical simulation model being used to simulate traffic flow. There are two primary ways in which conflicting traffic streams can be managed: spatial and temporal separation. Temporal separation involves the application of traffic control techniques to enable conflicting traffic flows to share the same road space, on a timesharing basis.

Spatial separation involves the construction of grade-separated facilities incorporating asphalt, concrete, and steel to separate conflicting traffic flows in space. This latter approach is expensive, requiring that high traffic volumes are present to justify the level of investment. Urban traffic intersections employee time separation using traffic signal control technology. Time separation is used for both vehicle-to-vehicle conflicts and vehicle-to-pedestrian conflicts. Traffic signal aspects or lights are mounted on poles and displayed to the drivers of vehicles approaching the intersection from different directions. While there are variations around the globe, the most common traffic signal designs make use of three primary traffic signal colors: green, amber, and red. Green means go if the way is clear, amber means caution and red means stop. The three lights or aspects are mounted in a vertical stack, or sometimes in a horizontal row. The elimination of the lakes is controlled by a traffic signal control located at one of the corners of the intersection. In some cases, the traffic signal controllers can be connected using wireline or wireless communications back to a traffic or transportation management center. There are many variations and design permutations possible for traffic signal control systems. The intention here is to provide an overview sufficient

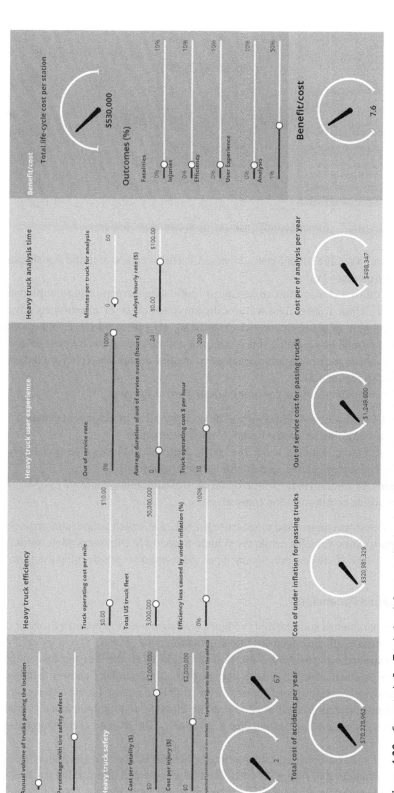

Figure 4.20 Case study 2 – Truck tire defects monitoring benefit and cost analysis, dashboard.

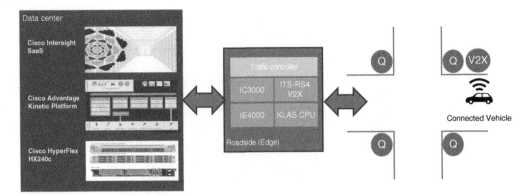

Figure 4.21 Case study 3 – Advanced traffic management, conceptual design.

for the understanding of the benefit-cost analysis. A useful overview of traffic signals can be found in the following reference [16].

The example utilized comes from an advanced traffic management system implementation in Las Vegas Nevada, United States. This features the implementation of an extremely advanced traffic signal control system along the corridor consisting of approximately 25 intersections. Each intersection is controlled by a traffic signal controller, which in turn is interconnected and connected to an advanced transportation management control center. The conceptual design for the system is shown in Figure 4.21.

The essential components of the conceptual design are as follows:

- A centralized data center contained within the transportation management center.
- A traffic signal controller situated adjacent to each intersection.
- Additional equipment installed inside the traffic controller includes high-speed wireless communications, connected vehicle, and LIDAR sensing equipment.
- At the roadside, in addition to the traffic signal control equipment, there are LIDAR sensors, and vehicle-to-roadside communication transceivers.

The LIDAR detectors can be used to measure the speed at which pedestrians cross the road, in addition to approaching vehicles, while the vehicle-to-roadside communication transceivers enable a two-way dialogue between suitably equipped vehicles and roadside infrastructure.

4.8.1 Evaluation Parameters

The evaluation parameters summarized in Figure 4.22 were used to develop a lifecycle cost analysis for the advanced traffic management implementation:

A series of expected outcomes were also provided as input to the lifecycle cost analysis. These were based on practical experience with very conservative numbers chosen to prove that the benefit-cost analysis was significantly positive even with low assumed values. These expected outcome assumptions are represented as sliders in the lifecycle cost analysis model, enabling a full discussion and debate regarding the outcome assumptions with the end client. Outcomes are summarized in Figure 4.23.

The data flow for the process used to determine the benefits for the next-generation traffic management system is illustrated in Figure 4.24. The data flow is divided into parallel streams numbered from 3 to 10 to facilitate tracking of the parameters to the methodology flow. Streams 3 relate to

No.	Parameter	Notes
1	Total population	This is the value for the analysis and was chosen as a published figure with little scope for disagreement
2	Vehicle miles traveled per person per year	
3.1	Number of fatal crashes per head of population	
4.1	Number of injury crashes per head of population	
5.1	Emissions cost per vehicle mile traveled (car)	
6.1	Emissions cost per vehicle mile traveled (truck)	
7.1	Operating cost per vehicle mile traveled (car)	
8.1	Operating cost per vehicle mile traveled (truck)	
9.1	Proportion of travel time expended on business travel purposes	
10.1	Proportion of travel time expended on personal travel purposes	
11	Proportion of vehicle miles traveled attributed to cars	
12	Proportion of vehicle miles traveled attributed to trucks	
13	The value of business time	
14	The value of personal time	
3.2	Assumed societal cost of a fatal road crash	
4.2	Assumed societal cost of an injury road crash	

Figure 4.22 Case study 3 – Advanced traffic management, parameters.

safety benefits, 5 through 8 relate to efficiency benefits and streams 9 and 10 relate to user experience benefits.

The dashboard that supports the benefit-cost evaluation and exploration is shown in Figure 4.25.

With the assumptions shown, which are very conservative, a benefit-cost ratio of 1.6 is forecast by the benefit-cost model.

3.3	**Expected portion of fatalities saved**
4.3	Expected proportion of injuries saved
5.3	Expected reduction in car emissions
6.3	Expected reduction in truck emissions
7.3	Expected reduction in car operating cost
8.3	Expected reduction in truck operating cost
9.3	Expected reduction in business travel time
10.3	Expected reduction in personal travel time

Figure 4.23 Case study 3 – Advanced traffic management, expected outcomes.

4.9 Developing a Comprehensive Benefits Analysis Framework

The four case studies featured in this chapter provide some guidance on how to approach benefit-cost analysis for specific smart mobility applications. It is up to the reader to use these as models to develop a comprehensive benefits analysis framework for specific smart mobility implementations within cities. The principles explained can be applied to a wider range of smart mobility services including mobility as a service, electronic payment systems, and connected and autonomous vehicles. Beginning with the development of a catalog of services to be delivered in the city, the principles can then be applied to estimate benefits and costs associated with the service.

4.10 Summary

This chapter explains the difference between features, benefits, and values related to smart mobility applications. It provides an overview of the challenges and opportunities presented by benefit-cost evaluation for smart mobility applications. To address the opportunities and challenges, a sketch planning framework for the estimation of benefits and costs associated with smart mobility implementations is introduced. The framework is explained by describing four specific examples relating to:

1) Roadside infrastructure sensors
2) Transit data analysis
3) Truck tire defects monitoring
4) Advanced traffic management

In the first example, a focus is placed on the estimation of relative lifecycle costs for different roadside infrastructure-based sensors. This illustrates the use of the cost estimation framework linked to a visual dashboard. The other examples extend the approach to include the estimation of benefits and costs. Collectively, these illustrate how to define the categories of benefits that can be delivered by smart mobility. The examples provide a practical explanation of how to apply these techniques to specific smart mobility applications and also to a wider review of a range of smart

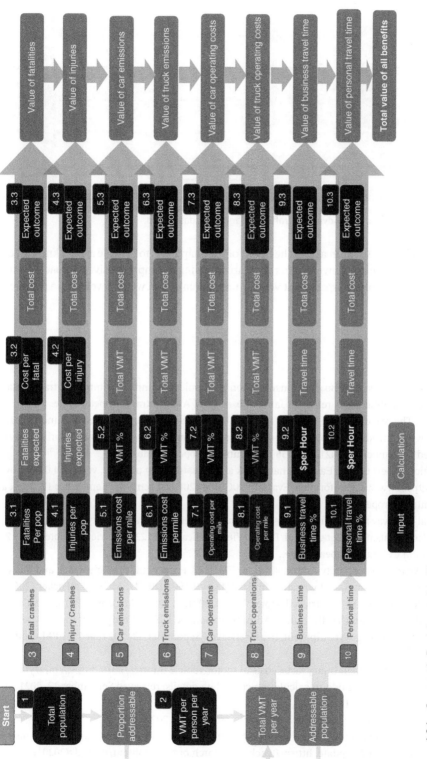

Figure 4.24 Case study 3 – Advanced traffic management, workflow.

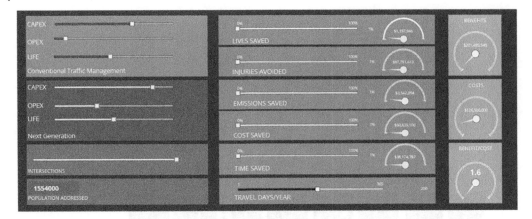

Figure 4.25 Case study 3 – Advanced traffic management, dashboard.

mobility applications within a city. The intent of the chapters is to provide a clear knowledge of the principles and techniques involved, rather than providing a comprehensive approach to any single city. Each city has unique aspects that will require customization when developing a comprehensive approach.

References

1 Transportation Research Board. Transportation economics committee, committee, transportation benefit-cost analysis page, retrieved January 6, at 11:41 AM Central European Time.
2 Felker, B. The real difference between features, benefits, and value, published on June 15, 2017. https://www.linkedin.com/pulse/real-difference-between-features-benefits-value-brian-felker, retrieved September 12, 2021, at 12:56 PM Central European Time.
3 Evaluation Expert Group. EasyWay handbook an evaluation best practice, version 4.0, 18 November 2012, https://www.its-platform.eu/EvalLib%20Guide%20Docs, retrieved September 12, 2021, at 6:02 PM Central European Time.
4 National Highway and Traffic Safety Administration (NHTSA). Traffic safety facts, 2014 data. https://crashstats.nhtsa.dot.gov/Api/Public/ViewPublication/812279, retrieved September 12, 2021, at 12:54 PM Central European Time.
5 Federal Motor Carrier Safety Administration (FMCSA). Pocket guide to large truck and bus statistics 2019. https://www.fmcsa.dot.gov/sites/fmcsa.dot.gov/files/2020-01/FMCSA%20Pocket%20Guide%20 2019-FINAL-1-9-2020.pdf, retrieved September 12, 2021, at 12:59 PM Central European Time.
6 National Highway and Traffic Safety Administration (NHTSA). Equipment topics, tires. https://www.nhtsa.gov/equipment/tires, retrieved September 12, 2021, at 1:03 PM Central European Time.
7 National Safety Council. Injury facts, costs of motor vehicle injuries, 2019. https://injuryfacts.nsc. org/all-injuries/costs/guide-to-calculating-costs/data-details, retrieved September 12, 2021, at 1:01 PM Central European Time.
8 International Road Dynamics. City of Cordelia truck study, 2020.
9 Nathan, W.; Daniel, M.; American Transportation Research Institute. An analysis of the operational cost of trucking: 2020 update. https://truckingresearch.org/wp-content/uploads/2020/11/ ATRI-Operational-Costs-of-Trucking-2020.pdf, retrieved September 12, 2021, at 1:12 PM Central European Time.

10 Pressure Guard Tire Inflation Systems. Industry information on the effects of improper tire inflation. https://www.ultra-seal.eu/images/pdf_bestanden/Industry_Info_on_Effects3.pdf, retrieved September 12, 2021, at 4:56 PM Central European Time.

11 Federal Motor Carrier Safety Administration. Roadside inspections. https://ai.fmcsa.dot.gov/SafetyProgram/RoadsideInspections.aspx, retrieved September 12, 2021, at 5:15 PM Central European Time.

12 Federal Motor Carrier Safety Administration. Roadside inspection out of service rates. https://ai.fmcsa.dot.gov/SafetyProgram/spRptRoadside.aspx?rpt=RDOOS, retrieved September 12, 2021, at 5:13 PM Central European Time.

13 Big Road. A fleet complete company, out of service costs, retrieved September 12, 2021, at 5:19 PM Central European Time.

14 The Texas Tribune. Government salaries explorer, July 1, 2021, Texas Department of Transportation, retrieved September 12, 2021, at 6:24 PM Central European Time.

15 Texas Department of Transportation. Construction and material tips, labor burden, fourth quarter 2007. https://ftp.txdot.gov/pub/txdot-info/cst/tips/labor-burden.pdf, retrieved September 12, 2021, at 6:32 PM Central European Time.

16 Wyoming Department of Transportation. Quick facts traffic signals. https://www.dot.state.wy.us/files/live/sites/wydot/files/shared/Traffic%20data/Traffic%20Signals.pdf, retrieved October 2, 2021, at 6:19 PM Central European Time.

5

Smart Mobility Progress Around the World

5.1 Informational Objectives

After reading this chapter, you should be able to:

- Describe smart mobility
- Illustrate the importance of smart mobility with examples
- Provide an overview of the adoption and implementation of smart mobility solutions worldwide
- Define the major cities leading the way in smart mobility initiatives
- Explain lessons learned
- Provide a brief introduction to smart mobility technologies, applications, and services
- Confirm that the smart mobility revolution is underway around the globe

5.2 Introduction

In Chapters 1–4, an overview of smart city activities, a smart mobility problem statement, and the value and benefits of smart mobility were described. This provides the context within which smart mobility is being deployed within smart cities around the world. Building on these previous chapters, this chapter provides an overview of smart mobility progress in leading cities around the globe. There are multiple objectives. The first is to provide the reader with an overview of the smart mobility services being deployed by leading cities around the world. This is a high-level overview with a focus on the projects that are being implemented. The second objective is to provide an initial introduction to the applications and technologies that can be deployed as part of a smart mobility initiative. The third and final reason is to provide evidence to the reader that the smart mobility revolution is underway, and that smart mobility applications and solutions are being deployed at a significant level in major cities around the world.

5.3 A Definition of Smart Mobility

For this book, smart mobility is defined as follows:

   The application of advanced technologies, data analytics, and innovative solutions to enhance the efficiency, safety, equity, sustainability, and user experience associated with transportation systems

Smart Mobility: Using Technology to Improve Transportation in Smart Cities, First Edition. Bob McQueen, Ammar Safi, and Shafia Alkheyaili.
© 2024 The Institute of Electrical and Electronics Engineers, Inc. Published 2024 by John Wiley & Sons, Inc.

in urban areas. This is achieved by leveraging digital connectivity, automation, and intelligent infrastructure to optimize the movement of people and goods, while minimizing undesirable side effects such as crashes, congestion, environmental impact, and excessive energy consumption. Smart mobility encompasses various aspects including intelligent transportation systems, electric and automated vehicles, shared mobility services, and multimodal transportation management.

Since the focus of the book is on the application of smart mobility within smart cities, the definition is limited to the application of smart mobility solutions and services in urban areas.

5.4 The Importance of Smart Mobility

Figure 5.1 provides a summary of the importance of smart mobility in global transportation.

To summarize the importance concisely, smart mobility enables the attainment of significant objectives associated with transportation, while enabling the avoidance or management of undesirable side effects.

5.5 Review of Key Cities Leading the Way in Smart Mobility

A sample of 10 leading cities was reviewed to illustrate progress made in smart mobility deployment and identify some of the lessons learned. While there are many more than 10 cities currently deploying smart mobility initiatives, solutions, and projects, these 10 were chosen to provide global representation across North America, Europe, the Middle East, and the Asia-Pacific region. They are also among the leading cities for which detailed information is currently available. Figure 5.2 provides a summary of the smart mobility highlights from each city.

Figure 5.3 provides a map of the 10 cities included in the study, showing that there is a good geographic spread, supporting a representative sample of global progress.

Please note that these cities are just examples, and there are many other cities around the world making significant strides in smart mobility. More information and specific details about each city's initiatives and projects can be found through the provided references. For each of the cities identified in Figure 5.3, a brief overview was prepared explaining the arrangements for managing transportation within the metropolitan area, the essential elements of the smart mobility implementations, and a summary of the practical experience and lessons learned.

5.6 Amsterdam, the Netherlands

Figure 5.4 indicates the position of Amsterdam on the globe.

5.6.1 Overview

Amsterdam is the capital city of the Netherlands and a prominent city in Europe with one of the busiest ports in the world. Several agencies are responsible for planning, regulation, development, and maintenance of the transportation network. Figure 5.5 provides an overview of the metropolitan region surrounding Amsterdam. At its core is the municipality of Amsterdam surrounded by the metropolitan region of Amsterdam.

#	Impact	Explanation
1	Efficiency	Intelligent traffic management systems real-time data analytics can improve the efficiency of transportation networks by optimizing roots, reducing congestion, and minimizing travel times and travel time variability
2	Sustainability	Promotes the use of eco-friendly transportation options such as electric vehicles, bike sharing systems, and public transportation This reduces carbon emissions, air pollution and dependence on fossil fuels, contributing to a more sustainable future
3	Safety	Collision avoidance systems, vehicle to vehicle communication, and advanced driver assistance systems enhance road safety by detecting potential hazards, provide warnings, and assisting drivers in making informed decisions Decision support systems can support a higher level of situational awareness enabling a greater understanding of crash mechanisms
4	Accessibility	Enabling accessible transportation options for all individuals, including people with disabilities are limited mobility This includes features like wheelchair accessibility, real-time transit information and on-demand transport services Smart mobility can be the basis for a new approach to mobility design based on an understanding of the needs of the individual
5	Connectivity	Enabling the integration of different modes of transportation including public transit, ridesharing, and micro mobility services through seamless connectivity This improves the overall travel experience and encourages multimodal transportation options for users
6	Data-driven decision-making	Smart mobility can provide a rich data stream, support effective conversion of the data to information, enable the extraction of insight from the information using analytics and provide decision support to ensure that the appropriate action is taken
7	Economic growth	Drives economic growth by enhancing connectivity, attracting investments, and creating new job opportunities in sectors related to technology, infrastructure development, and transportation services
8	User experience	Providing convenient, personalize, and on-demand transportation options This includes features like mobile apps of trip planning, seamless payment systems, and real-time updates on transportation services
9	Resilience	Enabling resilient transport networks that can adapt to unforeseen circumstances such as natural disasters, or emergencies Utilizing real-time data and intelligent routing algorithms that support a higher level of situational awareness
10	Equity	Addressing transportation inequalities by ensuring all communities of access to affordable and efficient transportation options This is regardless of their location, capabilities, or social economic status

Figure 5.1 Importance of smart mobility.

The following agencies have a responsibility for managing transportation in the greater Amsterdam metropolitan region.

5.6.1.1 Ministry of Infrastructure and Water Management [1]
This agency is responsible for planning, maintenance, and operation of the main transportation infrastructure in the Netherlands. This includes roads, waterways, and airports. The agency has a specific responsibility for the highway network surrounding the Amsterdam metropolitan area.

	City	Highlights			
1	Amsterdam	extensive cycling infrastructure and bike sharing programs	Electric vehicle charging infrastructure and incentives	Smart parking systems with real-time availability information	Intelligent traffic management systems and adaptive traffic signals
2	Barcelona	extensive bike sharing networks and pedestrian friendly initiatives	Electric vehicle charging infrastructure and incentives	Smart traffic management systems with real-time monitoring and control	Integrated transportation apps for multimodal journey planning
3	Berlin	sustainable transportation	Extensive public transportation network	Bike sharing programs	Initiatives to promote electric mobility
4	Dubai	Smart traffic signals	Autonomous shuttles	Mobile app for integrated transport planning	Advanced air mobility
5	Helsinki	mobility as a service platforms integrating various modes	Real-time transportation information and ticketing systems	Electric and shared mobility options for dedicated lanes and charging infrastructure	Smart traffic management and predictive analytics for congestion
6	New York	bike sharing	Car sharing	Digital payment options for public transportation	Smart traffic management using probe vehicle data
7	Oslo	electric public transportation	Incentives for electric vehicle adoption	Smart parking systems	
8	Seoul	advanced public transportation systems	Electric vehicle adoption	By electing automated vehicle technologies	Large-scale electronic payment for transit and services
9	Singapore	Extensive use of autonomous vehicles and raging services	Intelligent transportation systems for traffic management and public transit	Integrated payment systems for multiple modes of transportation	Dedicated cycling infrastructure and bike sharing programs
10	Stockholm	efficient public transportation system with real-time updates	Electric vehicle incentives and charging infrastructure	Smart traffic management systems and congestion pricing	Integrated mobility apps for multimodal transportation options

Figure 5.2 Summary of smart mobility highlights form each global study city.

5.6.1.2 Amsterdam Department of Transportation and Public Works [2]
This agency is responsible for transportation planning, policy development, and maintenance of public infrastructure, including roads and bicycle paths. The working close coordination with other transportation agencies.

5.6.1.3 Nederlandse Spoorwegen (NS) [3]
This is the main passenger railway operator in the Netherlands. They operate services connecting Amsterdam to other cities within the Netherlands and internationally central station operated by NS is a critical hub in Amsterdam's transportation infrastructure.

5.6.1.4 ProRail [4]
This agency is responsible for maintaining the railway infrastructure throughout the Netherlands, with an important role in Amsterdam.

5.6.1.5 GVB [5]
GVB is responsible for the operation of public transportation networks and around the city this includes 5 metal lines, 14 tram routes, 46 bus routes, and 10 ferry routes.

5.6.2 Smart Mobility Implementations
Amsterdam has long been known for its cycling infrastructure. This includes dedicated cycle lanes, special parking facilities, and traffic signal priority for cycles. The city has the policy of promoting cycling as the primary choice of travel mode. In the city, more than 60% of all trips are made by

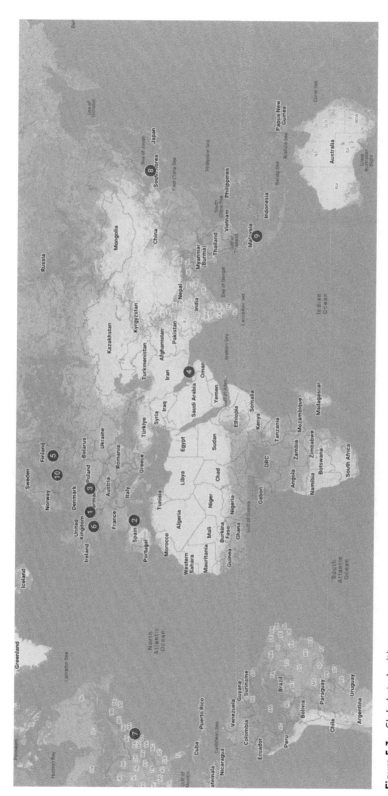

Figure 5.3 Global study cities.

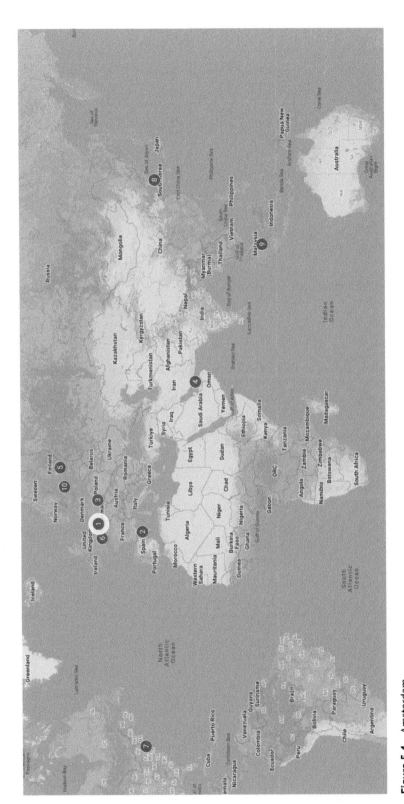

Figure 5.4 Amsterdam.

Figure 5.5 Amsterdam metropolitan region. *Source:* Amsterdam Tourist Information / https://www. dutchamsterdam.nl/2121-amsterdam-metropolitan-area / last accessed September 23, 2023.

bicycle. Amsterdam has made great progress in promoting the use of electric vehicles. Charging infrastructure has been implemented throughout the city enabling residents and visitors alike to have access to electricity supply required to charge electric vehicles. The approach also included incentives to make use of electric vehicles and coordinated awareness campaigns regarding the availability and benefits of electric vehicles. Amsterdam's approach to electric vehicles also includes extensive use of electric vehicles such as taxis. Subsequently, the city has unparalleled experience regarding the impact of electric vehicle use on the energy grid.

Amsterdam also supports car-sharing initiatives, providing residents with access to shared vehicles on an on-demand basis. This helps to reduce private car ownership, promoting more sustainable and efficient use of vehicles.

The city has also implemented extensive intelligent transportation systems designed to enhance traffic management and improve overall transportation efficiency. These systems include the capability to adjust traffic signal timings according to variations in demand and traffic conditions. Information can be delivered to the traveler using dynamic message signs to provide up-to-date information. Information delivery and traffic management are supported by an extensive traffic monitoring system, enabling data-driven decision-making.

The city has embraced the concept of mobility as a service. This enables travelers to make use of a single app to obtain information regarding the various travel mode choices available for a specific journey at a specific time. Travelers can also use the app to make a single reservation and pay for the traveler electronically. The city delivers mobility as a service in partnership with private sector companies who operate and maintain the app and the required back-office platform.

As part of a holistic approach to transportation management, Amsterdam has also implemented smart parking solutions. These use sensors and information delivery devices to provide parking availability information. The system also supports the application of dynamic pricing to discourage peaktime parking and encourage off-peak parking use. Users are offered electronic payment for speed and convenience. The total capability of the smart parking system is designed to optimize parking space utilization and reduce traffic congestion caused by drivers searching for available spaces.

The city has also implemented smart mobility for transit. This includes fleet management and passenger information systems along with an electronic payment system that makes use of contactless smart card technology. This enables travelers to pay for various modes of public transit with a single payment mechanism. Data from the electronic payment system is also used to support integrated real-time traveler information for apps and digital signage.

5.6.3 Key Lessons Learned

Among the many lessons learned from the implementation of smart mobility technologies, the following lessons stand out as being particularly important.

5.6.3.1 Collaboration
The city has learned that the successful application of smart mobility technologies requires partnerships and collaboration between government agencies, transportation operators, solution providers, citizens, and visitors.

5.6.3.2 Adopting a User Centric Approach
Amsterdam has placed an emphasis on design based on a full understanding of the user. Understanding the needs and preferences of the user has enabled the city to develop and implement services that are specifically tailored to travel requirements. This supports an increased level of adoption and user satisfaction.

5.6.3.3 Integrating Transportation Modes
Amsterdam has placed a particular emphasis on integrating the available modes of transportation. This includes the use of private cars, public transit, and nonmotorized, active travel modes such as cycling and walking. These are combined with the availability of shared mobility services across all modes. The city understands that it is vital to deliver a seamless and sustainable range of mobility options which balance transportation supply, reduce dependence on private cars, and improve overall mobility and accessibility.

5.6.3.4 Driving Decisions with Data
The city has learned the importance of making full use of data in decision-making and optimization of transportation systems. The city uses the various smart mobility implementations as data collection sources, combining this with data from additional sources such as infrastructure-based sensors, mobile applications, and transportation operators. This enables the city to extract important insight into the current demand for transportation and variations in travel conditions. This enables more efficient planning and better allocation of resources.

5.6.3.5 Piloting and Experimentation
The city has learned the importance of piloting and experimentation within an overall framework of smart mobility implementation. Especially with new and emerging technologies, it has been

defined valuable to test new concepts and new technologies at a smaller scale. The city has found that it is important to allow for an iterative and agile implementation process, supporting a learning process based on successes and failures prior to full-scale implementation.

5.6.3.6 Public–Private Partnership

Public–private collaboration has been crucial for Amsterdam smart mobility efforts. The ability to harness private sector resources and motivation has enabled Amsterdam to accelerate the implementation of innovative services and solutions. The city has found great value in leveraging the expertise and resources of both sectors.

5.6.3.7 Sustainable and Inclusive Mobility

During its experience, Amsterdam has understood the need to prioritize sustainability and inclusivity in smart mobility initiatives. The city has developed policies and objectives that include the reduction of carbon emissions, the promotion of electric and shared mobility options, and ensuring equitable access to transportation for all residents and visitors. They have found it important to include all residents and visitors, regardless of their socioeconomic background and characteristics.

5.6.3.8 Suitable Regulatory Framework

Amsterdam's experience has highlighted the need for appropriate and supportive regulatory frameworks. These are designed to protect the user and preserve the achievement of public sector objectives, while also fostering smart mobility innovations. Amsterdam has been proactive in adapting existing regulations and policies while developing new ones as required. The city has found it important that such regulations and policies ensure safety, privacy, fair competition, and innovation.

5.6.3.9 Continuous Learning and Adaption

Amsterdam's experience has also brought the understanding that the implementation of smart mobility is not a one-off initiative. Smart mobility is a fast-evolving field requiring effective continuous learning and adaptation. To support this, the city has found it essential to seek feedback from users and stakeholders, while monitoring the effects of smart mobility initiatives. The city has also understood the need to make appropriate adjustments to improve the effectiveness and efficiency of its smart mobility initiatives and the overall smart mobility ecosystem.

5.7 Barcelona

Figure 5.6 indicates the position of Barcelona on the globe. Figure 5.7 shows a tourist map of the Barcelona metropolitan region. This gives an impression of the size of the urban area and extensive transportation infrastructure.

5.7.1 Overview

Barcelona has emerged as a world leader in smart mobility by leveraging technology to improve efficiency, sustainability, and convenience. The city's bus network and public transport systems form a crucial part of the smart mobility effort, which includes the following major initiatives.

Figure 5.6 Barcelona.

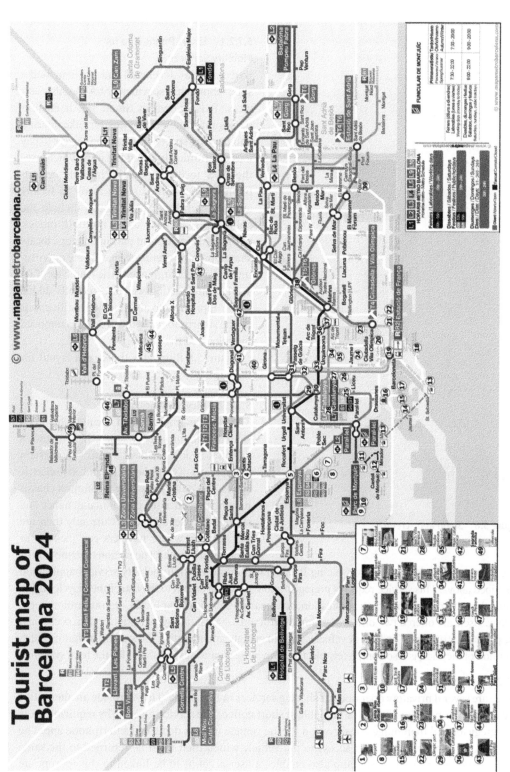

Figure 5.7 Map of Barcelona metropolitan region. *Source:* Flickr/David Beach.

Figure 5.8 Smou app screen.

5.7.2 Smart Mobility Initiatives

5.7.2.1 Smart Phone App for Mobility Services

Barcelona has developed and implemented a smartphone app specifically for mobility services. It is known as Smou [6]. It encompasses information services for bike sharing and parking spaces. It also allows users to pay fees electronically using the app, which uses the GPS technology built into the smartphone to locate the user to focus on locally available services. Services also include the ability to plan trips using multiple modes, pay for parking time at parking meters, locate shared mobility vehicles, and locate opportunities for recharging electric vehicles. Taxi service information was also recently added to the app. The app also enables notifications on the smartphone if a tow truck has removed the relocated the vehicle.

Figure 5.8 illustrates the screen layout for the app.

5.7.2.2 Bicycle Sharing

The city has implemented Bicing, which is a shared ride bicycle system which operates 24 hours a day, 365 days a year. The system incorporates 4000 pedal cycles and 3000 electric cycles. All 7000 cycles have a common chassis helping to reduce the cost of maintenance and simplifying user operation. Currently, there are 519 stations where travelers can pick up or drop off the cycle. The stations feature a mixture of pedal and electric bicycles. Bicing is integrated into the Smou mobility services application described above. This service requires preregistration, enabling access to the use of the bicycles using contactless smart card, the Smou application, or a smartphone with Near Field Communications technology. Users are offered two different payment structures – a flat rate subscription and an occasional use payment structure. The fleet of vehicles used to operate and maintain the system, including redistribution of bicycles, is also completely electric, providing greater sustainability for the total system.

5.7.2.3 Car Sharing

There are several privately operated car sharing services in Barcelona. These provide an alternative to car ownership, making vehicles available at short notice. The services operate by requiring users to preregister and then make a reservation for the vehicle using a dedicated smartphone app. The smartphone app is used to gain access to the vehicle which then must be returned to the same parking spot after the trip. The app also enables a user account to be linked to the appropriate means of payment, making it easy for the user to pay. It is estimated that each car-sharing vehicle replaces 20 private cars and the fleet includes lower emissions and zero missions models to improve sustainability. Many of the companies operating car sharing in Barcelona also make vehicles available in other major cities in Spain.

5.7.2.4   Automated Vehicles [7]

The port of Barcelona is piloting the use of a fully automated shuttle to transport port workers to and from a parking lot at the port the Barcelona and the World Trade Center, where the port authority offices are located. Figure 5.9 shows one of the shuttles in operation. A proof-of-concept trial for automated vehicles is also underway at Barcelona International Airport. Automated vehicles are being used to move cargo between the warehouse and the airport apron [8]. TMB the city's transit operator also makes extensive use of automated technology for the Metro [9].

5.7.2.5   Smart Traffic Signals [10]

The city has installed a network of smart traffic signals in various locations around the city. This incorporates emergency vehicle priority to speed the arrival of emergency services and incorporate adaptive control technology to reduce congestion. The city is in the process of converting traffic signals to the use of LED technology to reduce operating costs and increase sustainability. Some traffic signal installations have been converted to self-powered units which make use of a combination of solar and wind power energy generation.

5.7.2.6   Bus Network (TMB) [11]

The city has redesigned the bus network for greater efficiency and incorporated technology for real-time root planning and scheduling. Buses are equipped with IOT devices enabling tracking and other data to be transmitted to a central location. Smart mobility services available to the traveling public include the use of electronic payment for ticketing, trip planning services, and service status information. Technology applied to the bus network has also been applied to the Metro network.

Figure 5.9   Automated shuttle, Barcelona.

5.7.2.7 Superblocks

This initiative involves the application of a sophisticated planning approach, rather than the deployment of technology. It is an urban planning initiative implemented by the city. The objective is to transform city streets and public spaces to prioritize pedestrians and sustainable transportation. The concept is centered around the concept of superblocks which are clusters of city blocks that are reconfigured to create car-free or car-limited zones. The implementation area is divided into a grid of superblocks, typically consisting of nine existing blocks clustered together. Each superblock is around 400 × 400 m in size covering an area of approximately 16 ha [12].

Although this is not a direct application of smart mobility technologies and services, the superblock concept provides an excellent focus for combining such services and technologies in support of reduced private car use, increase sustainability, and improved accessibility. Figure 5.10 shows an aerial photograph of a superblocks zone in Barcelona. Superblocks also incorporate mobility hubs at strategic locations within the superblocks. These provide bike-sharing stations, electric vehicle charging points, and other facilities to encourage residents and visitors to use alternative modes of transportation.

5.7.3 Lessons Learned

5.7.3.1 Adoption of an Integrated Approach

One of the keys to success in Barcelona when implementing smart mobility is the adoption of an integrated approach. The city has focused on integrating various modes of transportation including public transit, bicycles, electric vehicles, and shared mobility services. By creating a seamless and interconnected transportation network, Barcelona has made it easier for travelers to choose sustainable, smart, and efficient transportation options.

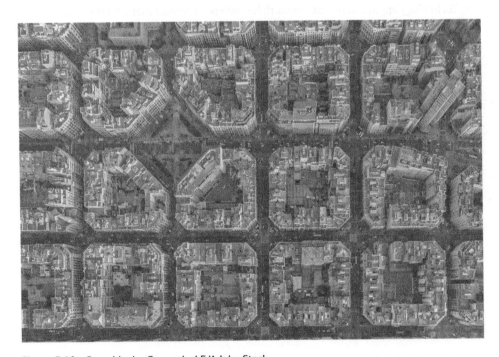

Figure 5.10 Superblocks. *Source:* JackF/Adobe Stock.

5.7.3.2 Effective Use of Public–Private Partnerships

Effective collaboration between both the public and the private sector has been a crucial aspect of rapid progress for Barcelona smart mobility initiatives. The city is actively partnered with technology companies, transportation providers, and startup companies to develop and implement innovative solutions. Alignment of public and private sector objectives and resources has enabled the deployment of new technologies and services in a rapid manner.

5.7.3.3 Driving Decisions Through Data

The city collects and analyzes data from various sources including sensors, smartphone apps, and various transportation providers. This has enabled the city to gain valuable insight into travel patterns, transportation conditions, and variations in the demand for transportation. Such insight has enabled the city to optimize its transportation system by making informed decisions about service operations and infrastructure investments.

5.7.3.4 Focusing on Sustainability and Efficiency

The city saw the need to place a strong emphasis on sustainability and efficiency in the deployment of smart mobility services. This includes measures to reduce emissions and improve the quality of using electric vehicles and promoting active travel modes. This focus on sustainability and efficiency is also supported by an investment in smart traffic management systems to optimize the flow of vehicles and reduce congestion.

5.7.3.5 Multimodal Integration

The city recognized the importance of multimodal transportation management. This involves a seamless combination of different modes of transportation for a single trip. Multimodal integration can be fostered by providing integrated ticketing systems in smartphone apps that allow access to several different modes.

5.7.3.6 Effective Citizen Engagement

Barcelona has actively involved citizens in the planning and implementation of smart mobility. This has included surveys, workshops, and digital platforms for public feedback. Effective citizen engagement enables valuable insights for the design and deployment process, builds trust with the traveler, and makes sure that the smart mobility services deployed are designed to meet the actual needs of residents. Building on the effective citizen engagement described above, Barcelona has understood the need to prioritize a detailed understanding of the needs and preferences of residents and visitors. User-centric design principles are extremely valuable in achieving this. Focusing on user experience also encourages greater adoption of sustainable transportation options.

5.8 Berlin

Figure 5.11 indicates the position of Berlin on the globe. The figure provides an overview of the Berlin-Brandenberg Metropolitan Region (Figure 5.12).

5.8.1 Overview

Berlin is the largest city in Germany and the nation's capital. It is a major hub for international travel and trade, with an extensive transportation system. Transportation modes include roadways, railways, airways, and waterways. The transportation system is administered by several public agencies.

Figure 5.11 Berlin.

5.8.1.1 Senate Department for the Environment, Transport, and Climate Protection
This agency oversees transportation policy in Berlin. This includes the management of road, rail, air, and waterway transportation. The primary functions of the agency include planning, implementation, and maintenance of transportation infrastructure.

5.8.1.1.1 *The City of Berlin [13]*
The city has an overall strategic plan that encompasses a range of smart mobility initiatives. Led by the mayor, the city is engaged in a range of smart city initiatives, within which smart mobility sits.

5.8.1.2 Verkehrsverbund Berlin–Brandenburg (VBB)
This agency is responsible for public transportation in the Berlin metropolitan area, including the surrounding Brandenburg region. The authority manages several transportation modes including buses, trams, underground trains, suburban trains, and ferries.

5.8.1.3 Berliner Verkehrsbetriebe (BVG)
This agency operates Berlin's public transportation system including rail, tram, bus, and ferries. It works under the auspices of the VBB as described above.

5.8.1.4 Deutsche Bahn (DB)
This is Germany stay on Railway Company. It provides regional intercity and international real services to and from Berlin. It also operates a longer distance S-bahn network in the greater Berlin metropolitan area.

Figure 5.12 illustrates the Berlin Metropolitan Region including the surrounding Brandenburg area.

5.8.2 Smart Mobility Initiatives

Berlin is emerging as one of the global leaders in the application of smart mobility to cities. With a focus on sustainability, innovation, and technological advancement, the city is developing a robust system of transportation. The city smart mobility initiatives cover a broad range of technologies, along with policies designed to create a more efficient, sustainable, and user-friendly transportation system. The initiatives are founded on a combination of digital infrastructure, comprehensive policy reform, and innovative technologies.

5.8.2.1 Integrated Mobility Solutions
The objective of the city is to create an integrated and seamless transportation network that spans a range of transportation modes. This range includes public transit, cycling, car-sharing, and ride-sharing services. Significant elements in this integrated mobility solutions approach include the unified ticketing and pricing system. This is coordinated by the Berlin Brandenburg Transport Association (VPP). This enables a single ticket to be used across multiple modes of transportation within specified zones. It is supported by the BPG app [14] which enables mobile ticket purchase and provides real-time travel information. Berlin has also implemented several major transportation hubs for multimodal interchange The intended outcome is to provide citizens and visitors with convenient and flexible options for mobility around the city.

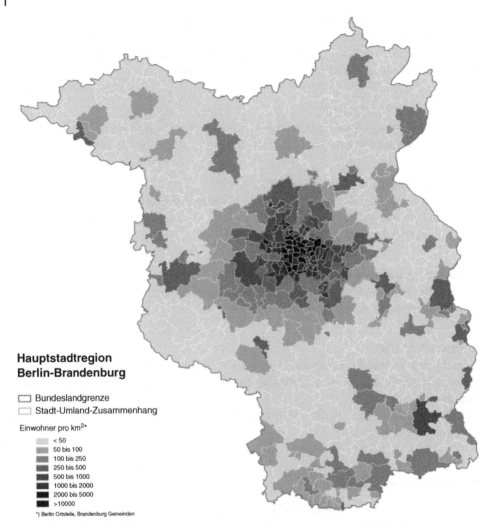

Hauptstadtregion Berlin-Brandenburg

☐ Bundeslandgrenze
☐ Stadt-Umland-Zusammenhang

Einwohner pro km²*

▫ < 50
▫ 50 bis 100
▪ 100 bis 250
▪ 250 bis 500
▪ 500 bis 1000
▪ 1000 bis 2000
▪ 2000 bis 5000
▪ >10000

*) Berlin Ortsteile, Brandenburg Gemeinden

Figure 5.12 Berlin Metropolitan Region. *Source:* https://en.wikipedia.org/wiki/Berlin/Brandenburg_Metropolitan_Region.

5.8.2.2 Digital Infrastructure

Digital infrastructure such as sensors, telecommunication transmission systems, and data management systems have been developed to support smart mobility initiatives. This also includes the implementation of intelligent traffic management systems, real-time data collection and analysis, and the use of mobile apps and platforms to deliver traveler information.

5.8.2.3 Public Transit Enhancements

Berlin is known for its extensive public transit system comprised of trams, buses, U-Bahn (rapid transit subway system, and S-Bahn [urban rail network]) train networks. In addition to infrastructure improvements to expand and enhance the transit system, some of the digital infrastructure described above is deployed to enhance usability and approve accessibility, while integrating with other modes of transportation.

5.8.2.4 Cycling Infrastructure

Building on its extensive cycling infrastructure including bike lanes, Berlin has implemented bike-sharing programs and supports bicycle use that are the implementation of bicycle-friendly policies. The city has a continuing program to enhance and expand cycling infrastructure including digital initiatives.

5.8.2.5 Electric Mobility

The city promotes the use of electric vehicles to reduce emissions and support decarbonization of mobility. Electric vehicle charging infrastructure has been deployed throughout the city [15]. Electric vehicle use is also encouraged by offering a range of incentives for electric vehicle owners at the national level [16]. Electric vehicles are also incorporated into car sharing and rental services [17].

5.8.2.6 Mobility as a Service

The city of Berlin has become a European focal point for mobility as a service. That are numerous startup and technology companies along with transportation providers partnering to deliver mobility as a service. Important private sector service providers include Jelbi [18] and Trafi. [19]. These companies deliver services via a single smartphone app that enables access to public transit, car sharing, scooter sharing, and bike sharing apps. Figure 5.13 shows the Jelbi app.

Figure 5.13 Jelbi app. *Source:* https://www.jelbi.de.

5.8.2.7 Smart Parking

These are technology-driven solutions that apply advanced technology along with data analysis to optimize parking management. Systems are comprised of sensors, real-time data collection, and smartphone apps to provide users with real-time parking information [20] drivers are provided with information on the closest available spaces, leading to a reduction in the amount of time and distance expended in looking for parking spaces. Figure 5.14 illustrates the concept.

5.8.3 Lessons Learned

5.8.3.1 The Importance of Transportation Integration

Berlin has understood the importance of a comprehensive approach to mobility. This requires that various modes are coordinated with services designed to complement each other. This enhances conductivity, increases efficiency, provides travelers with greater choice, and leads to higher usage of sustainable transportation options.

5.8.3.2 The Need to Embrace Data-driven Solutions

The city has taken advantage of the growing volume of data available from different sources and harnesses this to improve mobility services. The city has understood the importance of leveraging data from multiple sources including both public and private sector. These include roadside sensors, mobile apps, GPS devices, connected vehicles, and probe vehicle data. In addition to the ability to optimize routes based on effective situational awareness, the data has enabled the city to understand transportation demand patterns and user preferences. The use of the data has enabled Berlin to optimize transportation services and allocate resources efficiently. It has also supported better-informed decisions for infrastructure investment and for travelers.

Figure 5.14 Smart parking. *Source:* Akarat Phasura/Adobe Stock.

5.8.3.3   The Important Role of Public–Private Partnerships

Berlin has purposefully fostered collaboration between the public and private sectors. This has been vital to driving innovation and accelerating the attainment of benefits from smart mobility. The city has found that close cooperation with startups, technology companies, and mobility service providers enables access to a wide range of expertise and resources. Public–private partnerships have been extremely valuable in accelerating the deployment will be scooter sharing, right-hailing services, and mobility as a service platform.

5.8.3.4   User Focused Design with Feedback Loops

The city has understood the need to give priority to user needs and listening to feedback in the planning, design, and implementation of smart mobility services. It has been found that this leads to more effective services and better results. This requires active engagement with residents and visitors through opinion surveys and public consultations. They should be designed to understand customer characteristics, preferences, and challenges. This can also provide a way to understand opportunities for new services and better ways to do things. The incorporation of our customer focus along with support for user feedback leads to a more effective development process, which in turn leads to more user-friendly and efficient mobility services.

5.8.3.5   The Importance of a Focus on Sustainability

The city has understood the need to focus smart mobility initiatives with a strong emphasis on sustainability and managing the environmental impact of mobility services. Such a focus includes promotion of the use of electric vehicles, active travel modes, mobility as a service, shared ride, and reducing the dependence on the private car. The city also integrated smart mobility solutions within wider environmental policies.

5.8.3.6   The Need to Be Flexible and Adaptable

Berlin has understood the need for flexibility and adaptability when dealing with new and emerging technologies. Smart mobility also addresses the dynamic situation with variations in the demand for transport and mobility trends. The city has considerable value in adopting a pilot or experimental approach to enable effects and outcomes to be understood in a practical way before widespread implementation. The city's approach is specifically designed to learn practical lessons from successes and failures, paving the way for better strategies at scale.

5.9   Dubai

The figure provides an overview of the metropolitan area around Dubai (Figure 5.15).

5.9.1   Overview

The city of Dubai is the most populous city in the United Arab Emirates (UAE) and the capital of the Emirate of Dubai. It serves as a major hub for international travel and trade and has a sophisticated transportation network. Dubai is the most populated of the seven minutes that comprise the UAE. The city is known for its technological advancement which includes leadership in smart mobility initiatives. The city has been at the forefront of testing and evaluating smart mobility technologies including air taxis, automated buses and taxis, and mobility services.

Figure 5.16 shows the area around Dubai with an indication of congestion hotspots.

Figure 5.15 Dubai.

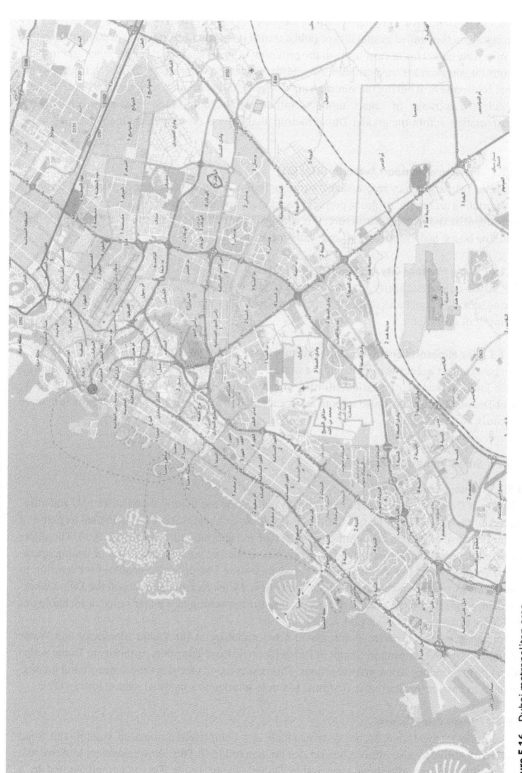

Figure 5.16 Dubai metropolitan area.

The city has launched a program of smart mobility initiatives driven by the need to manage traffic congestion, the limited availability of public transit infrastructure, and high dependency on the use of private cars. This in turn leads to the provision of insufficient parking facilities and limited options for sustainable transportation. An additional factor of considerable concern in the city of Dubai is extreme heat during the summer, making some transportation options infeasible. The city has adopted a range of smart mobility initiatives aimed at addressing these challenges. Transportation within the greater Dubai metropolitan area is the responsibility of several public agencies.

5.9.1.1 Roads and Transport Authority (RTA) [21]
This is the primary agency responsible for transportation in Dubai. The agency plans and executes transportation and traffic projects, prepares legislation, and develops strategic plans. It also develops residential and road infrastructure. It is responsible for all modes of land-based transportation including buses, taxis, metro, tram, and water buses.

5.9.1.2 Dubai Maritime City Authority (DMCA) [22]
This is the official authority that regulates, coordinates, and supervises all aspects of Dubai's Marine sector. This includes managing operations in the Emirates waters, licensing, and regulating marine craft operations.

5.9.1.3 The Ministry of Energy and Infrastructure Development (MoEI) [23]
This government agency is responsible for strategic planning and development of energy, mining, water, land, and sea transportation, utilities, housing, building, and construction within the emirate of Dubai. This agency is also responsible for regulation and licensing including vehicle operating cards.

5.9.2 Smart Mobility Initiatives

5.9.2.1 Automated Vehicles
The city of Dubai automated vehicle program includes the use of automated buses [24] and automated air taxis. A series of pilots of automated vehicle technology has been launched as part of a wider objective of achieving automation of 25% of transportation in the city by 2030. The pilots involve the operation of automated bus services on a specific bus route as a way of testing operational feasibility and the effectiveness of safety measures. The pilot also enables user reaction to automated buses to be ascertained. In a related effort, Dubai recently announced the Dubai World Challenge for self-driving transport. This enables further testing of a wider range of technologies and more extensive field trials [25].

Travelers can also experience automated bus technology at the Dubai Electricity and Water Authority (DEWA) Innovation Center in Dubai [26]. The Road Transport Authority in Dubai is also planning to introduce autonomous air taxis. This makes use of electric vertical takeoff and landing vehicles to provide an alternative first mile last mile solution to congested road networks [27].

5.9.2.2 Smart Public Transit
In addition to the Dubai Metro system which is a completely automated system, the Road Transport Authority also offers a bus service on demand [28]. This allows travelers to move in a smart, efficient, and affordable way within Dubai's major zones. This is complemented by a smart card-based ticketing system that enables payment across various transportation modes

with a single tap of the card when readers are installed in transit vehicles. The card can be used in Dubai Metro, buses, Dubai Tram, and Marine transport modes. The payment service can also be used to pay for parking [27].

5.9.2.3   Smart Traffic Systems

The Road Transport Authority (RTA) has implemented within the city of Dubai one of the largest and most sophisticated traffic control centers in the world. The traffic control center is linked to variable message signs and closed-circuit TV cameras along with traffic sensors and road weather information systems. An extensive network of fiber optics is used to support two-way communications between the field and the control center. After an initial phase which improved instant monitoring by 63%, shortened emergency response times by 30%, and cut travel times by an average of 20%, it was decided to enhance and expand the initial system. [29].

5.9.2.4   Smart Mobility Services

Building on its modern transportation infrastructure, the city of Dubai has supported the implementation of several ridesharing services. The two major services are Uber and Careem. The latter was acquired by Uber in 2019. Both services involve the use of a specific smartphone app to enable travelers to access on-demand services using cars, taxis, bicycles, and rental cars. Careem also offers specialized school rights packages enabling parents to purchase a package of school rides to transport children to school. The service uses its real-time tracking and notification capabilities to update parents on the child's journey. A smartphone app allows the user to choose a specific ride, track it and real time, and pay securely with a choice of electronic payment methods. The vision for Careem includes a continued expansion toward an "everything app" providing access to services beyond mobility. In addition to the digital services, an innovative fuel delivery service is also operational in Dubai. Registered users can request fuel delivery at home, on an emergency fuel delivery enroute in less than 20 minutes. This is available 24 hours a day, seven days a week. The services have also been expanded to include additional vehicle services including carwash, tires, battery, and engine oil delivery. Online services also include the ability to purchase car insurance [30].

5.9.3   Lessons Learned

5.9.3.1   Taking an Integrated Approach Beyond Individual Technologies or Solutions

One of the challenges identified by the city is a need for focused projects to move forward quickly, while also ensuring the investments are conducted within a wider coordinated framework. There is a need to balance focused, rapid progress, was coherent integration within a bigger picture.

5.9.3.2   Leveraging the Private Sector

The city is particularly adept at leveraging the resources and motivation of the private sector. In many cases, the private sector takes a lead in the implementation of smart mobility services. The thinking within public agencies focuses on how to achieve public sector objectives by being a good partner to the private sector. Public agencies also consider the possibility of future transportation systems that have little or no involvement from the public sector, other than regulation and monitoring.

5.9.3.3   Aligning Decision-making with Data Collection

The city has placed a heavy emphasis on developing an appropriate level of situational awareness to enable long-term investments in short-term operational management to be guided by a good understanding of demand and operating conditions. In addition to the installation of infrastructure-based sensors, the city understands the importance of combining public and private sector data sources to create a better understanding of the transportation network.

5.9.3.4 Designing for People, Citizen Engagement

Dubai has prioritized public awareness and engagement to ensure the successful adoption of smart mobility solutions. It also makes use of the dialogue with citizens and visitors to ensure that solutions and services are well aligned with the needs and characteristics of the user.

5.9.3.5 Scalability Driven by Pilot Project Experience

Dubai has excelled in the establishment and operation of pilot projects involving smart mobility technology. This gives the city important practical experience regarding technological capabilities and constraints. The city has understood the need to use this practical experience to guide larger-scale implementations of solutions and services.

5.9.3.6 Development of Enabling Regulatory Framework

Since the city is at the forefront of testing and deploying new smart mobility technologies such as automated vehicles, the city has understood the need for policy and regulation to keep pace with technology implementation. The road transport authority has issued guidelines for testing automated vehicles on public roads. These include provisions for minimum safety standards, arrangements for remote monitoring, and requirements for incident and anomaly reporting. Similar guidelines and policies have also been put in place for electric vehicles and electric vehicles charging infrastructure. Policies of also been implemented to protect data privacy and security under the auspices of law number 26 of the 2015 regulating data dissemination and exchange in the Emirate of Dubai [31].

5.10   Helsinki

5.10.1   Overview

Figure 5.17 indicates the position of Helsinki on the globe. The figure provides an overview of the Helsinki metropolitan area, highlighting the major public transit facilities (Figure 5.18).

This figure shows a map of tram routes and the Metro system showing the extent of the transportation network in the Helsinki metropolitan area. Source: https://welcome.hel.fi/getting-around-helsinki/public-transport-in-helsinki-using-buses-trains-trams-and-the-metro-in-helsinki.

The following major partner agencies are involved in developing and delivering transportation services in the Helsinki metropolitan area:

Helsinki region transport authority (HSL): this is the agency responsible for operating public transportation in the Helsinki metropolitan area.

Helsinki city transport (HPL): this agency operates metro and tram services in Helsinki under the authority of HSL.

Fintraffic: a special assignment group operating under the auspices of the Ministry of Transport and Communications. Responsible for traffic control and management services.

City of Helsinki urban environment division, traffic, and transport planning: responsible for the planning and maintenance of traffic and transportation within the city.

Finnish transport communication agency (Traficom): national government agency responsible for services for what, transport system services, digital connections, and the national

Figure 5.17 Helsinki.

Figure 5.18 Helsinki metropolitan area.

cybersecurity center. The agency is also tasked with planning and implementing activities related to new European Union digital and data laws.

Finnish transport infrastructure agency (FTIA): responsible for the development of Finland's road, rail, and waterway networks.

The port of Helsinki: the port is one of the busiest passenger ports in Europe, serving both international and domestic ferry routes as well as a significant amount of cargo traffic.

5.10.2 Smart Mobility Initiatives

Helsinki, the capital of Finland has adopted a progressive approach to urban development and sustainability. The objective is to create a livable and environmentally friendly city. The city has become a leader in the application of smart mobility solutions to address several challenges. These include population growth, transportation-related omissions, traffic congestion, efficient public transit, and the dependence on fossil-fueled vehicles.

5.10.2.1 Public Transportation Modernization

The city has invested in initiatives to integrate public transportation including buses, trams, trains, and the Metro system. Helsinki region transport (HSL) is a joint regional authority responsible for public transportation planning. Integration is assisted by the introduction of comprehensive route planning, coordination, ticket prices, marketing, and passenger information. HSL is also responsible for ticket sales and ticketing enforcement. HSL has introduced a smartphone app that allows travelers to access real-time data across modes enhancing the convenience and reliability of public transportation. Using the HSL app [32] travelers can buy tickets for single journeys and multiple journeys. A variety of payment options is available including season tickets. The app also offers a journey planner which allows travelers to identify the best routes for the particular trip and time and provides information to show where the transit vehicles are in real-time.

5.10.2.2 Bicycle Sharing and micro Mobility

Source: https://mobilitylab.hel.fi/helsinki-ecosystem/.

The city has introduced city sponsored bicycle sharing programs. These provide a fleet of icicles stationed at various docking stations across the city. Travelers can register to access the bicycles using a smartphone app. Several subscription plans are available to pay for their use [33]. In addition to the city-sponsored bicycle-sharing program, there are several private bicycle-sharing companies operating in Helsinki. These again use a smartphone app to register to use the vehicles. The apps also provide information to locate available bicycles and unlock them. In a similar manner, electric scooters and bicycles are also available across the city [34]. The city has worked to integrate bicycle sharing and micromobility closely with public transportation, to provide first mile, last-mile solutions to multimodal trips [35]. This has involved close cooperation with private sector micromobility providers to ensure proper parking and safety regulations adherence for electric scooters and bicycles [36].

5.10.2.3 Intelligent Traffic Management

The management of traffic on land, sea, and air is the responsibility of Fintraffic. This is a special-purpose company under the auspices of the Ministry of Transport and Communications. The objective is to develop and deliver the required traffic control services to make transportation across Finland more efficient. The company makes use of smart traffic signals, data collection sensors, and traffic management centers to control urban roads, tunnels, and rural highways. In urban areas, the company works with a network of partners responsible for city transportation. In addition to the deployment of smart traffic signals, the company has focused on the establishment and operation of a traffic data ecosystem. This contains data for all motor transportation within the auspices and open access business model [37].

5.10.2.4 Automated Vehicles

Finland has put a lot of effort into creating a positive legal and regulatory environment for automated vehicles. The entire national road network is available for field trials [38].

In addition to operating automated vehicles on public highways, the pilot projects have also involved the deployment of an on-demand mobile phone application and the remote-control center for automated vehicles. Automated vehicle operation is also enhanced by the establishment of mobility hubs [39] automated vehicle deployment has also included the world's first all-weather automated bus in partnership between a Japanese company and a Finland-based startup [40].

Source: https://www.sick.com/br/en/sick-sensor-blog/autonomous-buses-an-uncommon-sight-on-finnish-streets/w/blog-gacha-autonomous-bus-finland.

5.10.2.5 Data-driven Decision-Making

Recognizing the possibilities associated with knowledge management and digitalization, the city has promoted an open information environment that can be the foundation for better decision-making and new smart mobility services. The city has a proactive stance toward the use of data to drive productivity and improve quality of smart mobility services [41].

5.10.2.6 Mobility as a Service

Helsinki has become a world leader in the practical application of mobility as a service. The system provides travelers with a user-friendly interface using either a mobile home phone app or a web portal. Through these interfaces, travelers can access and interact with the mobility as a service platform. The interfaces allow users to plan, reserve, and pay electronically for trips across multiple modes of transportation. It can justifiably be claimed that the Mobility as a Service concept was invented in Helsinki. The city has made it a key component in national transportation policy. The implementation of mobility as a service in Helsinki is widely recognized as an innovation that has the possibility to change the entire smart mobility universe [42].

Source: https://www.g-mark.org/zh-Hant/gallery/winners/9e0cab6f-803d-11ed-af7e-0242ac130002?years=2019.

One of the keys to successful implementation of mobility as a service in Helsinki is an inclusive approach that includes multiple transportation modes including public transport, taxi, shared ride bicycles, car rental, and micromobility e-scooters.

The mobility as a service system coordinates with different mobility service providers such as public transportation agencies and micromobility companies. A series of application programming interfaces supports data exchange, enabling integration across the platforms. In addition to providing travel services, the MaaS system collects and manages vast amounts of data relating to user preferences, trip patterns, and transportation conditions. Analytical tools are applied to data to derive insights and enable data-driven decision-making. This leads to improvements in the transportation system efficiency, including optimized writing and personalized recommendations for travelers. The mobility as a service system also supports seamless ticketing and fair integration across different modes of transportation, enabling a single electronic reservation for multimodal trips. All this is made possible by robust backend infrastructure to handle processing, storage, and retrieval of data.

Helsinki uses this mobility as a service implementation to show other cities around the world how it can be deployed successfully. It uses an organization known as MaaS Global to achieve this [43].

5.10.2.7 Electric Vehicles and Alternative Fuels

The use of ethanol as an alternative fuel for transportation vehicles is already well-established in Scandinavia. Building on this the city has promoted the use of electric vehicles by ensuring that adequate electric vehicle charging infrastructure is available. The national government has also reinforced the promotion of the use of alternative fuel vehicles by offering registration tax and ownership tax benefits for both individuals and companies. This also includes purchase incentives for households to purchase or lease a new battery electric vehicle, and companies to purchase electric vans and electric trucks. The government supports alternative fuel-charging infrastructure by offering a subsidy for the construction of these [44].

Figure 5.19 illustrates the geographical coverage of electric vehicle charging infrastructure in the Helsinki region [45].

The city has taken a comprehensive approach to the design and development of the electric vehicle charging infrastructure. In addition to encouraging residents to install at-home charging facilities, the city has promoted electric vehicle charging infrastructure at residential parking and at major destinations including park-and-ride sites will use of the electric vehicle can interface with public transportation. The infrastructure incorporates fast charging facilities which have the capacity to fully charge electric vehicles in less than 30 minutes. In addition to support for better decision-making, the insight obtained from electric vehicle charging data is also used to ensure that mobility solutions match the needs and characteristics of residents and visitors.

5.10.3 Lessons Learned

5.10.3.1 The Value of Data

Both the city and the country have recognized the value of data and adopted a centralized approach to data management and open access to the data. The city has recognized that data can feel better decision-making and serve as a solid foundation for innovative new smart mobility services. The mobility as a service application makes full use of its ability to collect data as well as deliver travel information services and centralize reservations.The Need for Comprehensive Coverage of Modes in the Application of Mobility as a Service

Helsinki is a world leader in the application of mobility as a service and has understood the need for the system to include as many modes of transportation as possible. It is also understood that integration between smart mobility services and existing public transportation enhances the value of both. Many of the smart mobility implementations in the city feature

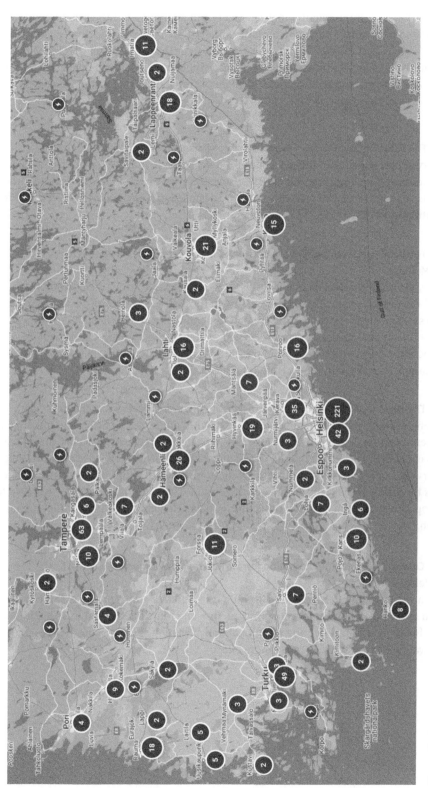

Figure 5.19 Geographical coverage of electric vehicle charging infrastructure in the Helsinki region. *Source:* https://kartta.lataus.helen.fi.

close integration with infrastructure improvements and alignment of infrastructure investments with smart mobility services. Smart mobility integration is also supported by an integrated ticketing system called the Helsinki regional transport authority (HSL) card. This allows users to access multiple modes of transportation using a single payment mechanism. Ticketing integration enables travelers to choose the best mode of transportation for their needs, characteristics, and time of the trip.

5.10.3.2 The Value of Coordinated Advanced Traffic Management

The country adopts a well-coordinated, centralized approach to multiple modes of transportation management including land, sea, and air. There are strong partnerships in place between the various agencies at national and local levels. These are designed to support a higher level of coordination for both transportation objectives and operations.

5.10.3.3 The Value of Close Cooperation Between the Public and the Private Sector in Shared Mobility

In Helsinki, the application of shared mobility services features a strong cooperation between the public and the private sector. The core shared mobility service from the city is complemented by additional services driven by the private sector. Cooperation includes alignment on infrastructure and facility investment alongside private sector investment in delivering smart mobility services. Close cooperation between the public and private sectors also extends to continuous experimentation, evaluation, and adaptation through pilot projects.

5.10.3.4 The Need for Automated Vehicles to Operate in Severe Weather Conditions

The implementation of automated vehicles in Helsinki addresses a particular challenge relating to severe weather. The system deployed in pilot testing has been selected for all-weather operation capabilities. The city has also recognized an important role automated vehicles can play in serving as the link between original origin, ultimate destination, boarding, and alighting points for public transportation. Experimentation with automated vehicles has also led to the development of effective regulations and policies that protect safety while encouraging innovation.

5.10.3.5 Combining Public Transportation Infrastructure Improvements with Smart Mobility Yields Better Results

In addition to applying smart mobility techniques to public transportation, the city has also invested in significant improvements to infrastructure and route planning. It has been understood that a combination of innovative and traditional transportation solutions, involving both the public and private sector, can have the maximum effect.

5.11 London

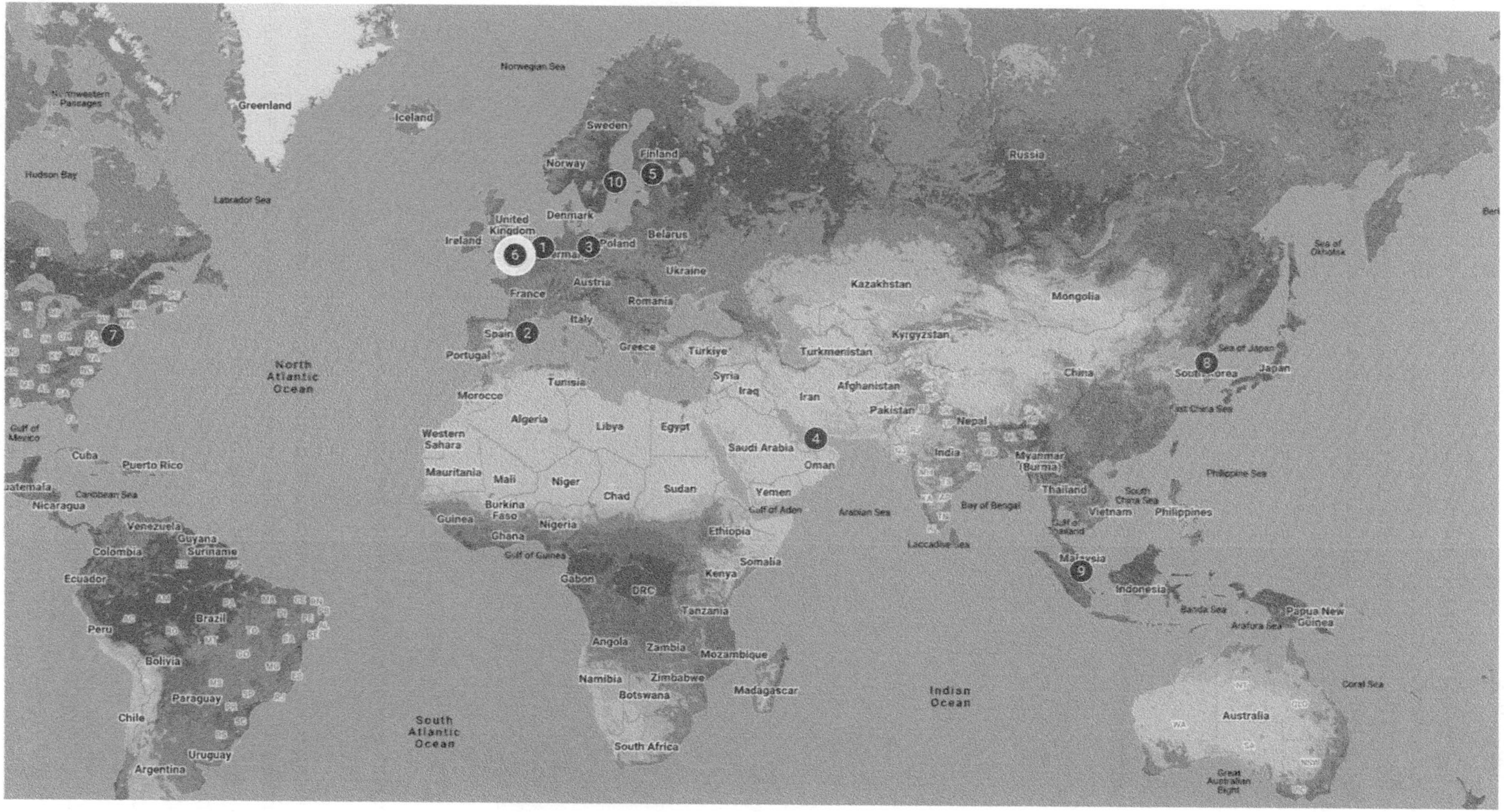

Greenland
Iceland
Norwegian Sea
Hudson Bay
Labrador Sea
Sweden
Norway
Finland
Russia
United Kingdom
Ireland
Denmark
Poland
Belarus
Ukraine
Austria
France
Romania
Italy
Spain
Portugal
Greece
Türkiye
Kazakhstan
Mongolia
Kyrgyzstan
Turkmenistan
Syria
Iraq
Iran
Afghanistan
Pakistan
Nepal
India
China
Sea of Japan
Japan
North Atlantic Ocean
Tunisia
Morocco
Algeria
Libya
Egypt
Saudi Arabia
Oman
Western Sahara
Mauritania
Mali
Niger
Chad
Sudan
Yemen
Gulf of Aden
Arabian Sea
Bay of Bengal
Myanmar (Burma)
Thailand
Vietnam
Philippines
Philippine Sea
Cuba
Puerto Rico
Caribbean Sea
Nicaragua
Burkina Faso
Guinea
Ghana
Nigeria
Gulf of Guinea
Ethiopia
Somalia
Kenya
Venezuela
Guyana
Suriname
Colombia
Ecuador
Gabon
DRC
Tanzania
Malaysia
Indonesia
Brazil
Peru
Bolivia
Angola
Zambia
Mozambique
Banda Sea
Papua New Guinea
Namibia
Zimbabwe
Botswana
Madagascar
Indian Ocean
Coral Sea
Australia
Paraguay
Chile
South Atlantic Ocean
South Africa
Uruguay
Argentina
1
2
3
4
5
6
7
8
9
10

5.11.1 Overview

London is a global economic and cultural hub. London's transportation system is complex and integrated conveying travelers on a large scale every day. The greater London metropolitan area. Consists of 32 London boroughs, and the City of London. Note that the city of Westminster is both a city and a Borough. The greater London region faces several transportation challenges that impact both residents and visitors.

Source: https://londonmap360.com/london-boroughs-map.

Challenges include traffic congestion, inadequate public transportation, air quality, accessibility and inclusivity, funding and financial constraints, and future planning and urban development challenges. The city also faces the challenge of multimodal integration and effective inclusion of stakeholders and the traveling public. London has implemented a series of smart mobility initiatives designed to address these challenges.

Agencies responsible for transportation in London include the following:

5.11.1.1 Transport for London (TfL)

This agency is responsible for most of the transportation systems within the greater London metropolitan area. It reports to the Greater London Authority. The agency is responsible for fares, grants, and investments. Modal responsibilities include the London underground, buses, trams, the Documents Right railway, overgrown railway services, river services, and the London Congestion Charging Program.

5.11.1.2 Greater London Authority (GLA)
This is the strategic regional authority for London. It is led by the mayor and the London assembly. Strategy and budget for transport for London fall under the remit of the Greater London Authority. It also sets transportation policy for London.

5.11.2 Smart Mobility Initiatives

5.11.2.1 Intelligent Transportation Systems
These include adaptive traffic signal control, the application of connected vehicle technologies, and real-time traffic management. With respect to intelligent transportation systems, London has placed a particular emphasis on advanced traffic signal control, or advanced traffic management systems. This includes the implementation of advanced microprocessor-optimized traffic signal controllers that can collect real-time traffic data to dynamically adjust signal timings. These are complemented by a range of detectors and sensors which are used to monitor traffic conditions and provide input to the traffic signal controllers. Sensors being used include inductive loops, video cameras, and specialist pedestrian and cycle detectors. An extensive, robust telecommunication network is used to enable data transmission between the traffic signal controllers, the traffic management center, and other systems. The telecommunications infrastructure allows real-time monitoring and coordination of traffic signals. The telecommunication infrastructure includes copper wire, fiber optics, wireless, and combinations of these telecommunication technologies. Several traffic management centers provide centralized monitoring and command-and-control for the entire traffic signal system. Data is received from sensors and used to support decision-making on signal timing and coordination. The traffic management centers also coordinate with other transportation management systems for incident management, public transportation, and emergency response. The traffic signal system is integrated with a wider intelligent transportation system. This uses a range of technologies and applications to enable more effective traffic management. The system utilizes dynamic message signs, traffic information distribution, traffic data analysis, and adaptive signal control techniques.

5.11.2.2 Smart Public Transportation
London has the advantage of a single unitary transportation agency addressing all modes of transportation in the city. This includes public transportation, traffic management, and maritime. An essential element of London's approach to smart public transit has been the introduction and successful operation of a contactless payment system known as the oyster card. In addition to providing access to multiple modes of transportation, the system generates valuable data regarding traveler characteristics, operating conditions, and trip patterns. In parallel with improving public transportation, transport for London, the responsible agency has implemented an effective congestion charging program, designed to manage the demand for private cars and delivery vehicles within central London. The congestion charging program is also used to administer and urban lower missions on (ULEZ).

5.11.2.3 Electric Vehicle Infrastructure
The city has implemented a comprehensive electric vehicle charging infrastructure, reinforced with incentives for electric vehicle adoption. The combination of both initiatives encourages

residents and businesses to transition from conventional vehicles to electric ones. At the time of writing London was on track to reach the 40,000–60,000 charging points needed by 2030 to match the demand for electric vehicles. London already has close to 13,000 charging points, representing an average of one charging point for every four registered electric vehicles [46] public investment in charging points has been complemented by private-sector implementation, often guided by the public sector.

5.11.2.4   Ridesharing Services

There are multiple private-sector ridesharing services available in London (https://ride.guru/cities/london-united-kingdom). These services cover multiple modes including traditional taxes, private car rideshare shared ride shuttles, Uber, and Lyft. Most of these use smartphone apps to enable travelers to register and access services and then pay for them electronically.

5.11.2.5   Data-driven Analytics

Transport for London makes extensive use of data-driven analytics to create investments and optimize service. A good example of transport for London's data-driven approach is the use of predictive maintenance to reduce costs and improve services [47]. Large-scale data collection and data analytics are being used to predict when components associated with the London Underground might fail, since around half of all delays in the London Underground are associated with assets, this is an extremely important application. The analytics approach features a heat map which indicates the impact of multiple factors on every system and component. This information is provided to specialist maintenance and operations staff, who use the input to make better decisions. Remote condition monitoring data is collected and analyzed to enable predictions.

5.11.3   Lessons Learned

Lessons learned from London's implementation of smart mobility initiatives include the following:

5.11.3.1   The Value of Adopting an Integrated, Multimodal Approach

London has recognized the importance of integrating multiple modes of transportation. This includes public transportation, cycling, walking, and ride sharing. This enables the creation of a comprehensive mobility ecosystem that promotes seamless connectivity while offering the traveler multiple options. This also enables London to encourage people to choose sustainable and efficient modes of transportation.

5.11.3.2   The Importance of Data-driven Decision-making

London has embraced the power of data to inform decision-making processes in smart mobility. Data is collected from a wide variety of sources such as sensors, GPS devices, and mobile apps. This provides greater insight into traveler patterns, congestion hotspots, and the variation in the demand for different transportation modes. Policymakers and enable to optimize existing infrastructure, plans, and define targeted interventions when necessary.

5.11.3.3 Effective Arrangements to Embrace Technology

London has defined arrangements to leverage new and emerging technologies that can be applied to London's transportation system. London is a leader in the practical application of technology such as contactless payment systems, congestion pricing, and advanced traffic management. Emerging technologies such as electric, autonomous vehicles, and mobility as a service platforms have been applied to take London to the forefront of smart mobility innovation.

5.11.3.4 Collaboration and Partnerships

Strong collaboration and partnerships between the public sector, the private sector, research institutions, and community organizations have been a significant feature of London success in smart mobility. The various partnerships and collaborations are designed to facilitate knowledge sharing, resource pooling, and joint problem-solving. It has been understood that resource pooling can have a significant effect on accelerating the benefits attained from smart mobility by leveraging private sector resources and motivation.

5.11.3.5 Active Involvement of End Users in the Design Process

London has learned that prioritizing the needs and preferences of residents and visitors is essential. The London approach is designed to actively involve end users in the design process and incorporate their feedback. This has resulted in mobility services that are more inclusive, accessible, and user-friendly. This has helped to increase the adoption rate of smart mobility services by ensuring that they are relevant and effective.

5.11.3.6 Integration of Sustainable and Environmental Considerations into Regulatory and Policy Frameworks

London experience is that balanced, robust policies, and regulatory frameworks are essential to success. It is important to establish clear rules, guidelines, and standards while providing an environment that fosters innovation. It has been found that there is an important balance between safeguarding public safety, privacy, and security while also adapting regulations to align with emerging technologies and business models. London has found integrating sustainability and environmental considerations into policies and regulatory strategies is key to success. Many of the smart mobility initiatives in London directly support sustainability and environmental considerations. These include the ultra-low emissions zone, congestion charging, and integrated transportation ticketing.

5.12 New York

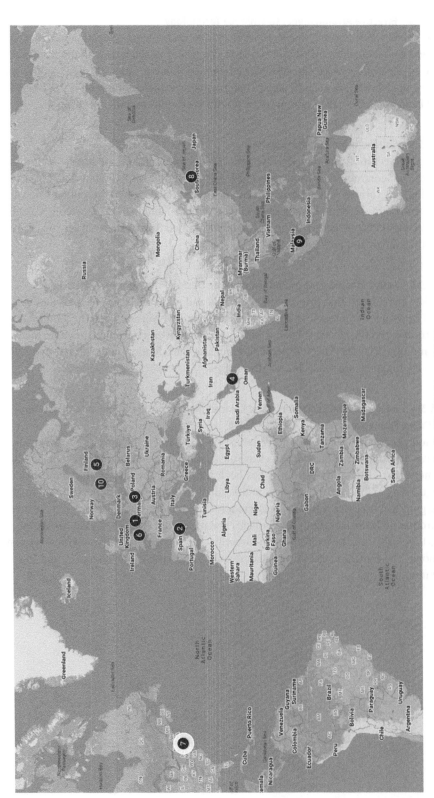

Source: https://en.wikipedia.org/wiki/New_York_metropolitan_area#/media/File:NY_Metro_Divisions.png.

5.12.1 Overview

The city of New York is comprised of a wide region encompassing multiple jurisdictions and multiple transportation agencies. The extent of the New York City metropolitan region is illustrated in Figure 5.20 and is home to 23.5 million people and 10.4 million jobs. The major public transportation agencies operating in the city include the following:

5.12.1.1 The Metropolitan Transportation Authority (MTA)

MTA operates the largest transportation network in the United States with 12 commuter rail lines, 3 subway lines, and 7 bus networks. The agency operates in a 5000 mi.^{2} traveled region around New York City.

1. NEW YORK (MANHATTAN)
2. BRONX (THE BRONX)
3. QUEENS
4. KINGS (BROOKLYN)
5. RICHMOND (STATEN ISLAND)
6. HUDSON

Figure 5.20 New York metropolitan region.

5.12.1.2 The Port Authority of New York and New Jersey (PA NY NJ)

This agency is a joint venture between the states of New York and New Jersey. It operates a large proportion of the Metropolitan region transportation infrastructure, including bridges, tunnels, airports, seaports, and commuter rail.

5.12.1.3 New York State Department of Transportation (NYS DOT)

This agency manages the city's transportation infrastructure including fairies, 12,000 miles of roads, highways, and pedestrian paths, and approximately 800 bridges. It also implements traffic safety measures and promotes the use of sustainable modes of transportation.

5.12.1.4 Amtrak

Amtrak, a railway operator, works nationally and plays a significant role in connecting New York with other major cities. Amtrak operates services from several stations in New York.

5.12.2 Smart Mobility Initiatives

As one of the world's most populous and dynamic cities, New York faces significant transportation challenges. These include traffic congestion, limited infrastructure, and significant environmental impact from mobility activities. The city has implemented several smart mobility initiatives to support the city's ambition to revolutionize urban transportation using technology and data-driven solutions.

5.12.2.1 Intelligent Transportation Systems

New York has implemented an integrated combination of sensors, telecommunication networks, and advanced analytics capability to provide a complete approach to transportation system management. This includes the generation of real-time travel information, parking availability information, and public transportation service bulletins. The data is used to optimize transportation operations and provide decision-quality information to travelers to support better decision-making.

5.12.2.2 Connected and Automated Vehicles

NYSDOT is at the forefront of connected and automated vehicle implementation. Pilot projects, regulatory actions, and the development of infrastructure support were part of the activities. The state of New York is also working on the implementation of connected and automated vehicle technologies including several pilot projects and initiatives. Automated vehicle testing is allowed on public roads under strict guidelines and with a trained operator in the vehicle. The city has also led the way by equipping thousands of vehicles including taxis, buses, and private cars with vehicle-to-vehicle and vehicle-to-infrastructure technology. This has enabled the city to gain daily experience and how to incorporate the technology to achieve improvements in traffic safety and efficiency. The city has hosted automated vehicle tests in several regions enabling the collection of vital data to help shape policies and infrastructure planning for the future. One of the most significant of these is the Brooklyn Navy Yard autonomous vehicle pilot project. [48]. This features a fleet of automated shuttle vehicles operating on a 1.1-mile loop within the yard 300-acre industrial park. The loop includes various types of network features including hills, intersections, and areas with heavy pedestrian traffic. The shuttles connect the New York City ferry stop to other points within the industrial park. It is open to the public, enabling it to serve as a demonstration showcase for automated vehicle technology.

5.12.2.3 Electric Vehicles and Charging Infrastructure

NYC DOT is creating a comprehensive network of publicly accessible electric vehicle charging stations [49]. These will allow electric vehicle owners to charge their vehicles while parked at home, at work, or curbside. This is in response that many people parking at the curb within the city do not have access to a home charger. To promote the use of electric vehicles, NYC DOT has partnered with a local energy company to install charging infrastructure at curbside locations across the city for a period of four years. The intention is for this to serve as a demonstration project. Operation of the infrastructure will be managed by a private-sector company.

To complement this initiative, NYC DOT is also building a network of fast charger hubs across the city. These will be in city-owned public parking lots and garages.

Source: https://www.nyc.gov/html/dot/html/motorist/electric-vehicles.shtml#/find/nearest.

In addition to electric vehicle charging infrastructure, the state is also promoting the use of electric vehicles by offering financial incentives. [50] This provides a rebate of up to $2000 for an electric vehicle purchase. This is also supplemented by a tax credit of up to $7500 for eligible buyers of qualified clean vehicles [51].

5.12.2.4 Mobility Hubs

New York has implemented several mobility hub initiatives to combine investments in infrastructure with those in smart mobility. These are innovative solutions to enable smooth interchange between modes improving the efficiency of multimodal trips, enhancing transportation connectivity, and promoting the use of sustainable transportation modes. The three most significant mobility hub projects are Brooklyn Navy Yard, Grand Central Terminal, and Hudson Yards. The Brooklyn Navy Yard [52], as previously described in the section on Connected and Automated Vehicles, features shuttle buses, New York City ferry services, car-sharing services, and bike-sharing facilities. These are combined with automated vehicles as discussed previously.

Grand Central terminal [53] is one of the largest train stations in the world and serves as a significant mobility hub which integrates train lines, subway lines, and buses. Ongoing development will also connect the Long Island Railroad to the station, further enhancing multimodal interchange possibilities.

Source: Oleksandr Dibrova/Adobe Stock.

Hudson yards [54] is Manhattan's newest neighborhood and integrates seven subway lines with ferries, buses, and bike sharing stations. The surrounding neighborhood was designed and active modes of travel.

Source: https://en.wikipedia.org/wiki/Hudson_Yards,_Manhattan#/media/File:Midtown_Manhattan_from_Weehawken_September_2021_HDR.jpg.

5.12.2.5 Smart Public Transportation

The city has a heavy reliance on public transportation and has invested in advanced technologies such as Artificial Intelligence, the Internet of Things, big data, and blockchain to enhance transfer operations and services. Key elements of smart public transportation in the city include real-time information services, automated payment systems, smart routing, and predictive maintenance. One of the artificial intelligence applications detects fare invasion on the transit system [55]. The system compares the number of paid fares with the actual number of people on the transit vehicles. Initial results suggest that the evasion rate was around 16%. The system enables peaks in the evasion affairs to be pinpointed by time of day and by station location.

5.12.2.6 Shared Mobility Services

The city defines shared mobility as the shared use of a private car, taxi, shuttle, carpooling, and van pooling. It also includes public transit, bicycles, and scooters. The term micromobility is reserved for bicycle and scooter sharing [56]. Shared mobility systems deployed in New York include docked bike share systems for the bicycle must be collected from and returned to one of a network of bicycle docks, and dockless systems where the bicycle can be left at any location at the discretion of the user.

5.12.2.7 Data Analytics and Integration

The city has invested in systems and arrangements to collect mobility data from a wide variety of sources. This data is analyzed to form the basis for future planning decisions and operational management support. The New York Metropolitan Transportation Council is responsible for a significant amount of data collection and analysis [57]. The New York State Department of Transportation also has significant activity in data acquisition and analysis [57].

5.12.3 Lessons Learned

5.12.3.1 The Importance of Bicycle Sharing and Mobility Hubs

New York has learned that the implementation of bicycle-sharing services and the creation of mobility hubs is of vital importance in promoting sustainable mode choice. Making shared bicycles accessible and integrating the service into the overall transportation network is crucial. Modal interchange through the creation of physical mobility hubs to complement the digital ridesharing services, delivers convenience and choice for travelers.

5.12.3.2 Embracing Multimodal Transportation

While the various smart mobility initiatives are focused on specific modes, it is important to embrace overall multimodal transportation. It is necessary to ensure the appropriate level of conductivity to allow different modes of transportation to work together and offer travelers multimodal trip possibilities. It is also important to include as many modes of transportation within overall multimodal management as possible. This should include public transportation, cycling, walking, and ride-sharing services. It is important to create a seamless experience for users by making transfers between modes and services as efficient and user-friendly as possible. In addition to the physical conductivity required, digital services that deliver real-time information on service availability and status are also crucial in delivering a multimodal transportation experience.

5.12.3.3 Prioritizing Data-driven Decision-making

New York has learned that smart mobility initiatives rely heavily on effective data collection and efficient analysis to turn the data into insight. It is necessary to leverage data from a wide variety of sources including public agencies and third-party private sector data providers. The insight acquired from data enables the identification of travel patterns, variations in the demand for transportation, and changes in transportation system operating conditions. In addition to improving operational management, the insight also informs policy decisions and enables effective future investments.

5.12.3.4 Fostering Public–Private Partnerships

The city has successfully collaborated with other government agencies and the private sector to accelerate the introduction of new technology. Collaboration has taken the form of pilot projects and testing initiatives for innovative solutions. An effective partnership approach can provide access to resources and funding that can achieve public sector objectives while creating market opportunities for the private sector.

5.12.3.5 Prioritizing Equity and Accessibility

The city has understood that an important aspect of smart mobility solutions involves a detailed understanding of the needs and characteristics of travelers. These are both residents and visitors. It is important to ensure that the application of smart mobility solutions benefit all travelers including underserved communities and vulnerable populations. This requires that a holistic approach is taken to the delivery of the service including consideration of appropriate fare options and that areas with limited transportation services are addressed.

5.12.3.6 Testing and Iteration

New York is understood that due to the nature of emerging technologies, the adoption of a test and data rate approach is most effective. The city has engaged in numerous pilots, demonstrations, and experiments to evaluate the effectiveness of different technologies and a small scale, prior to large-scale implementation. Deliberate arrangements to learn from both successes and failures have led the city to adopt a continuous improvement strategy, ensuring that smart mobility initiatives are as effective as possible.

5.12.3.7 Community Engagement

New York is involved in residents community organizations and advocacy groups as part of a focused effort to gather input and feedback. Effective community engagement has involved the clear communication of the smart mobility initiatives to the various groups and understanding their feedback based on practical use of the technology. The city has found that effective community engagement helps build trust, increases adoption rates for the services, and ensures that solutions align with local needs.

5.12.3.8 Considering the Broader Urban Context

The city has understood that for maximum effectiveness smart mobility should be seen as part of a broader urban planning strategy. Smart mobility initiatives have been reinforced by coordinating land-use planning and investing in pedestrian-friendly infrastructure. This includes policies to reduce congestion and pollution. It is vital to take a holistic approach to urban development ensuring that smart mobility is not implemented in isolation but as part of a coordinated effort.

5.13 Seoul

5.13.1 Overview

Seoul is one of the leading megacities in Asia. It has a high population density with over 25 million residents in the metropolitan area. Major transportation challenges include continuing organization, air quality, and traffic congestion. The wide-ranging extent of the metropolitan area is illustrated in Figure 5.21.

The city experiences heavy traffic congestion especially during peak periods, leading to substantial travel times and lowering the city's productivity. Transportation in Seoul is the responsibility of the following public agencies.

5.13.1.1 Seoul Metropolitan Government (SMG) [58]

This agency plays a central role and policymaking and overall management of transportation within the city. It oversees urban planning, transportation infrastructure development, and public transportation services. It is also responsible for managing the bus services using private contractors.

5.13.1.2 Seoul Transportation Corporation

This agency is responsible for the operation and management of subway lines within the Seoul metropolitan area. It was formed in 2017 when the Seoul Metro Corporation and the Seoul Metropolitan Rapid Transit Corporation merged to improve efficiency of management on the subway lines.

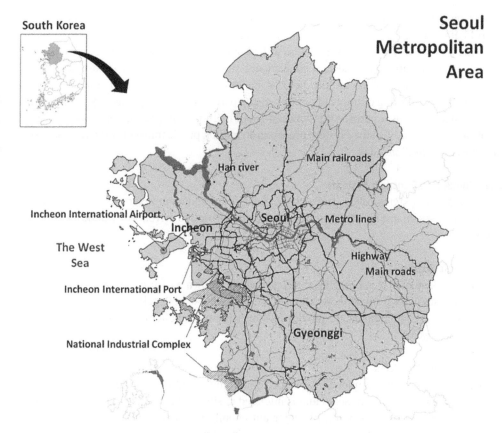

Figure 5.21 Seoul metropolitan region. *Source:* https://www.researchgate.net/figure/Seoul-Metropolitan-Area_fig1_318505374.

5.13.1.3 Seoul Metropolitan Infrastructure Headquarters

This is a Department of Seoul Metropolitan Government focusing on construction and maintenance of infrastructure including roads, bridges, and tunnels within the Seoul metropolitan area.

5.13.1.4 Seoul Transport Operation and Information Services (TOPIS)

This agency collects and analyzes transportation data in real-time, representing, and integrated transport operation Center for Seoul. It provides real-time information on traffic, manages traffic signals, and supports public transportation operation.

5.13.1.5 Korea Transport Institute (KOTI)

In addition to conducting research, this agency provides policy recommendations for transportation within the Seoul metropolitan region and other parts of the country.

5.13.1.6 Korea Rail Network Authority (KRNA)

This is a national agency that plays an essential role in planning construction and maintenance of railway networks that service the Seoul metropolitan area in addition to national networks.

5.13.2 Smart Mobility Initiatives

5.13.2.1 Mobile Payment System

Seoul Metropolitan Government has implemented one of the leading electronic payment systems in the world. It is known as T-money [59]. The contactless smart card and cell phone-based system allow payment for subway and train services, as well as buses. The card can be loaded and reloaded at subway and train stations and several convenience stores. Some taxi services also accept the money as payment. The system can also be used for low-value transactions in restaurants and cafés. The money is a leading example of public–private partnership, it is operated by the T money company limited which is a partnership between Seoul metropolitan government, LG CNS, and the credit card union. The T money system processes 40 million public transportation transactions, with a value of $17 million a day.

5.13.2.2 Intelligent Transport Systems

Intelligent transport systems collect and analyze transportation data in real time. They also manage the city's traffic signals and offer real-time traffic information to drivers and public transportation users.

5.13.2.3 Bicycle Sharing

The city has implemented a public bike-sharing program known as Ttareungi [60]. The bicycle-sharing system uses its own smartphone app and the Discover Seoul Pass [61] to manage access to bicycles. Bicycle parking facilities are provided all over the metropolitan area to enable users to rent and return bicycles.

5.13.2.4 Automated Vehicles

The Korean government has implemented a special automated vehicle testing facility known as K-City. It is located 20 miles from Seoul and is intended to simulate real-life conditions for testing automated vehicles. This has stimulated major automobile manufacturers and several startup companies to develop automated vehicles and car-sharing services. The city is also implementing automated vehicle support infrastructure around the city [62].

5.13.3 Lessons Learned

5.13.3.1 The Importance of an Integrated Approach

The city has learned that integration is required on multiple levels, including public agencies, public, and private sector cooperation, and coordinated management of multiple modes. Integration has been supported by the introduction of a multimodal transportation payment system and the travel information system that provides information to the users for multiple modes of transportation.

5.13.3.2 Data-driven Decision-making

Seoul relies heavily on data collection and analysis using advanced technologies. The insight extracted from the data is used to guide decision-making on transportation planning and transportation operations. The data is collected from a wide range of sources including the public sector and private sector. The city has understood that major systems designed to deliver services such as electronic payment, can also be major generators of important data.

5.13.3.3 Planning to Embrace Technology and Innovation

The Seoul metropolitan government has been proactive in developing initiatives to adopt emerging technologies. These include test sites and major implementations from which practical lessons are learned regarding the capabilities and limitations of technologies. By embracing technology and innovation Seoul has been able to improve the traveler experience, increase accessibility and provide a greater sense sustainability in transportation.

5.13.3.4 Public–Private Partnerships

Seoul is home to one of the leading public–private partnerships for smart mobility in the world. The T money system is owned and operated by a combination of public agencies and private company partners. This not only delivered a high capability system but also proved a business model of combining public resources and motivation with private sector energy. The system was implemented at an impressive pace and manages large volumes of financial transactions every day.

5.13.3.5 People Focused Approach to Design, Continuous Evaluation, and Improvement

Seoul has placed a focus on understanding and prioritizing the needs and preferences of travelers within the metropolitan area. They understood the importance of actively seeking feedback from citizens and visitors. This enables the city to tailor smart mobility services to the specific needs and characteristics of different user groups, raising the level of inclusivity and improving the overall user experience. This process also includes continuous evaluation and improvement on the effectiveness of the smart mobility initiatives. The city readily monitors and evaluates the performance of the smart mobility initiatives and the overall transportation system.

5.13.3.6 Developing Sustainable and Eco-friendly Solutions

The city has placed a strong emphasis on sustainable and eco-friendly smart mobility initiatives. This includes the promotion of electric vehicles and investing in electric vehicle charging infrastructure, and bicycle-sharing systems. These encourage travelers to make better modal choice decisions leading to a higher level of sustainability, minimizing the impact of the transportation process and the environment.

5.14 Singapore

5.14.1 Overview

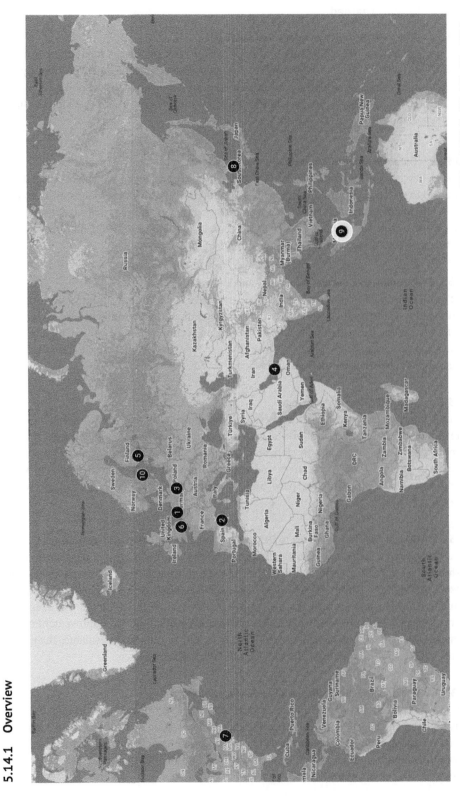

Source: https://www.google.com/maps/place/Singapore/@0.7709072199,3520017,6z/data=!4m6!3m5!1s0x31da11238a8b9375:0x887869cf52abf5c4!8m2!3d
1.3520831!4d103.8198361!6zL20vMDZ0MnQ?entry=ttu.

Singapore is a city-state in Southeast Asia on the southern tip of the Malay Peninsula. It is a thriving metropolis with world-class infrastructure and fully integrated islandwide transportation network.

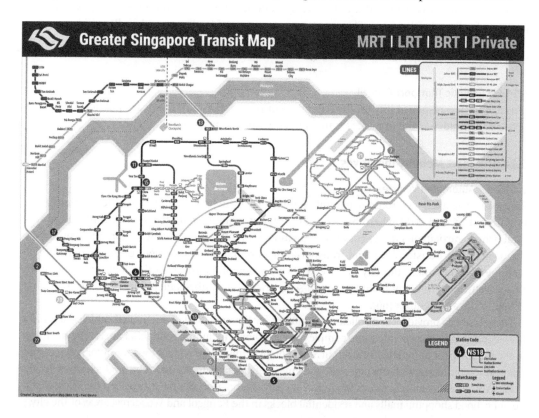

The following agencies of responsibility for transportation in Singapore.

5.14.1.1 Ministry of Transport

This agency has overall responsibility for connectivity of land, air, and sea in Singapore. While holding ultimate responsibility for setting policy, the implementation of policies and day-to-day operations are managed by several statutory boards under the auspices of the Ministry of Transport. These include the Land Transport Authority Maritime and Port Authority of Singapore public transport counsel, and the Civil Aviation Authority of Singapore.

5.14.1.2 The Civil Aviation Authority of Singapore [63]

This agency is responsible for developing and operating the air hub and aviation industry in Singapore. This includes the development of Singapore as a global air hub and ensuring safe and efficient air traffic flows. The agency has responsibility for aviation safety and developing industry. The agency represents Singapore when contributing to international aviation and has responsibility for contingency planning and crisis management.

5.14.1.3 Land Transport Authority [64]

This agency leads land transportation activities in Singapore, with responsibilities for planning, designing, building, and operating Singapore's land transport infrastructure and systems.

5.14.1.4 Maritime and Port Authority of Singapore [65]

This agency has responsibility for developing Singapore as a global hub and international maritime center. It manages Port and Maritain development activities maritime import regulations and planning for Port development. The agency manages several facilities that handle a wide range of cargo including containers and conventional bulk cargo. The agency manages terminal operators from the private sector.

5.14.1.5 Public Transport Council

This agency has responsibility for the regulation of public transportation fears and ticket payment services. Also advises the minister for transport on public transportation matters.

5.14.2 Smart Mobility Initiatives

5.14.2.1 Automated Vehicles

Singapore has taken a structured, careful approach to the deployment of automated vehicles. The initiative started in 2014 when the committee on autonomous road transport for Singapore was established. Participants in the committee include both public and private sector enterprises with the objective of transforming land transportation through the deployment of automated vehicle technology. Since then, the city has taken an incremental approach to the deployment of automated vehicle technology including small-scale trials and testbeds including one North, Jurong Island, NUS, and Sentosa [66].

These testbeds provide a controlled environment for the operation of automated vehicles and have been expanded to incorporate more complex environments including residential areas. The next step in the roadmap for automated vehicles in Singapore is limited deployment for commuter services and selected towns, along with operational deployment of truck platoons and utility vehicles in selected areas. Autonomous mobile robots for delivering parcels and groceries have also been trialed between neighborhood shopping malls and residential blocks in the Punggol and Woodcrest areas of the city.

*Source:*https://www.straitstimes.com/singapore/over-50-driverless-vehicles-approved-for-trials-on-spores-roads-and-public-paths-in-last-5-years.

5.14.2.2 Digital Payment Solutions

The transport authority has implemented an account-based ticketing system called SimplyGo. This allows several different payment mechanisms to be used to pay for transportation in the city including contactless smart card and cell phones using near-field communication technology. The payment system can be used on buses and trains. Building on this initial success, trials are now underway to use radio frequency identification and bluetooth for hands-free payment for services.

5.14.2.3 Smart Traffic Management

Within the auspices of an overall intelligent transport system, the city has implemented an advanced traffic management system. At the heart of the system lie two hubs, one for data and one for operational control. The data hub is known as I transport that consumes all ITS data to be integrated into a unified platform for distribution to the ITS operations control center. This lateral control center works 24 seven to monitor and manage traffic and incidents on expressways and road tunnels. Incident management includes the coordination of recovery crews and traffic marshals. The center works in close association with other agencies responsible for transport in the city. Information is also shared by the land transport authority through the LTA data model which enables distribution to the private sector.

5.14.2.4 Shared Mobility

Shared mobility services facilitating the use of a variety of vehicles on a temporal basis are operated by several private sector companies in Singapore. The leading company, grab, offers carpooling services, private taxis, and coaches utilizing a variety of ridesharing models. Services also include special vehicles equipped with Pepsi covers, enabling pets and service animals to travel with the traveler. This also extends to special services from selected hospitals enabling wheelchair-bound travelers to use the vehicle while remaining in the wheelchair. The company has also expanded to include food delivery and financial services.

5.14.2.5 Mobility as a Service

The city has experimented with various mobility as a service apps in close cooperation with the private sector. These have not been commercially sustainable for several reasons including the Covid-19 pandemic. There is also some concern that mobility as a service is not attractive to travelers in Singapore due to the high population density and compact transportation network. City continues to study the effective implementation of mobility as a service.

5.14.2.6 Cycling Path Networks

A network of second passes in the development to promote the use of the bicycle as a mode of transportation. A national level that a 525 km cycling pass which will be expanded to approximately 1300 km × 2030. The intention is to connect commuters between their homes and the SMRT stations, enhancing the value of bicycle use as a first mile, last-mile mobility strategy.

5.14.2.7 Smart Parking

The city has implemented a parking guidance system that delivers information through 29 roadside variable message signs. This informs drivers of the location of parking lots with available parking spaces, reducing the amount of time spent looking for a space.

5.14.2.8 Congestion Pricing [67]

Singapore was one of the pioneers of urban congestion pricing. This was introduced as a traffic management and emissions reduction strategy in 1975. It was operated within a two-square-mile central business district area. This required vehicles entering the store through 28 entry points required to display a prepurchase daily or monthly windshield license. These were available at retail outlets including banks, convenience stores, and gas stations. The system was operated on a manual basis using roadside enforcers with citations sent by mail to vehicle owners. The city converted the system to electronic road pricing after extensive testing from 1995 to 1997. The electronic system was introduced in 1998 and takes a fully automated approach to collecting charges for entering the zone. The system operates over the busiest time of the day from 7 a.m. to 7 p.m. on weekdays. The system uses overhead gantries to enable the electronics and enforcement system to be mounted over the road.

5.14.3 Lessons Learned

5.14.3.1 Comprehensive Planning

Like many cities around the world, Singapore has understood the need for a comprehensive planning approach to smart mobility. In addition to the islands of smart mobility deployed, there is a need to integrate them and coordinate the deployments but long-range transportation planning and land use planning within the city. The integration of smart mobility with conventional technologies such as road signs and road marks is also vital to ensure consistency and to ensure that the highest values obtained from sunk investments. This also extends to integration of modes within a multimodal transportation planning framework.

5.14.3.2 Data-driven Decision-making
Perhaps because of its rather unique position as a unitary Transportation Authority within an island, city-state, the land transport authority has become a leader in the effective collection, information processing, and in-state extraction process for smart mobility data. The city has understood the need to collect and analyze a wide variety of data and other to raise situational awareness and provides full support for better decision-making for investments and operations.

5.14.3.3 The Importance of Public–Private Partnerships
The city has been very effective in harnessing the motivation and resources of the private sector, working in close coordination to ensure the achievement of public sector objectives, while also fostering markets for the private sector. This has enabled the city to accelerate the deployment of smart mobility and achieve early benefits.

5.14.3.4 Adopting a User Centric Approach
The city is focused on using smart mobility technologies to deliver a detailed understanding of the characteristics and preferences of all users within the city. Insight is used to guide the design process and to inform future investments. Operational management is considerably enhanced by the insight delivered due to effective data collection from multiple sources and data unification.

5.14.3.5 Multimodal Electronic Payment Systems
In addition to the convenience of avoiding cash payment and payment for multiple modes in one transaction, the city is also understood that contactless payment mechanisms can have a significant impact on accessibility to public transportation for citizens of all abilities.

5.14.3.6 Public Awareness and Education Including the Needs for Behavior Change Award Programs
In addition to the need to raise public awareness regarding the capabilities of new smart mobility technologies, it has also been understood that a compelling case must be made for travelers to change behavior. This should include consideration of appropriate reward programs for behavioral change.

5.14.3.7 Shared Mobility as the Foundation for Mobility as a Service
The city has experimented with shared mobility solutions and is home to several private sector companies offering commercially viable services. The city also understood that shared mobility and shared mobility platforms could provide an effective foundation for future mobility as a service. The city has realized that she had mobility services could be an important step that will facilitate the creation and use of smart mobility services in the city.

5.15 Stockholm

5.15.1 Overview [68]

Stockholm is a city built on 14 islands and connected by 57 bridges. This represents a unique challenge with respect to transportation. It is a capital city of Sweden and is the country's largest city with over 1 million inhabitants. It is a growing metropolis featuring increasing population and economic growth. This leads to significant vehicular density and traffic congestion.

Agencies responsible for transportation and Stockholm
Swedish transport administration
Stockholm public transport administration

5.15.2 Smart Mobility Initiatives

5.15.2.1 Integrated Transport Network

Stockholm has developed an integrated transport network. This combines buses, trains, trams, bicycles, and ferries, to offer a seamless experience for multimodal travel. An integrated electronic ticketing system and timetable coordination reinforce the connectivity across multiple modes. The Stockholm Metro can be considered the backbone of the integrated transportation system, complemented by buses trams, ferries, and bicycles. The digital information system including displays, mobile apps, and online portals also provides real-time information across modes and trip planning facilities.

5.15.2.2 Smart Traffic Management [69]

The city already has an advanced traffic signal management system and is building on this to make it even better. This includes incorporation of sensor-based monitoring techniques and the integration of data from multiple sources. The city has switched from underground inductive loops to aboveground smart cameras for vehicle detection which reduces costs and improves the capability of the detection system. The traffic management system ingests data from multiple sources and makes it available for traffic management, transportation planning, and to the private sector. The current Smart Traffic project has the objective to test and evaluate new ways of planning and managing the traffic in the city by incorporating the use of real-time traffic data and multiple data sources. Other notable features of the Smart Traffic project include dynamic capacity adaptation smart pushbuttons and traffic signals, and the development of sophisticated time-of-day traffic plans. The Smart Traffic project works in close cooperation with Project IoT Stockholm. This project is establishing a central IoT platform to enable data collection in real time, while managing and controlling sensors.

Source: univrses / https://univrses.com/press-releases/univrses-leads-ground-breaking-smart-parking-project-funded-by-vinnova-and-ffi/ / last accessed September 23, 2023.

Smart parking [70].

The project involves the installation of suitably equipped smartphones on the Stockholm taxi fleet. These were used to collect probe vehicle data for various transportation uses including mapping on street parking availability in real-time. The probe vehicle data was collected using machine vision algorithms, utilizing the cameras on board the smartphones.

5.15.2.3 Bicycle Friendly Infrastructure Development

An important feature of Stockholm's approach to bicycle friendly infrastructure has been the development of a network of cycle pathways, designed from the beginning to operate as a single system. It incorporates bicycle lanes across the city and neighborhood lanes for short-distance trips. Because of the extensive network, the travelers are offered door-to-door availability of cycling facilities. Reserved cycle lanes of differentiated from roads and sidewalks. This bicycle-friendly infrastructure is complemented by a range of micromobility providers in the city. These private sector companies are working in coordination with the city to offer e-scooter and e-bicycle rental on an on-demand basis. The city has recently conducted a pilot project on data sharing with these micromobility providers.

5.15.2.4 Electric Mobility

Under the auspices of the eRoad Arlanda project an electrified road test track is being used to charge electric vehicles with no need to stop. A test track is located on a 10 km section of road 893 between the Arlanda cargo terminal and the Rosesberg logistics area. 2 km of the road has been equipped with a charging system. This directly addresses electric vehicle users concerns regarding the availability of charging infrastructure and the possibility of running out of electricity. The system uses an electric rail installed on the road to power and recharge vehicles during the trip. The focus is on charging heavy goods vehicles. The project is managed by the Swedish Transportation Administration and involves a consortium of public and private sector enterprises.

5.15.2.5 Electric Vehicles [71]

While trains and buses in Stockholm have been using 100% renewable energy since 2017, the city is still working toward the development of the most sustainable public transportation system in the world. Goals include fossil-free maritime traffic, emissions reduction, and using 15% less energy per passenger kilometer. The Stockholm public transport administration has been studying the effects of introducing electrically powered buses [72]. As part of this effort, the city of Norrtalge, [73] in northern Stockholm, has replaced the entire bus transportation system with 110 electrically powered vehicles. The buses run on renewable electricity, offering more sustainable and quieter public transportation. With respect to electric cars, the city has had considerable success. About 36% of all new vehicle purchases and electric. This has been achieved by investing in a network of public charging facilities and providing incentives for electric vehicle adoption. Sweden provides grants of up to 25% of electric cars cost, available to both individuals and businesses. There are other grants in place for electric buses and other public vehicles as well as a considerable tax deduction for electric vehicles. In addition, car owners who subscribe to parking spaces in Stockholm can get free charging for electric vehicles as part of the parking fee.

5.15.2.6 Mobility as a Service [74]

The pioneering service for mass in Stockholm is known as UbiGo. The service provides access to public transportation, car rental, car sharing, taxis, and bicycles. Users preregister and pay a single monthly subscription. A single account can be shared by all members of the household, with payment for services linked to this account. The app features significant integration with public transportation services.

5.15.2.7 Congestion Pricing

This project, sometimes known as the Stockholm congestion tax is a leading example of a successful congestion pricing program. It was designed to reduce traffic congestion and promote the use of alternative sustainable transportation. There are several key components including infrastructure, cameras, sensors, and a backend system. Overhead gantries are equipped with cameras and sensors to capture and collect data about vehicles entering and exiting the congestion pricing zone. Initially, RFID tags were used as a primary means of vehicle identification, supplemented with automated number plate recognition. However, the accuracy of the ANPR was found to be high enough to rely on these without the RFID tags, saving a considerable amount of capital and operating costs, by obviating the need to equip vehicles. The automated number plate recognition system is designed to collect data regarding the owner of the vehicle and identify individual vehicles. An extensive back-end system was developed that automatically extracts and recognizes license plate numbers from the images captured by the ANPR system. The ANPR data along with other data collected at the roadside is transmitted to a centralized backend system which conducts the transaction processing for the congestion pricing. Payment date along with the vehicle ID is combined with license plate and vehicle registration data to identify the vehicle owner, the entry point, or exit point, and the specific time of day. The backend system then calculates the appropriate charge based on time of day, vehicle type, and whether the vehicle meets specific exemption criteria. Vehicle owners are notified of their congestion tax obligations with various payment methods available such as online, mobile apps, or physical payment points. Because the system is deemed to be a taxation system and all existing legislation about the enforcement of taxation can be applied to this system also. A public–private partnership has been implemented to provide customer service and support at public agency offices and a range of convenience stores. The backend system features robust data privacy and security measures due to the sensitive nature of the data collected. This protects the personal information of vehicle owners and ensures the overall integrity of the system.

5.15.3 Lessons Learned

5.15.3.1 The Need for an Integrated Approach Focused on Sustainability

The city is focused on combining different modes of transportation including public transportation, cycling, and walking to create a comprehensive and interconnected mobility network. The city has embraced the need for sustainability and has understood that an integrated approach ensures that residents are encouraged to use the best mode and right for the trip with consideration of sustainability.

5.15.3.2 Effective Public and Private Sector Collaboration

Perhaps the most significant example of public and private sector collaboration in Stockholm was the implementation of the congestion pricing system. The entire system was designed, developed, and implemented by a private-sector company under the auspices of a public contract. This ensured that the system implementation was guided by the needs of the public sector, while effectively harnessing the resources of the private sector, and expertise obtained from other locations around the world. This public–private collaboration also extends to micromobility and radiating services as well as electric vehicle charging infrastructure.

5.15.3.3 Data-driven Decision-making

Stockholm has understood the need to collect data from a wide variety of data sources from both the public and private sector. The city has put considerable effort into the combination and analysis of the data to make it available for a wide variety of applications. Using the insight obtained from the data of the city can determine traffic patterns, identify congestion hotspots, and optimize public transportation routes. This enables data-driven insights to ensure effective allocation of resources and support evidence-based policy decisions.

5.15.3.4 People Focused Design

The city actively involved the residents in the planning and design process. The insight obtained has been incorporated into plans and designs that take full account of user needs and preferences. The resulting user-friendly designs, encourage travelers to use new services and take ownership in the resulting outcomes.

5.15.3.5 Integrating Scalability and Flexibility

Struggles approach to smart mobility fully supports scalability and flexibility. In most cases, the city adopted an incremental approach to technology implementation featuring pilots and demonstrations. Arrangements are in place to learn practical lessons and gain experience that can be used to guide large-scale deployment of the technology and technology-based services. This approach extends to business models as well as technologies. The city can accommodate new and emerging technologies because of the flexibility offered by the incremental approach.

5.15.3.6 Appropriate Policy and Regulatory Support

The city has understood the need to create supportive policy and regulatory framework for smart mobility initiatives. While it is essential to address and preserve the needs of the public, the city has understood that there is also a need to foster innovation startup companies, and other private sector efforts to establish and sustain new markets related to smart mobility. The city has obtained a balance between safety and efficiency by integrating smart mobility into the policy and regulation frameworks that also enable innovation.

5.15.3.7 Continuous Monitoring and Incremental Evaluation

Aligned with the incremental approach involving pilots and demos, the city has understood the importance of continuous monitoring and incremental evaluations of the effects of the smart mobility initiatives. This enables course corrections to be made in the business case created for widescale deployment. The information from continuous monitoring and evaluation also provides raw material to explain to the traveling public the importance of smart mobility services.

5.16 Summary

There is a large amount of activity with respect to smart mobility initiatives in the leading cities around the globe. This chapter scratches the surface of the amount of activity and investment at this current time. It is not a comprehensive catalog of all an activity but provides some highlights that indicate the scale of investment and the successes being achieved. The global situation is dynamic so providing a comprehensive catalog would not add much value as the picture would change even before the book was published. The smart mobility revolution is real and moving forward. Figure 5.22 provides a summary of the smart mobility implementation highlights from each of the 10 cities studied.

While each city has a unique set of characteristics and different starting points depending on prior investments. There is a degree of commonality among the smart mobility initiatives deployed. There is a strong focus on promoting transportation modes that offer an alternative to the private car, improving sustainability through reducing emissions and energy. A considerable amount of investment has been placed in electric vehicle infrastructure and incentives to encourage car drivers and businesses to adopt electric vehicles. There is a common understanding that an integrated and comprehensive approach is required for successful smart mobility implementation in cities. This requires that several different modes are managed as a single entity and that the transportation system is managed as a network, rather than individual pieces. While smart mobility initiatives and others to be efficiently managed and implemented tend to be conducted on a project basis, results of such projects must fit together to form an overall framework or architecture. Another common thread is the investigation of the practical application of autonomous vehicles. While these are still in their infancy, considerable effort has been placed around the world to implement pilots and demonstrations leading to a better understanding of the capabilities and limitations of this technology. Another major theme across the globe is the application of mobility as a service. This can be viewed as the tip of the iceberg, one-eighth of the total picture that is visible. It has been recognized that to deliver comprehensive mobility as a service, with travelers able to access information and make reservations across all modes of transportation in an urban area, then make a single electronic payment, a large amount of conductivity, coordination, and consolidation is required. Practitioners around the world have always understood that the goal is a completely integrated urban transportation system from hardware, software, data, and people perspectives. This has been difficult to achieve, so it is interesting that mobility as a service may act as the leading edge for the work required to achieve this. Shared mobility services and micromobility services have also been extensively deployed worldwide. The practical application has led to the understanding that these two can form the foundation for a bigger integrated approach. In fact, shared mobility and micromobility can provide the foundation for mobility as a service.

Finally, what could be considered as the original smart mobility, intelligent transportation systems are also heavily deployed in these major cities. This includes the installation of data collection capabilities using sensors and probe vehicle data, a centralized traffic management or transportation management center and distribution channels for providing information to the traveler and to

	City	Highlights			
1	Amsterdam	extensive cycling infrastructure and bike sharing programs	Electric vehicle charging infrastructure and incentives	Smart parking systems with real-time availability information	Intelligent traffic management systems and adaptive traffic signals
2	Barcelona	extensive bike sharing networks and pedestrian friendly initiatives	Electric vehicle charging infrastructure and incentives	Smart traffic management systems with real-time monitoring and control	Integrated transportation apps for multimodal journey planning
3	Berlin	sustainable transportation	Extensive public transportation network	Bike sharing programs	Initiatives to promote electric mobility
4	Dubai	Smart traffic signals	Autonomous shuttles	Mobile app for integrated transport planning	Advanced air mobility
5	Helsinki	Mobility As a Service platform integrating various modes	Real-time transportation information and ticketing systems	Electric and shared mobility options for dedicated lanes and charging infrastructure	Smart traffic management and predictive analytics for congestion
6	London	the need for an integrated and multimodal approach	Suitable policy and regulatory frameworks	Embracing new technology	Taking a user centric approach to smart mobility design
7	New York	bike sharing	Car sharing	Digital payment options for public transportation	Smart traffic management using probe vehicle data
8	Seoul	advanced public transportation systems	Electric vehicle adoption	Electric and automated vehicle technologies	Large-scale electronic payment for transit and services
9	Singapore	Extensive use of autonomous vehicles and mobility sharing services	Intelligent transportation systems for traffic management and public transit	Integrated payment systems for multiple modes of transportation	Dedicated cycling infrastructure and bike sharing programs
10	Stockholm	Efficient public transportation system with real-time updates	Electric vehicle incentives and charging infrastructure	Smart traffic management systems and congestion pricing	Integrated mobility apps for multimodal transportation options

Figure 5.22 Implementation highlights.

other systems. It has also been understood that there is considerable scope to improve the operation of traffic and transportation management centers using new technologies such as artificial intelligence, location intelligence, and probe vehicle data collection techniques. The sunk

#	Lesson	Explanation
1	Integration	Smart mobility solutions are better when there integrated into a system approach to transportation, this includes modal integration, coordination and integration with existing infrastructure and facilities and urban planning initiatives
2	Data-driven decision-making	Efficient data collection and analysis is vital to enabling effective decision-support Use of multiple traditional and innovative data sources
3	User centric approach	Understanding and focusing on the needs and preferences of users, including user experience, accessibility, and affordability Continuous feedback from the user is also important
4	Public-private partnerships	Leveraging the expertise and resources of both sectors, taking full advantage of public sector resources and investments can develop more effective and sustainable solutions
5	Sustainability and inclusive	Reducing the environmental impact of transportation, or the undesirable side effects This includes promoting electric vehicles and prioritizing low carbon modes of transportation Equitable access to services for all residents and visitors
6	Scalability and adaptability	Scalable and adaptable to changing needs and technologies, this should include planning for future growth and the use of flexible infrastructure designs to allow cities to expand and upgrade their transportation systems as new technologies emerge
7	Public awareness and education	Educating the public about the benefits of smart mobility, encouraging behavioural changes, raising awareness of our availability and advantages of transportation alternatives and teaching users on how to get the best from the wide range of options
8	Collaboration	Among various takeovers, government agencies, transportation operators, technology providers, and citizens
9	Piloting and experimentation	Testing new mobility concepts and technologies, taking an iterative and agile implementation approach
10	Regulatory framework	A supportive regulatory framework to enable smart mobility innovations, while also ensuring safety, privacy, unfair competition
11	Continuous learning and adaptation	User feedback, monitoring the impact of initiatives, making necessary adjustments
12	Focus on cycling infrastructure	An extensive cycling infrastructure is required to encourage cycling as a mother transportation This includes bike lanes, bridges, and parking facilities This also includes bike sharing facilities and mobility hubs
13	Pedestrian friendly urban planning	Creating vibrant and walkable streets, widening sidewalks, pedestrian zones, and improve crossings
14	Political commitment and long-term vision	Policymakers prioritize sustainable mobility with strong political commitment and long-term vision for sustainable transportation
15	Electric vehicle infrastructure	Charging stations and infrastructure to support the adoption of electric vehicles
16	Mobility as a service	Multiple applications supporting better decision-making by travellers enabling them to plan, reserve and pay for multiple modes of transportation

Figure 5.23 Lessons summary.

investment in the establishment and operation of these management centers can be leveraged through the adoption of emerging technologies.

The use of electronic ticketing across all modes of transportation and for small-value retail transactions is also a common feature. It has been understood that such implementations while delivering value and benefit to the traveler, can also serve as the source for a considerable volume of data that can provide insight into the variation of transportation demand and operating conditions in an urban area. Electronic payment is seen as another unifying factor to enable the achievement of multimodal transportation management.

Some of the cities have also deployed congestion pricing to provide a more direct means of managing private car traffic in urban areas. These include London, Singapore, and Stockholm. The cities have had considerable success in applying this technique while also making parallel improvements in alternative modes.

Figure 5.23 provides a summary of the most important lessons learned in the study of the 10 cities.

References

1 Government of the Netherlands. Ministry of Infrastructure and Water Management website, August 7, 2023. https://www.government.nl/ministries/ministry-of-infrastructure-and-water-management, retrieved August 07, 2023, at 9:12 PM Central European Time.

2 City of Amsterdam website August 07, 2023. https://www.amsterdam.nl/en, retrieved August 07, 2023, at 9:24 PM Central European Time.

3 NS website, August 07, 2023. https://www.ns.nl/en, retrieved August 07, 2023, at 9:26 PM Central European Time.

4 pro rata website, August 07, 2023. https://www.prorail.nl, retrieved August 07, 2023, at 9:26 PM Central European Time.

5 Wikipedia. GDB Amsterdam, August 07, 2023. https://en.wikipedia.org/wiki/GVB_(Amsterdam)#:~:text=GVB%20(often%20informally%20called%20de,the%20metropolitan%20area%20of%20Amsterdam, retrieved August 07, 2023, at 9:27 PM Central European Time.

6 smou website. August 7, 2023. https://www.smou.cat/en, retrieved August 07, 2023, at 9:28 PM Central European Time.

7 Bax and Company website. Bringing autonomous vehicles to Barcelona with Pendel mobility, August 7, 2023. https://baxcompany.com/insights/bringing-autonomous-vehicles-to-barcelona-with-pendel-mobility, retrieved August 07, 2023, at 9:30 PM Central European Time.

8 Airport Industry News. WF S and Aena trial autonomous cargo transport at Barcelona, August 07, 2023 airport. http://airportindustry-news.com/wfs-and-aena-trial-autonomous-cargo-transport-at-barcelona-airport, retrieved August 07, 2023, at 9:31 PM Central European Time.

9 TMB website. Automated metro, August 07, 2023. https://www.tmb.cat/en/about-tmb/transport-network-improvements/automated-metro, retrieved August 07, 2023, at 9:32 PM Central European Time.

10 Metropolitan Barcelona website. Smart city Barcelona. A city of the future? August 07, 2023. https://www.barcelona-metropolitan.com/features/smart-city-Barcelona/#:~:text=Smart%20traffic%20lights%20help%20emergencyare%20transitioning%20to%20LED%20technology, retrieved August 07, 2023, at 9:35 PM Central European Time.

11 TMB website. Journey planner, August 07, 2023. https://www.tmb.cat/en/home, retrieved August 07, 2023, at 9:36 PM Central European Time.

12 Cities forum website. Superblock Barcelona – a city redefined, May 31, 2021. https://www.citiesforum.org/news/superblock-superilla-barcelona-a-city-redefined, retrieved August 07, 2023, at 9:37 PM Central European Time.

13 City of Berlin website. Public transportation Berlin, August 07, 2023. https://www.berlin.de/en/public-transportation, retrieved August 07, 2223, at 9:38 PM Central European Time.

14 BVG website. Our BVG apps at a glance, August 07, 2023. https://www.bvg.de/en/subscriptions-and-tickets/all-apps, retrieved August 07, 2023, at 9:39 PM Central European Time.

15 Stromnetz Berlin website. Are you looking for a connection, August 07, 2023. https://www.stromnetz.berlin/en/electric-mobility, retrieved August 07, 2023, at 9:40 PM Central European Time.

16 Wallbox website. The ultimate guide to EV incentives in Germany, August 07, 2023. https://blog.wallbox.com/ev-incentives-germany, retrieved August 07, 2023, at 9:41 PM Central European Time.

17 Rhino car website. Electric car hire Berlin, August 07, 2023. https://www.rhinocarhire.com/Germany/Berlin-Downtown-Car-Hire/Electric.aspx, retrieved August 07 2023, at 9:42 PM Central European Time.

18 BVG Jelbi website. Jelbi one for all. https://www.jelbi.de, retrieved August 07, 2023, at 9:43 PM Central European Time.

19 Trafi website. We are Trafi, August 07, 2023. www.trafi.com, retrieved August 07, 2023, at 9:43 PM Central European Time.

20 Search medium website. Intelligent parking: a tale of five cities, October 12, 2018. https://medium.com/predict/intelligent-parking-a-tale-of-five-cities-31b14056261 retrieved August 07, 2023, at 9:51 PM Central European Time.

21 Government of Dubai. Portal website, August 07, 2023. https://www.rta.ae/wps/portal/rta/ae/home?lang=en, retrieved August 07, 2023, at 9:53 PM Central European Time.

22 Dubai pulse. Dubai Maritime City Authority, August 07, 2023. www.dubaipulse.gov.ae/organisation/dubai-maritime-city-authority-dmca#:~:text=Dubai%20Maritime%20City%20Authority%20(DMCA)%20monitors%2C%20develops%20and%20promotes,infrastructure%2C%20operations%20and%20logistics%20services, retrieved August 07, 2023, at 9:54 PM Central European Time.

23 United Arab Emirates Ministry of Energy and Infrastructure website. August 07, 2023. www.moei.gov.ae/en/about-ministry/TMB.aspx, retrieved August 07, 2023, at 9:55 PM Central European Time.

24 Volvo bus is global website. Volvo buses electoral mobility solution trials in Dubai, 17th November 2021. https://www.volvobuses.com/en/news/2021/nov/Dubai-trials-electric-solution.html, retrieved August 07, 2023, at 9:56 PM Central European Time.

25 Arabian Business. Dubai announces $2.3 million self-driving bus challenge tests, July 13, 2023. https://www.arabianbusiness.com/industries/transport/dubai-announces-2-3m-self-driving-bus-challenge-tests, retrieved August 07, 2023, at 9:58 PM Central European Time.

26 Golf News. UAE, now in Dubai: ride a driverless bus for free, February 17, 2022. https://gulfnews.com/uae/now-in-dubai-ride-a-driverless-bus-for-free-1.85802078, retrieved August 07, 2023, at 9:59 PM Central European Time.

27 Government of the United Arab Emirates website smart mobility solutions, August 07, 2023. https://u.ae/en/information-and-services/transportation/smart-mobility-solutions, retrieved August 07, 2023, at 10 PM Central European Time.

28 Government of Dubai, Road Transport Authority website. Dubai bus on demand, August 07, 2023. https://www.rta.ae/wps/portal/rta/ae/home/rta-services/service-details?serviceId=11170105, retrieved August 07, 2023, at 10:01 PM Central European Time.

29 The National News. Dubai's main road network to be covered by congestion busting smart traffic system, August 07, 2023. https://www.thenationalnews.com/uae/2022/10/23/dubais-main-road-network-to-be-covered-by-congestion-busting-smart-traffic-system, retrieved August 07, 2023, at 10:06 PM Central European Time.

30 CAFU website. 24/7 emergency fuel delivery less than 20 minutes, August 07, 2023. https://www.cafu.com/emergency-fuel-service, retrieved August 07, 2023, at 10:08 PM Central European Time.

31 Dubai Government. Law number 26 of 2015 regulating data dissemination and exchanging the Emirate of Dubai, 2015. https://dlp.dubai.gov.ae/Legislation%20Reference/2015/Law%20No.%20(26)%20of%202015.pdf, retrieved August 07, 2023, at 10:09 PM Central European Time.

32 HSL website. HSL app, August 07, 2023. https://www.hsl.fi/en/tickets-and-fares/hsl-app, retrieved August 07, 2023, at 10:10 PM Central European Time.

33 HSL HRT website. Helsinki and Espoo, August 07, 2023. https://www.hsl.fi/en/citybikes/helsinki, retrieved August 07, 2023, at 10:11 PM Central European Time.

34 CGI website. CGI Move360, August 07, 2023. https://www.cgi.com/fi/fi/tuoteratkaisut/cgi-move360, retrieved August 07, 2023, at 10:12 PM Central European Time.

35 whim website. write e-scooters with whim, August 07, 2023. https://whimapp.com/helsinki/en/mode/e-scooters, retrieved August 07, 2023, at 10:13 PM Central European Time.

36 Bird cities blog. Bird e-scooters launch in Helsinki: a guide for how to ride safely, April 1, 2022. https://www.bird.co/blog/bird-e-scooters-launch-helsinki-guide-how-to-ride-safely, retrieved August 07, 2023, at 10:14 PM Central European Time.

37 Fintraffic website. The traffic data ecosystem in brief, August 07, 2023. https://www.fintraffic.fi/fi/fintraffic/liikenteen-dataekosysteemi-lyhyesti, retrieved August 07, 2023, at 10:15 PM Central European Time.

38 KPMG. 2019 autonomous vehicles readiness index, 2019. https://assets.kpmg.com/content/dam/kpmg/xx/pdf/2019/02/2019-autonomous-vehicles-readiness-index.pdf, retrieved August 07, 2023, at 9:45 PM Central European Time.

39 Fabulos. Helsinki pilot, August 2023. https://fabulos.eu/helsinki-pilot, retrieved August 07, 2023, at 9:46 PM Central European Time.

40 Helsinki Partners website. MUJI and finish startup Sensible 4 take self-driving innovation from Finland to Japan, June 16, 2022. https://www.helsinkipartners.com/case/muji-and-finnish-startup-sensible-4-take-self-driving-innovation-from-finland-to-japan, retrieved August 07, 2023, at 9:47 PM Central European Time.

41 City of Helsinki website. Data and digitalization help run a smart city, August 2023, City. https://www.hel.fi/en/decision-making/strategia-ja-talous/kaupunkistrategian-ja-talouden-seuranta/alykasta-helsinkia-johdetaan-tiedolla-ja-digitalisaatiota-hyodyntaen, retrieved August 07, 2023, at 9:48 PM Central European Time.

42 Whim website Helsinki. How to get there, August 07, 2023. https://whimapp.com/helsinki/en, retrieved August 07, 2023, at 10:16 PM Central European Time.

43 Whim Maas global website. Maas global, August 07, 2023. https://whimapp.com/maas-global, retrieved August 07, 2023, at 10:18 PM Central European Time.

44 European Commission. European alternative fuels observatory, August 07, 2023. https://alternative-fuels-observatory.ec.europa.eu/transport-mode/road/finland/incentives-legislations, retrieved August 07, 2023, at 10:19 PM Central European Time.

45 Helen website. Find charging points, August 07, 2023. https://kartta.lataus.helen.fi, retrieved August 07, 2023, at 10:20 PM Central European Time.

46 London government website. Mayor of London and London councils and incisors more electric charge points across the capital, April 11, 2023. www.london.gov.uk/Mayor%20of%20London%20

and%20London%20Councils%20announce%20thousands%20more%20electric%20charge%20 points%20across%20the%20capital#:~:text=%E2%80%9CThere%20is%20a%20comprehensive%20 charging,per%20cent%20increase%20from%202019, retrieved August 07 2023, at 10:21 PM Central European Time.

47 Computerworld. How Transport for London is using predictive analytics to keep the Underground moving, February 2, 2018. https://www.computerworld.com/article/3427521/how-tfl-is-using-predictive-analytics-to-keep-the-underground-moving.html, retrieved August 07, 2023, at 10:24 PM Central European Time.

48 Curbed magazine New York. Self-driving cars debut at the Brooklyn Navy Yard, August 6, 2019. https://ny.curbed.com/2019/8/6/20756830/autonomous-vehicle-self-driving-car-brooklyn-nyc, retrieved August 07, 2023, at 10:29 PM Central European Time.

49 New York City Department of Transportation website. Electric vehicles, August 07, 2023. https://www.nyc.gov/html/dot/html/motorist/electric-vehicles.shtml#/find/nearest, retrieved August 07, 2023, at 10:29 PM Central European Time.

50 New York State website. Drive clean rebate for electric cars, August 07, 2023. https://www.nyserda.ny.gov/All-Programs/Drive-Clean-Rebate-For-Electric-Cars-Program, retrieved August 07, 2023, at 10:30 PM Central European Time.

51 New York State DOT website. Inflation reduction act: vehicles, August 07, 2023. https://www.nyserda.ny.gov/All-Programs/Inflation-Reduction-Act/Vehicles, retrieved August 07, 2023, at 10:30 PM Central European Time.

52 Brooklyn Navy Yard website. A place to build your history, August 07, 2023. https://brooklynnavyyard.org, retrieved August 07, 2023, at 10:31 PM Central European Time.

53 Grand Central terminal website. Always moving, Grand Central terminal, August 07, 2023. https://www.grandcentralterminal.com, retrieved August 07, 2023, at 10:32 PM Central European Time.

54 Hudson Yards website. All Roads Lead Here, August 07, 2023. https://www.hudsonyardsnewyork.com/getting-here, retrieved August 07, 2023, at 10:33 PM Central European Time.

55 Statescoop website. NYC using AI to track subway fare evasion, report claims, July 24, 2023. https://statescoop.com/nyc-ai-subway-fare-evasion/#:~:text=Seven%20undisclosed%20subway%20 stations%20infrom%20the%20city's%20transportation%20authority, retrieved August 07, 2023, at 10:34 PM Central European Time.

56 New York State MPO website. Shared mobility, August 07, 2023. https://www.nysmpos.org/sharedmobility, retrieved August 07, 2023, at 10:35 PM Central European Time.

57 New York Metropolitan transportation Council website. Transportation data and statistics, August 07, 2023. https://www.nymtc.org/en-us/DATA-AND-MODELING/Transportation-Data-and-Statistics, retrieved August 07, 2023, at 10:35 PM Central European Time

58 So Metropolitan Government website. Service and policy, August 07, 2023. https://english.seoul.go.kr, retrieved August 07, 2023, at 10:37 PM Central European Time

59 Tmoney website. Overview, August 07, 2023. https://eng.tmoney.co.kr/en/aeb/aboutUs/overview/overview.dev;jsessionid=Ul9qf5nfv51farHWMMkiaZ4waZTmuJjkBAIf7RMOegNA3a2K7IsQ0jYFj32jJLqf.czzw02ip_servlet_ksccweb, retrieved August 07, 2023, at 10:37 PM Central European Time.

60 Seoul bike website. August 07, 2023. https://www.bikeseoul.com, retrieved August 07, 2023, at 10:39 PM Central European Time

61 Discover Seoulpass website. About pass, August 07, 2023. https://www.discoverseoulpass.com/app/guide/index/how/bikeseoul, retrieved August 07, 2023, at 10:40 PM Central European Time.

62 Seoul Z website. Top autonomous vehicle startups in Korea – Korean mobility startups, June 27, 2022. https://www.seoulz.com/top-autonomous-vehicle-startups-in-korea-korean-mobility-startups, retrieved August 07, 2023, at 10:41 PM Central European Time.

63 Civil aviation Authority of Singapore website. August 07, 2023. www.caas.gov.sg retrieved August 07, 2023, at 10:41 PM Central European Time.

64 Land Transport Authority website. August 07, 2023. www.lta.gov.sg/content/ltagov/en.html, retrieved August 07, 2023, at 10:42 PM Central European Time.

65 MPA Singapore website. August 07, 2023. www.mpa.gov.sg/home, retrieved August 07, 2023, at 10:42 PM Central European Time.

66 Land Transport Authority website. Autonomous Vehicles, August 07, 2023. www.lta.gov.sg/content/ltagov/en/industry_innovations/technologies/autonomous_vehicles.html, retrieved August 07, 2023, at 10:43 PM Central European Time.

67 Federal Highway administration website. Lessons learned from international experience in congestion pricing, August 07, 2023. https://ops.fhwa.dot.gov/publications/fhwahop08047/02summ.htm#:~:text=Congestion%20pricing%20has%20been%20a9%3A30%20in%20the%20morning./, retrieved August 07, 2023, at 10:44 PM Central European Time.

68 City of Stockholm website. August 07, 2023. https://start.stockholm/other-languages/the-city-of-stockholm, retrieved August 07, 2023, at 10:44 PM Central European Time.

69 City of Stockholm. Smart Traffic Management, August 07, 2023. Https://Smartstad.Stockholm/Smart-Traffic, retrieved August 07, 2023, at 10:45 PM Central European Type.

70 Univrses website. Univrses leads groundbreaking smart parking project funded by Vinnova and January 14, 2020, univrses. https://univrses.com/press-releases/univrses-leads-ground-breaking-smart-parking-project-funded-by-vinnova-and-ffi/, last accessed September 23, 2023.

71 Your Living City website. How Stockholm is leading the world in electric vehicle usage, August 3, 2022. https://www.yourlivingcity.com/uncategorized/how-stockholm-is-leading-the-world-in-electric-vehicle-usage, retrieved August 07, 2023, at 10:48 PM Central European Time.

72 Smart city Sweden. Sustainable public transport Stockholm, August 07, 2023. https://smartcitysweden.com/best-practice/368/sustainable-public-transport-in-stockholm, retrieved August 07, 2023, at 10:49 PM Central European Time.

73 Transdev website. Transdev brings new claimant smart and high-performance buses to northern Stockholm, September 9, 2022. https://www.transdev.com/en/success/transdev-brings-new-climate-smart-and-high-performance-buses-to-northern-stockholm, retrieved August 07, 2023, at 10:50 PM Central European Time.

74 Fluidtime website. The MaaS project UbiGo became the starting point for the new mobility paradigm that is becoming a reality in cities and regions around the world, August 07, 2023. https://www.fluidtime.com/en/ubigo, retrieved August 07, 2023, at 10:52 PM Central European Time.

6

Planning for Smart Mobility

6.1 Informational Objectives

After you read this chapter, you should be able to:

- Define planning
- Define the objectives of smart mobility planning
- Explain a framework for approaching it
- Explain the starting point
- Describe the elements of a roadmap
- Describe the characteristics of a vision
- Explain how to implement a rapid focused approach to planning
- Describe a way to approach smart mobility planning based on practical experience

6.2 Introduction

First, a formal definition of planning:

> Planning is deciding in advance what to do, how to do it, when to do it, and who should do it. It bridges the gap from where the organization is to where it wants to be. The planning function involves establishing goals and arranging them in logical order. [1]

Essentially, it is all about building a bridge from the current situation to the desired future vision. The authors have discovered that in addition to defining what, how, when, and who, it is valuable to also answer the questions of why and where. Using the opening part of the Rudyard Kipling poem "I keep six honest serving men" is a good way to remember this [2]:

> *I keep six honest serving men*
> *(They taught me all I knew).*
> *their names are what and why and when*
> *and how and where and who*

For the authors, this sums up the six fundamental questions to be answered in the planning process. Each question will be discussed in a bit more detail in the following sections.

Smart Mobility: Using Technology to Improve Transportation in Smart Cities, First Edition. Bob McQueen, Ammar Safi, and Shafia Alkheyaili.
© 2024 The Institute of Electrical and Electronics Engineers, Inc. Published 2024 by John Wiley & Sons, Inc.

6.2.1 What?

This can refer to the plan implementation of the anticipated outcomes. It can be a description of the initiative that is planned or the outcomes that are desired. These can also be described as goals. Over the course of experience, the authors have also found it very important to define what is not wanted as well as what is desired. A clear understanding of what is to be achieved can be complemented by a clear understanding of what is to be avoided, or if not avoidable then managed.

6.2.2 Why?

This question addresses the reasons that the plan is being developed and the implementation being considered. This will capture the desired values, benefits, and effects.

6.2.3 When?

The timescale over which the implementation will be spread in the order in which things will be done. This can also describe the order in which the goals are to be achieved. They are really the same thing from different perspectives.

6.2.4 How?

This can describe the technologies that will be applied, and the solutions that will be implemented in the services that will be delivered to achieve the goals. It can also describe the organizational arrangements that will be used to support the implementation of the business models that will be used to effectively engage the private sector.

6.2.5 Where?

This may seem like a trivial question, however by defining where the intended audience is to be addressed by the implementation and the boundaries of the zone of influence or area of effect are also defined.

6.2.6 Who?

The cast of characters, the list of players, the stakeholders, the organizations that will be involved. This question can also address the nature and characteristics of the end user. This would also involve the identification of those who are likely to benefit from the implementation and those who might not.

Note that in practice each of these questions is not completely independent and there is significant common ground or overlap between them. This is okay, for the intention is to be comprehensive in the answers to the questions while not constraining a complete answer. The more light that the answers to the six questions shed, the better the results of the planning process.

6.3 The Objectives of Planning

So, the objective of smart mobility planning is to answer the six questions while defining a starting point, the desired endpoint or vision, and a roadmap to get from where the city or region is to where it wants to go. Of course, part of the planning process is to ensure that appropriate agreement is achieved on all of these elements. Socializing the answers is a crucial part of setting the scene for moving forward effectively. This is particularly important within the context of smart mobility as it is likely that there will be a wide range of participants or stakeholders whose agreement to the overall direction will be important to success.

The authors also believe that successful planning requires a balance between the level of detail provided and studied and the need to move quickly through the planning process. It is of particular concern that most smart mobility plans take somewhere between nine and 12 months to create. Smart mobility involves the application of new and emerging technologies making it important to define the plan in at least some detail as quickly as possible. Speed, efficiency, and effectiveness in the smart mobility planning process can also deliver another extremely valuable tool. There is a chicken and egg situation in that the existence of the plan would make it easier to define and agree to the plan. In working with people from the information technology industry, the authors have become aware that this is a problem that they share and that their response to the problem is to adopt something called rapid evolutionary development. This has been employed in software development and information technology for at least the past 20 years, so it is a good model to use as it has been proven to be robust and effective. This takes advantage of new tools and capabilities to move very quickly from initial concept to working prototype.

The lessons learned in developing the prototype are then incorporated into more detailed and refined designs, using an iterative process. The prototypes are then used to communicate to the end user to seek agreement on the characteristics and features required in the final product or solution. It also fits well with experience that implementing something provides a richer stream of lessons learned and practical experiences than simply planning or documenting. Remember the old adage "a picture is worth a thousand words," it is also useful to add that "a demonstration is worth a thousand pictures." This is true when dealing with new technology or concepts that are relatively complex. In many cases, high-level technical knowledge is required to explain how technology works and what the effects will be when it is applied. It is often much more efficient to simply demonstrate. A detailed exposition of rapid evolutionary development techniques for information technology is not provided in this book, but that are several very good references that can be used to get more information [3–5].

6.4 The Essentials of an Effective Smart Mobility Planning Approach

In simple terms, an effective smart mobility planning approach will answer the six questions, define the starting point, define the endpoint, and prepare a roadmap that connects the two. This is a very high-level view of what has to be accomplished. There is considerable detail attached to each of the questions, the starting point, the endpoint, and the roadmap. This will be discussed in more detail later in this chapter. It is also important that what is meant by "objectives" is addressed. The objectives that were defined above for the smart mobility planning approach or process are in fact interim objectives or a means to an end. There is also a set of end objectives, which can be referred to as "goals" which describe the desired outcomes beyond the deliverables of the planning process. This addresses the destination end of the bridge by describing and defining where the city or region wants to go. Essentially, while the bridge is an extremely valuable output from the planning process, it is not the bridge itself which is the end goal it is where the bridge leads to, the destination, that is the goal.

6.5   Smart Mobility Planning Goals

There is a range of legitimate smart mobility planning goals that can be brought together as a goals framework. It is highly unlikely that a focus will be placed on achieving one goal to the detriment or exclusion of others. Therefore, the best approach is to develop a multi-criteria framework that takes account of the need to support multiple goals. Several years ago, one of the authors provided consulting services to a large (more than 500 miles of toll road network) toll road agency. In discussions, the senior client expressed a wish to make the toll road the safest road in the world. The author's rather flippant response was well let us just close it, because when there is no traffic there can be no crashes. Of course, the point was made that is not possible to focus exclusively on one goal. The toll road management had other goals, improving safety was a very important one. The toll road served as an important economic development tool by connecting different parts of the state, a crucial component in congestion management, and a major facility for improving the efficiency of the overall transportation service delivery in the region. All of these goals had to be achieved, while also preserving safety. So, the choice of goals is not a binary one, but rather one of balance and in some cases prioritization.

Based on extensive interaction with smart mobility practitioners in the course of consulting assignments, a list, or menu of possible goals for smart mobility, is offered. These are not in any particular order and are intended to represent a broad sweep across the mobility domain based on experience and perspective. It is not designed as a recipe for smart mobility goal definition, but rather a model which can guide the development of a specific customized list for the city or region being addressed and the program being planned. Also, this list can be considered as high-level headings which can be detailed with subheadings according to the needs of the program. A wide range of key performance indicators or performance measurements have been applied to mobility and these can be sliced and diced into many different configurations. The author's experience has been distilled over about 40 years in the advanced transportation technology business resulting in the following list. Again, it should be noted that this is designed as a model and a starting point and not as an off-the-shelf recipe for defining goals for the smart mobility program. Also, the names chosen for these goals are based on the authors preference for explaining it to clients and end users. Other names can be chosen, depending on what works best for the audience is being addressed.

6.5.1   Safety

Smart mobility will have a number of very important safety goals. These include reduction in fatalities, injuries, and damage caused by crashes. They also include the safety of transit systems and those facilities designed to transport goods. Over the course of experience, it has also been noted that it is very important to address the perception of safety as well as the actual safety levels. This is an outreach and communications issue that will be dealt with later in the book. Perception is important as it is the basis for end-user decision-making.

6.5.2   Efficiency

Efficiency relates to how much value is obtained from the amount of resource invested. It can be measured in terms of minimizing delay, stops, travel time, or travel time reliability. Using this definition, it is possible to achieve efficiency goals in excess of 100%, when the amount of resource invested is outweighed by the value of the benefits obtained. This may also be characterized as "return on investment." Equity and climate change are also included under the heading of this efficiency goal. Equity refers to ensuring that the effects of the smart mobility implementation are fair to all sectors of the community and did not unduly disadvantage a particular group. Climate change refers to the energy and emissions implications of the smart mobility implementation. In recent times both subjects have become of greater importance and are the focus of considerable activity. As stated earlier, the actual goals used will be designed to fit the needs and it may be necessary to call equity and climate change out as high-level goal headings, and other to communicate the importance of the subjects.

6.5.3   User Experience

As discussed above, perception is an important part of goal definition and goal attainment. User experience has an element of perception. This goal is focused on improving the end-user experience, where the end user includes the traveler and those responsible for operating transportation systems. User experience management is an important part of smart mobility implementations, as friends in the IT industry say, "the end user will have an experience whether you have designed it or not." Having accepted that user experience is inevitable, then the job becomes one of ensuring that the experience is positive. This can also be extended to happiness and well-being. In the United Arab Emirates, this became such an important focus that the post of minister of state for happiness was created in 2016. This was intended to harmonize all government plans, programs, and policies to achieve a happier society. In a cabinet reshuffle in July 2020, the portfolio was transferred to the Ministry of Community Development [6]. This visionary government understood the importance of focusing on people as a critical part of government activities. The authors believe that smart mobility must take the same approach. Smart mobility implementations have often been encountered, where the effort has been distracted by technologies and the focus on people has been lost.

6.5.4   Economic Development

Note that many of these are closely related to the problems described in Chapter 2 where a problem statement for smart mobility was provided. In many cases, the definition of goals can be considered as a mirror image of the definition of problems. After all, a rational approach to the definition and description of a problem is to determine a solution and seek the appropriate resources to implement the solution. This is also a good opportunity to conduct a check on the logic of both the problem statement and the goals. If a problem does not have an accompanying goal, then the needs have not been completely addressed. If a goal cannot be matched to a problem, then the usefulness of the goal must be questioned. So why are both problems and goals needed? While problems are stated in such a way that the focus on issues that must be addressed and problems that need to be fixed. Goals on the other hand have a positive perspective on what is planned to address the problems. They are very closely related, and it is important that they are; however, each brings its own perspective and adds information to the overall picture.

6.6   The Value of Planning

6.6.1   Defining What You Want (And What You Do Not Want)

One of the author's grandmothers was known for using proverbs to communicate knowledge. Short, memorable sayings were used to pass valuable information from one generation to another. While literacy was lower for that generation, making verbal communications important, this also proved to be an excellent and efficient way to communicate. In addition to the core information, this communication channel also benefits from being compelling and getting the attention of the person intended to receive the information. One of her sayings was

> *There are two things you need to do to get what you want. First, you have to decide what you want and second you have to ask for it.*

As usual, the proverb or saying seems to be very simple, almost trivial at first glance. However, the more you think about it, the more the depth and value of the information is revealed. Maybe this is why this particular method of communication is so powerful. It is simple enough to persist from home and enables reevaluation over time. Persistence also enables recollection and application to appropriate circumstances. This points to two of the essential values in planning. Deciding what is required and identifying the best way to obtain it. There is an unspoken part to the need for planning in this respect. It is also very important to have a clear definition of what is to be avoided, especially when dealing with multiple stakeholders and partners.

6.6.2 Getting What You Want

The definition of what you do not want also forms the basis for a robust approach to risk management in which to identify things that can be avoided and develop ways to mitigate things that cannot.

6.6.3 Finding the Best Partners

A successful approach to smart mobility planning will also enable the identification and engagement of the best partners for the initiative. These might be public-sector agencies or private-sector enterprises. The ability to communicate effectively with potential partners is a very valuable part of the smart mobility plan.

6.6.4 Getting Everyone to Agree

In addition to finding the best partners, it is also necessary to get all appropriate stakeholders to agree to the forward direction that is being proposed. Identifying the appropriate stakeholders is a considerable exercise. The identification of appropriate stakeholders must take account of sources of funding, levels of authority, and how much influence stakeholders have in future projects. It will also take account of the resources available within each stakeholder organization and how these can be used to assist and enable the successful implementation of the smart mobility initiative.

6.6.5 Providing the Basis for Progress

A successful smart mobility plan will also form the basis for progress. In addition to a generalized description of the proposal, it will also provide sufficient information to enable the plan to be applied in practical terms. It could be argued that the best plan is the one that gets implemented, regardless of its level of sophistication or glossy cover. Perhaps this is the overarching goal of the planning process stated in a very simple way. The purpose of planning is to get the things done that enable attainment of the goals.

6.6.6   Enabling Integration Between Transportation Service Delivery Activities

Planning is an important part of the spectrum of activities that are supported to deliver effective transportation. The high-level view is illustrated in Figure 6.1 which shows five primary activity areas for transportation service delivery: planning, design, delivery, operations, and maintenance. Figure 6.1 was first introduced in Chapter 1 and is reproduced here for convenience.

While each primary activity area is extremely focused and requires specialist expertise, as discussed in Chapter 1. There is a strong need for communication between those involved in each activity area. For example, planning can benefit from practical experience and design, delivery, operations, and maintenance. There are other areas that can also benefit from experience and lessons learned gained during planning activities.

6.6.7   A Cheaper Better Way to Go

One of the biggest values of planning lies in the effective use of resources. The best planning approaches will allow for what could be called a "waterski" approach. A large number of alternatives, options, and roadmaps are addressed and analyzed by the planning process. This wide and shallow approach allows resources to be managed across as wide a reach as possible. The actual reach can also be extended by learning lessons and gaining practical experiences from previous deployments. While the reach is long, and the depth is shallow it does not need to be indiscriminate. Gaining information and knowledge from prior deployment enables resources to be focused on promising approaches and enables dead-end approaches to be considered and ignored early.

6.7   Conventional Planning Approach

While there are several variations on a theme, the typical approach to smart mobility and transportation plan development involves the following steps:

- Development of an intelligent transportation systems vision
- Identification and agreement of goals and objectives

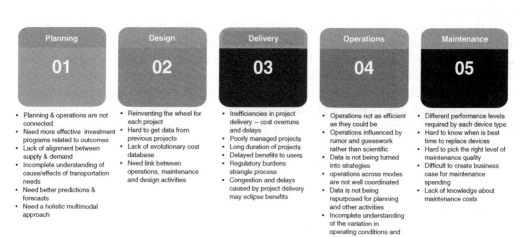

Figure 6.1   Transportation service delivery primary activities.

- Assessment of existing conditions and development of an infrastructure inventory
- Needs assessment
- Definition of appropriate smart mobility strategies
- Definition of a regional architecture
- Development of a concept of operations

These steps can be supported by a specialist consultant or conducted in-house by the appropriate planning agency. In either approach, there is typically a significant steering committee, task force, or working group to provide consultation and then put at regular points along the plan development process. This parallel development of appropriate solutions in close consultation with the client group can typically take anywhere from 9 to 18 months. It does result in a robust and effective plan that has evaluated most available options, considered the regional starting point, and defined a viable plan for moving forward. Having been involved in this type of project that adopts this parallel long-term approach to planning where aware that it provides a couple of challenges. As mentioned earlier, working with emerging technologies puts pressure on developing at least a high-level plan at the earliest possible opportunity to make sure that technology progress is captured. This approach also restricts the ability of those developing a plan to communicate and demonstrate technology solutions and anticipated effects to the client group at an early stage.

To get around this, there is a danger of expanding duplicate resources. At an early stage in the planning process, there is a need to educate and make the client group aware of the possibilities. This is usually done by defining some strawman solutions. After the client group is briefed and the needs of the city or region are fully understood, then further resources must be expended to define and communicate the proposed solution. The consensus development process which is crucial for a successful outcome to the plan can consume a lot of resources and take a significant amount of time. The need to support a long-time span dialogue of between 9 and 12 months with the client group also expends resources in the form of formal events and deliverables to communicate progress and expected results to the client group. Unless every member of the client group is fully aware of the possibilities and constraints of technological solutions, there can be a lack of agreement in the client group at this stage in the planning process. Another typical drawback is that private sector solution providers and potential investors are held at arm's length until the plan is complete.

This prevents the private sector from providing valuable input into what can be done and what has been successful in other places. Let us try to illustrate this with a quite silly analogy. Imagine that you are trying to sell tools to prehistoric cave-dwelling people. Having never seen these before, the client group needs to have a fundamental awareness of the characteristics of the tool and the results to be expected. Language is also a difficulty, so perhaps the best results can be achieved by demonstration and visual learning. It is not suggested that the client group can be compared to prehistoric cave-dwelling people. In fact, the authors believe quite the opposite, most client groups today have a more detailed awareness and understanding of technology capabilities than those from the past. The advent of the Internet, YouTube, and remote learning capabilities provides a multichannel source for people whose jobs are focused on transportation and mobility, to also gain a detailed understanding of information technology and other advanced technologies used in smart mobility. The point is that demonstration and visual learning are powerful tools in any era and for any planning process. If some sort of prototype or solution is planned anyway, then why not develop it sooner, and use it as a communication tool?

6.8   The "What How" Cycle

Earlier in this chapter, the six questions that need to be asked and answered in a planning approach, were discussed. It was also noted that there is an overlap between all six questions but that addressing all six makes sure that a complete answer is provided, and a full understanding of the situation is achieved. Two of these questions have a particular relationship: what and how?

Over time, it has been understood that there is a successful way to approach the definition of requirements. Requirements can be goals, objectives, or outcomes for smart mobility programs. Let us first talk about the way that is often used and does not always lead to success. The authors call this a linear sequential approach to the questions of what and how. It goes something like this. A planning consultant or solution provider responsible for providing the answer to how, asks the client group what they want. The solution provider then takes this information and develops a solution. However, in the course of defining the solution, the client group plans about new possibilities and new things that they did not know could be done. This results in a change in the definition of what is required. If the linear sequential approach has been adopted there could be significant investment in the solution, the answer to the how question, causing reluctance or resistance to make changes considering the new information. This is another reason for advocating the accelerated insights approach. That is some additional advice regarding the most effective approaches to planning for smart mobility; however, the best way to communicate it is to incorporate further advice into a description of the accelerated insights approach.

6.9   Accelerated Insights Approach

Before describing the proposed approach which the authors call "accelerated insights," it is important to be clear on one thing. It is not suggested that less resources be invested in planning for smart mobility projects. What is being advocated is the shifting of resources toward a sharper more focused initial planning approach. In fact, some of the major issues that have occurred in previous implementations have resulted from a lack of sufficient resources at the planning stage. This has also been a phenomenon that has been experienced in information technology, consider the Bohm study [6]. It is a few years old now but the results are evergreen. Every dollar spent in planning is worth $1000 at the operations stage of the initiative. This is because planning, if done correctly, will provide a degree of agility and support the cycle between what and how that will avoid significant changes later in the process when knock on inconsequential effects because the cost of alterations or corrections to be more expensive.

Information technology practitioners often referred to this as the "fail fast" approach. The concept is to move quickly and lightly through the material required to define the problems and solutions, to enable revisions to be made, supporting what the authors call the "what how cycle" Supporting an intensive effort at the beginning of the planning process, with the major acceleration of insight and understanding leading to an early smart mobility plan. The smart mobility plan can then be used as a vital tool in communicating what is going to be done and what anticipated values are at an early stage in the process. The accelerated insights approach is feasible if enough experience about what works and what does not work with regard to smart mobility technology is possessed by the workshop enablers.

There are numerous deployments of smart mobility technology in cities around the world and considerable practical lessons learned regarding technology, business models, and organizational

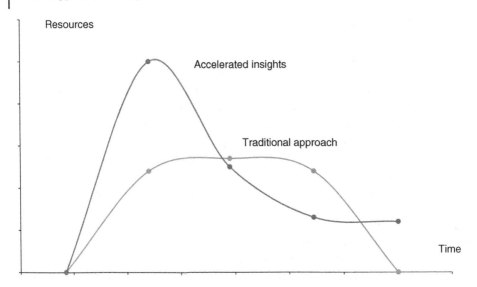

Figure 6.2 Expected resource expenditure for the accelerated insights and traditional approach.

arrangements required for success. By distilling this experience into a carefully engineered event, with the appropriate tools and resources, it is possible to support an accelerated insights experience within a condensed period of time. The high-level assessment of the resource expenditure expected for the accelerated insights compared to the traditional approach is shown in Figure 6.2. This indicates that while the accelerated insights approach is frontloaded, overtime a similar number of resources will be expended for both approaches.

The accelerated insights event is intended to be an intense starting point for a new approach to smart mobility planning. An agenda for an accelerated insights event is provided to illustrate the major elements of the event and provide a framework for providing further advice on successful smart mobility planning. The event agenda is a Compass It of previous experiences that the authors have accumulated in planning, delivering, and managing similar events for several smart mobility and transportation planning activities over the years. Please note that the event is planned over a period of four days in this description. This does not need to be a hard and fast constraint and as a custom version of the accelerated insights event is developed, feel free to change the duration to assert regional needs and to encompass all the activities required to set the scene for rapid progress.

The authors are big advocates of Instructional System Design as defined by Dr. Donald Kirkpatrick [7]. Based on the Kirkpatrick model and previous experience in organizing intensive events focused on smart mobility subjects, it is considered that the following objectives are crucial in the success of the event:

- Reaction: Did the participants enjoy the experience?
- Learning: Did they gain sufficient knowledge to be of practical application for their needs?
- Behavior: Is the information and knowledge practical and actionable?
- Results: Were ways to measure the success and outcomes from the plan defined and agreed?

Achieving all four objectives to the appropriate level requires a high degree of skill and experience, along with appropriate tools and relevant information regarding practical experience and lessons learned from prior deployments. In short, a successful event will require appropriate input, the relevant tools, the relevant expertise and experience, and a robust structure.

Figure 6.3 Accelerated insights three-stage approach.

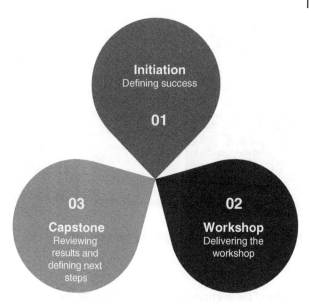

A three-stage approach to the accelerated insights event is proposed, as illustrated in Figure 6.3 each stage is summarized in the following sections.

Stage 01

Initial discussion and definition of a successful event outcome. A meeting is held with key leadership to define their perception of a successful outcome from the accelerated insights event. Input from this discussion is used to customize the accelerated insight approach.

Stage 02

Conduct of the accelerated insights event. An accelerated insights event is be designed and delivered.

Stage 03

A capstone event is designed, developed, and delivered. This will enable the accelerated insights plan to be delivered to the relevant stakeholders. During the capstone event, the essential elements of the plan are discussed and arrangements for moving the plan into action defined and agreed.

6.9.1   Summary of the Stage 02 Event

The following agenda is proposed, that offers a structured approach to delivering accelerated insights in support of smart mobility. The event is designed for a duration of four days with each day providing the foundation for the subsequent days.

On the first day, the focus is placed on defining and agreeing needs, issues, problems, and objectives related to smart mobility. A checklist of objectives is presented to the event participants to start a discussion on what is required for the city or region. This first day is also a time to develop a reasonably detailed picture of prior investments in smart mobility and transportation technology. This does need to be a detailed catalog but will provide sufficient detail to form the basis for a description of the smart mobility starting point. The second day will focus on defining and matching solutions to problems and objectives, by matching "what" to "how." On day three, the deployment of large-scale building blocks of solutions is defined and discussed. On day four,

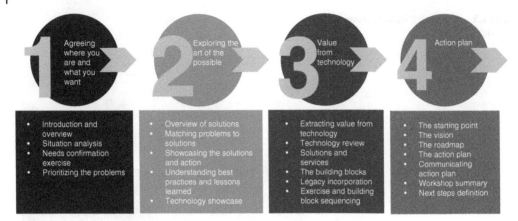

Figure 6.4 Accelerated insights agenda overview.

results of the event are brought together and the accelerated insights plan is outlined. A high-level summary of the event agenda is shown in Figure 6.4 and the following sections address each element in the agenda, step-by-step. In addition to explaining each element, a commentary is also provided containing advice and some rationale regarding why the element is included. Suggested inputs and recommended outputs for each day are also provided.

6.9.1.1   Day One: Needs and Starting Point
Preparatory materials and a scene-setting discussion will set the scene for the remainder of the event.

6.9.1.1.1   *Review the Accelerated Insights Event Objectives*
Following the Kirkpatrick model, the objectives for the event will be explained, discussed, and agreed. This will build on prior discussion with senior leadership regarding their intended outcomes for the event and ensure that all event participants have ownership of the intended work product.

6.9.1.1.2   *Discuss Information Sources and Tools to Be Used*
One of the major factors in engaging event participants is to provide a good overview of what to expect in the event and what approaches and tools will be used. This will also include information sources that will be tapped and a general overview of the communication approach to be taken.

6.9.1.1.3   *Define the Intended Work Product: Accelerated Insights Action Plan*
Part of the scene setting is to provide a reasonably clear picture of what the expected outcome from the accelerated insights event will be. One of the most tangible outcomes will of course be the accelerated insights plan that will be the highest level and earliest view of the steps to be taken in the period after the accelerated insights event. This is an important time to identify any concerns or questions that event participants have and to address them effectively. One of the reasons for using high-quality tools and information sources is to ensure that people are encouraged to ask questions. It has been understood over the course of experience in this area that cynicism and skepticism are great opportunities to provide information, but apathy is more difficult to deal with.

6.9.1.1.4 *Introductory Videos to Frame the Subject*

This is described as introductory videos to frame the subject. However, the media used does not necessarily need to be videos. It could be animations or visualizations, really anything that brings the subject material to life. At this point, a waterski approach will be taken to the subject that defines the boundaries of area of interest and provides a good overview of the ground it will be covered over the entire event. This is not the time for deep dives, which will come later in the event and the focus will be placed on ensuring that a complete picture is provided of smart mobility, rather than answering every question about what it is and how it works. One of the most important things in waterskiing is to keep moving. Lack of forward motion could cause submergence. This is a framing exercise intended to maximize the probability that focus is maintained and straying into the areas that are not the subject of the event, is avoided.

6.9.1.1.5 *Situation Analysis*

The following factors will be presented and discussed. Note that this is an example list of factors which will be customized to the specific region or city, it is a model not a recipe. The authors have learned that people are better at revising things than creating them, so this is intended to help make a start with a custom version and not as a substitute for that version.

Green Field or Rescue Planning Situation?   What is the context within which this exercise has been conducted? Is it designed in a positive light to show the way forward based on what has been achieved so far, or is it a rescue planning situation? Is the initiative that the result of a crisis or challenging situation in which the need for structured action has become apparent. It is likely that there will be a senior nontechnical decision-maker involved as the champion for the event and it is important to understand what they are trying to achieve and what they are trying to avoid. A delicate balance is required between a clear identification of the negative side of the equation represented by the challenges and issues and the positive side of the equation which is the possible solutions that can be applied to change and improve the situation. While clarity is required on the challenges to be addressed, it is also necessary to maintain a spirit of optimism by partnering the challenges with appropriate solutions and potential ways forward. It is quite likely that there will be a level of discontent and discomfort with the current situation. In fact, this may be what has driven the desire to hold this event and to make structured progress. Proper handling of this can provide excellent fuel for forward progress.

Equity   Equity aspects of smart mobility are currently trending as an emerging issue attracting considerable attention and investment. Effective management of equity is one of the "people" aspects of smart mobility planning and in fact an important part of the overall smart mobility service delivery spectrum. This is a complicated issue that requires a detailed understanding of the specific region or city to be addressed. Equity issues can include the following:

1) Identification of those groups of people that will be advantaged by the proposals and those who will be disadvantaged.
2) Definition of specific groups for which additional effort will be required due to past inequity.
3) Definition of the equity-related objectives that are desired.
4) Exploration of high equity effects will be measured.
5) Discussion on how any revenue generated by the smart mobility proposals will be distributed.

Note that the technologies and services involved in smart mobility can provide some new capabilities with regard to equity management. The availability of detailed data and analytics capability a much more high-resolution view of the impact of smart mobility on people, to be taken.

Environment   With respect to the environment and climate change is important to define desired future outcomes. It will be recognized that environmental objectives, in common with other smart mobility objectives, must be treated within an overall multivariate objectives analysis that includes trade-offs. Note that it could also be argued that environmental effects are also part of efficiency effects.

Economic Development   This is often an unstated objective within smart mobility initiatives. It is important to recognize that the successful application of smart mobility services will have a significant effect on the economic attractiveness of the region and workforce development within the region. This is something that has been defined to be very specific to the city or region involved. For example, if the city or region is a significant tourism destination, then the promotion of tourism and visitor experience management will be important here. If the city is suffering from a high rate of unemployment, then the focus may be placed on opportunities for job creation and accessibility to job opportunities that can be afforded by smart mobility.

Safety   This usually comes high in the list of objectives to be attained by the introduction of smart mobility. In addition to the expected goals of reducing fatalities, injuries, and property damage, this can also include improvements to the effectiveness and efficiency of the response process. It can also include ways to improve the measurement of safety and project the appropriate perception of safety to visitors and citizens alike.

Efficiency   Efficiency is all about saving time, lives, and money. Time savings are generated by a reduction in travel times and also an increase in the reliability of travel time predictions. The predictability of travel conditions is an important factor of efficiency and the ability to manage both travel conditions and the supply of travel services is an important aspect of success in this area. Efficiency will also take account of both people movement and goods movement and pay particular attention to individual modes efficiency and multimodal efficiency across the region.

Enhancing User Experience   This is often referred to as "customer service," but the authors like the term "user experience" as it places a strong focus on the people, the end users, the citizens, and visitors. It also sends a signal that the most important aspect is from the user's perspective and not from the organization delivering the service. As noted earlier, it is important to recognize the difference between perception and reality and ensure that any attempts to enhance user experience are properly perceived by the user. This may require considerable resources in terms of outreach and education to explain to users the values and benefits of the proposed smart mobility implementation. An important question here is "what are we trying to do for the benefit of the people?"

Getting the Best Out of Prior Investments   This requires discussion at an "under the hood level" rather than a "shop window level." What is meant by this is that prior implementations must be viewed from a realistic and practical point of view including everything that went right as well as everything that went wrong. This discussion will also look at what would be done differently based on the experience of other implementers around the globe. This may also require consideration of a different approach to investment planning or work program development, based on outcomes rather than a historical index-linked approach.

Automated Vehicles and Other Emerging Technologies   One of the major features of smart mobility is the use of new and emerging technologies to solve problems and address issues. Most new technologies generate positive and negative feelings. On the one hand the availability of new capabilities and the ability to do things faster, cheaper, and better create optimism about new technologies. On the other hand, a lack of understanding of the full effects of technology can cause apprehension and perhaps even fear of the unknown. At this situation analysis development stage of the process, it will be sufficient to gain a high-level understanding of how event participants perceived new technologies. The authors have specifically called out automated vehicles as this is a prominent example of an emerging technology with significant repercussions for transportation and society. A discussion on automated vehicles could shed considerable light on attitudes toward other emerging technologies such as artificial intelligence and machine learning.

6.9.1.1.6 Practical Exercise on Needs, Issues, Problems, and Objectives

Up to this point, day one has featured a considerable amount of talking with the audience. Therefore, i is important to balance the knowledge transfer approach by introducing a degree of interactivity. An interactive, firsthand exercise will be conducted to explore requirements for the accelerated insights plan and reach agreement on these. The exercise will be designed to define common ground in terms of agreed accomplishments and uncommon ground where there is no agreement currently regarding what has been done in the past and what will be done in the future. The exercise will also begin to explore questions that must be asked and identify possible answers. A detailed explanation or a design for this exercise will not be provided. It will be interactive and will enable event participants to actively engage in the exercise. In the past, it has been found that a good approach is to let expert consultants play a supervisory role but enable event participants to take a lead and work in small teams to define the current situation.

6.9.1.1.7 Day One Inputs

- Accelerated insights event objectives
- Kirkpatrick model
- Information sources
- Introductory videos or other media
- An initial overview of the current situation in terms of prior investments
- An initial menu of potential needs, issues, problems, and objectives

6.9.1.1.8 Day One Outputs

- Agreed list of prior investments
- Agreed needs, issues, problems, and objectives
- Agreed situation analysis

The primary output from this stage of the event is that all participants have a clear idea of what achievements are intended and how they will be achieved. Participants will also have a good understanding of the essential elements of smart mobility and have an overview-level grasp of the essentials.

6.9.1.2 Day Two: The Art of the Possible

The focus of this part of the event is to present the needs, issues, problems, and objectives defined in day one and start to match them to the capability of technology-driven services.

6.9.1.2.1 Prioritizing the Problems, Summarizing the Current Situation

This draws on activities and outputs from day one to summarize the needs, issues, problems, and objectives that were identified, attempting to prioritize them into at least a first-order list and providing the first few of the situation analysis, or starting point for the smart mobility planning initiative.

6.9.1.2.2 Matching Problems to Solutions

It is highly unlikely that most of the problems and challenges identified, if not all, have been addressed in prior implementations. Therefore, this activity involves taking all the problems and challenges that were identified on the first day and matching them to a predefined list or menu of potential solutions that has been developed based on prior experience and expertise around the globe. In addition to the solutions, they will incorporate best practices from a technological, business model, and organization perspectives. This requires access to an extensive and current body of knowledge that captures prior implementation experience, lessons learned, best practices, and advice. In the course of showing how the city or region's problems can be addressed by the potential solutions, a discussion will also be supported on how the particular solution might work within a specific city context. The objective here is to do more than sure that there are existing solutions for all the problems, but also to start the process of understanding the outcomes and effects that solutions might have been the specific city or region.

6.9.1.2.3 Services and Technologies

While it is important to highlight the importance of technology, it is also important to put the technology into perspective. As illustrated in Figure 6.5, technologies are the building blocks for products and solutions that ultimately support the delivery of service. It is those services that deliver outcomes, values, and benefits. Therefore, this part of the event transfers the focus from technology to the enabled services.

This is also a good time to start introducing the concept of benefit-cost analysis for each of the services. The approach is that information on the effects of services is drawn from an international

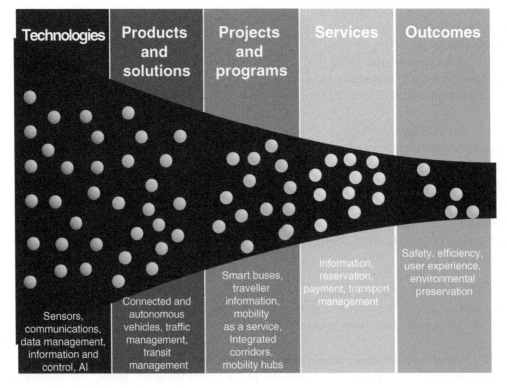

Figure 6.5 Technologies to outcomes.

experience base. The authors have witnessed what they call "global ping-pong." This is where one area of the world takes the lead with a particular technology and service and provides practical experience, best practices, and business models that can be transferred to another part of the world. Similarly, the other part of the world may have taken the lead in something that can be of value to the original region. The authors find this to be a particularly interesting effect as they have often reached a situation with client groups where there is a desire to be leaders in the field, accompanied by a strong desire to be reassured that the technology they will employ has been tested and is robust. By employing global ping-pong, it is possible for a public agency to be a leader in the region or nation, and yet take advantage of lessons learned from somewhere else. Being the leading edge and having robust approach could potentially be conflicting objectives unless this global ping-pong is utilized properly. The authors often joke with clients that they have a "roller coaster" effect. The psychology of roller coasters is interesting in that people want to be scared safely. Global ping-pong allows an approach to the leading edge while being reassured from a risk management point of view that what is being proposed is not just feasible but has been tried before somewhere else around the globe. With the appropriate lessons learned and practical experience acquired to minimize risk.

6.9.1.2.4 *What About the People?*

Earlier in the workshop, the focus was placed on understanding the people effects of smart mobility; in this section of the event, the expected impacts on both citizens and visitors are explored. This will take account of travelers and non-travelers. For example, even if people do not travel, the effects of smart mobility on the efficiency of the freight process can have an influence on the prices for essential goods. Hence this will impact non-travelers. Similarly, economic development effects of smart mobility will also have an effect on travelers and non-travelers alike. Again, drawing on global best practices and practical experiences, it is possible to provide at least an outline assessment of the people effects of the proposed technology-supported services. This would also be an appropriate time to discuss the plans for communicating the proposals to citizens and visitors in the form of an outreach plan.

One important aspect of this activity is to consider the definition of demographic groups or groups of citizens and visitors that have similar characteristics and needs. It may also be possible in these days of good availability of big data and strong analytics, to be able to move toward a "market segment of one approach," where the user groups are as small as possible, underwriting the possibility of a strong approach to equity.

Previous planning initiatives that the authors have been involved in I find particular success in communicating the proposals in terms of a game plan consisting of a series of big moves, or high-level strategies.

6.9.1.2.5 *Technology Showcase Exhibit*

This is where there is a strong role for the private sector in the workshop. It is also a way to break up the format to avoid the continual use of "talking heads." This involves the incorporation of a technology and solution showcase exhibit, within the overall framework of the event. This can be sponsored by a group of solution providers with a focus on highlighting capabilities and constraints and avoiding commercial sales pitches. The objective of this is to support a dialogue that enables event participants to visualize how the needs identified earlier in the event could be satisfied with the potential solutions on display. Where possible, the exhibits will focus on demonstrations and discussions on prior experiences around the globe. Sufficient time will be reserved to enable both formal and informal discussions on technological capabilities and constraints. It has often been defined useful to equip event participants with a menu of questions that they may wish to pose to technology solution providers.

6.9.1.2.6 Day Two Inputs

- A catalog of prior experiences and best practices from around the world. Incorporating lessons learned based on the view from the front, preferably utilizing multiple media formats and technologies.
- Solutions and services exhibition.
- A menu of needs, issues, challenges, and objectives drawn from global prior experience.

6.9.1.2.7 Day Two Outputs

An alignment between needs, issues, problems, and objectives and technology-supported services. In other words, a high-level catalog of proposed solutions that will completely address the needs.

6.9.1.3 Day Three: Harnessing Technologies for Value

6.9.1.3.1 Technology Review

Day three begins with a review of day two activities and outputs. This includes refreshing the information that was delivered and obtained from the technology showcase and he agreed set of needs, issues, problems, and objectives that were identified earlier in the day. The discussion will be focused on uncovering the practical consequences of applying the technology-driven solutions identified. It will also address how new possibilities and new solutions identified might fit with prior investments. The discussion will also continue to reinforce the relationship between technology as building blocks and services as value delivery mechanisms.

6.9.1.3.2 The Building Blocks

Based on the needs and technology-driven services discussion, this activity moves to the first few of which system engineers would: architecture. For the purposes of this event, the term architecture will not be used as it can lead to a lot of tangential discussion and confusion among a non-system engineering audience. In fact, it can also cause confusion among system engineers. The objective is simply to define a number of major building blocks that combine services and solutions based on previous experience around the globe. These building blocks will then be used like Lego bricks, to piece together an approximate future big picture for the city or region. The aim here is to be thorough and avoid being drawn into detailed discussions on how the building blocks fit together such as what telecommunications will be used. That's a job to be addressed later in design. Having defined a big picture using the building blocks, the fit with legacy processes and existing systems can be explored and defined. Maximization of the value of previous investments can be insured by incorporating them into the big picture. In addition to a high-level technology view, it will also be possible to begin the definition of the organizational capabilities required to support the solution and the likely business models that could work in supporting effective public and private cooperation. The discussion on organizational capabilities may lead to the identification of a need for a separate organizational capability assessment to identify strengths, challenges, and identify workforce gaps. An important part of the discussion will relate to the potential business models that could be employed to support the technology and organizational solutions. The big picture could be viewed as three separate layers – a technology or service layer, an organizational layer, and a business model or commercial layer. Again, these are all relatively high-level with detail delegated to a later design process.

6.9.1.3.3 Exercise on Building Block Sequencing

Having developed a big picture for the future, this is an interesting time to consider the order in which the building blocks will be implemented. This requires consideration of multiple factors including funding, benefits attainment, impact on people, and organizational capability.

An interactive exercise allowing event participants to get hands-on with the building blocks and come up with their own deployment sequence can yield a rich stream of information about what will work, what will not work, and what is suitable for city or regional needs. This building block sequence exercise will have a clear set of predefined objectives including the development of the deployment sequence, the definition of what to invest in, and when to do it.

The authors have found it useful to provide examples of a building block sequence or a big picture drawn from previous deployments around the globe. Here again, this supports an understanding that people are better at revising things than creating them, so giving them a starting point usually results in better outcomes. It also provides reassurance that the challenges are not being addressed for the first time and that there is considerable international experience in doing this successfully.

6.9.1.3.4 Day Three Inputs
- A series of predefined building blocks for smart mobility deployments based on international experience.
- Best practices on business models.
- Best practices on funding.
- Examples of complete architectures from previous deployments including technology, organizational, and business model solutions.

6.9.1.3.5 Day Three Outputs
- High-level big picture of a complete regional or citywide solution.
- An initial investment plan with sequencing of major building blocks.
- A list of challenges and opportunities regarding the architecture and investment plan.

6.9.1.4 Day Four: Finalizing the Action Plan
Results and insights from the event will be brought together to form the basis for an outline for the accelerated insights plan.

Day four of the event is designed to bring all the outputs from the previous three days together, setting the scene for rapid progress. Remember that any successful smart mobility planning approach has three essential components. A starting point definition, an endpoint or vision definition, and a roadmap defining how to get from the start to the end. So, at this point as the end of the event is approached, it is time to bring all the information together by defining the appropriate starting point, the intended endpoint or vision, and at least the first view of what the roadmap will look like.

6.9.1.4.1 The Starting Point
Enough information will be gathered from the first three days of the event to enable the definition of the current situation in a concise and clear manner. While some details may still be missing, the salient points regarding legacy systems and prior investments will have been captured and can be communicated. Based on extensive practical experience, the authors have understood that the best starting point for a city or region will have strong roots in prior investments. This is partly because the preservation of sunk investment is an important element of a successful plan and partly because those prior investments were justified by their own planning process. So not only investments in legacy infrastructure and systems being preserved, but also investment in previous planning efforts. It is also likely that the region or city will have considerable experience in design, delivery operations, and maintenance of these legacy systems making it easier to support new implementations and investments that are logical extensions to the current. Prior investments also take full account of the local factors and a strong understanding of the city and regional landscape

with respect to transportation, politics, and geography. The starting point can also define current organizational capabilities through the use of a capability assessment model.

6.9.1.4.2 The Vision

The vision for the smart mobility implementation is also the endpoint for the roadmap. It is a clear, concise, and agreed understanding of where the city or region is striving to go in the future. Note that sometimes things that are not incorporated in this vision are just as important as things that are because this indicates things that are not desirable and outcomes that are not sought.

Developing a vision and a detailed understanding of the desired endpoint is a bit like mental time travel. The group is trying to visualize and understand as much as possible, what the desired endpoint will look like from multiple perspectives. This can be difficult. Say for example our desired endpoint is 50 years in the future. There may be solutions available then that have not even been thought of now. Consider the Internet for example, is widely accepted that the official birthday for the Internet is January 1, 1983, well within the 50-year time horizon. If a plan was developed in 1980 for example, it may not have taken account of the capabilities of the Internet.

An important thing to notice is that the endpoint will remain static over time. Lessons learned in incremental deployment of smart mobility and changes in the needs and requirements of the city will cause the endpoint to move. It is important to recognize this in the planning approach and accommodate anticipated movement. Imagine being in a small rowing boat crossing a sea with strong currents. Because the boat is small, the currents are likely to cause the boat to drift off course. This is sometimes known as going with the flow. It is okay to go with the flow but it is also essential to ensure that eventually the desired destination is reached. This is accomplished by creating a vision that acts as a lighthouse. The little boat can be blown off course but because the lighthouse is always visible, course corrections can be made so that is still heading in the desired direction. This lighthouse effect is one of the essential features and functions of a smart mobility vision. Unlike a real lighthouse, the vision is going to shift a little over time. Therefore, an important aspect of vision design is to integrate sufficient flexibility into the planning approach, while also preparing thoroughly at the start of the journey to minimize the amount of movement or change in the vision. Change is inevitable as emerging technologies are involved, in a time when rapid advances and discoveries on how to do business will cause sudden and dramatic changes. After all that is what has been defined as the smart mobility revolution. So, it can be assumed that as the smart mobility planning approach is being developed that the future can be predicted on a linear sequential basis. It is not useful to work on the basis that the future can be extrapolated from the past. In fact, it is necessary to build a smart mobility planning approach to cope with the unknown and have the flexibility to take account of all of these changes. A thought at this point might be that as things are changing so much and will change a lot in the future then why bother to plan? The lighthouse effect makes planning crucial. One single point has been defined on which all can converge. This is one reason why the endpoint is known as a vision and not a detailed plan or design. It has sufficient detail to serve the lighthouse purpose but is also at a high enough level to accommodate changes and allow for the unknown.

6.9.1.4.3 The Roadmap

This is the third essential element of the accelerated insights plan, the roadmap, as its name suggests, is an outline definition of how to get from where the city or region is to where it wants to be. This begins with the definition of deployment building blocks, the development of a service deployment strategy, and an outline definition of programs and projects that will result in the delivery of the services. The roadmap will also show the relationship between the smart mobility planning process and other established planning processes related to land use and transport.

The initial perception of the essential features of the roadmap for smart mobility revolves around the need to define an overall program to get from the starting point to the vision, the definition of practical and realistic projects, and the delivery of tangible interim benefits from each incremental deployment. Further experience has taught the authors that a successful roadmap will go beyond these basic requirements and provide some excellent material for outreach, communications, and marketing. This is more than a technical exercise and is the foundation for a successful communications and outreach program to inform citizens and visitors about the proposed investments. The authors also believe that milestones are an important consideration in the definition of a successful roadmap. Milestones along the road will ideally be a sequence of breathtaking insights and glimpses of the future. Each milestone will also deliver its own standalone values and benefits in addition to the cumulative value and benefits of the entire smart mobility program. This is an essential part of a survival strategy to enable the end of the roadmap to be reached through the management of expectations and the delivery of results along the way. The roadmap will be punctuated by glorious stops along the way, that can serve as destinations in their own right but also clearly stepping stones to the intended future vision. If done successfully, then the roadmap will feature milestones that represent viewpoint opportunities that go beyond value delivery into perception-altering marketing.

6.9.1.4.4 Event Summary

As the conclusion of the accelerated insights event is approached, it is important to conduct a wrap-up session or an event summary. This describes the highlights of the event including the key lessons learned, the major features of the proposed approach, and an evaluation of the success of the event. This is a good time to reiterate the accelerated insights plan by defining the starting point, the desired future vision, and the definition of the roadmap as is currently known. An evaluation of the event will make use of the Kirkpatrick model and also a checklist of the initial objectives set at the beginning of the event.

6.9.1.4.5 Next Steps

Going beyond summary, this activity defines possible next steps now that the accelerated insights plan has been developed and agreed. During this section of the event, a possible timescale for implementation of the plan will be discussed. This will make use of the roadmap and identify significant milestones along the way from where the city or region is today, to where it wants to be tomorrow. It is also particularly valuable at this stage to identify challenges and opportunities that are currently perceived based on the information gathered and the results of the event. It is expected that challenges and opportunities will relate to at least one of the following:

- Funding
- Procurement
- Organizational capability
- Solution provider capability
- Political will

6.9.1.5 Beyond the Accelerated Insights Event

As the authors see it, the accelerated insights event will have generated considerable amount of potential energy. What is now required is to convert the potential to kinetic energy, by stimulating forward movement. This forward can be expected in the following areas.

- Procurement – using the accelerated insights plan to enable a practical discussion on how the smart mobility implementation can be procured.

- Organizational – a detailed assessment of organizational capabilities including the definition of roles and responsibilities for each of the stakeholder agencies, the private sector, and the definition of potential improvement programs.
- Business model and partnerships – the definition of roles and responsibilities for both public and private sector partners.
- Socializing the plan – taking the accelerated insights plan as the basis for an effective outreach and marketing program that will socialize the plan beyond event participants and to all those who must be consulted regarding the proposals.

6.9.1.6 Who Should Be There?

In addition to defining the kind of activities that will be supported in accelerated insights event, is also important to develop a clear view on who will participate.

Figure 6.6 provides an overview of the groups of people that will be invited to participate. Here again, this is based on the authors' knowledge and experience of prior implementations and may not be completely in alignment with what is required for the city or region. Note that there is one important group missing from the list of participants and that is the general public. While the authors believe that it is crucial to involve input and support consultation with the public, this will be done in a specific separate exercise after the accelerated insights plan has been prepared. This allows the use of the plan as a valuable communication tool to the general public and the end user of the smart mobility system. Please use this as a starting point, as a model for developing a custom list of invitees.

This question also leads to another question regarding why particular groups should participate. This has been addressed in Figure 6.7.

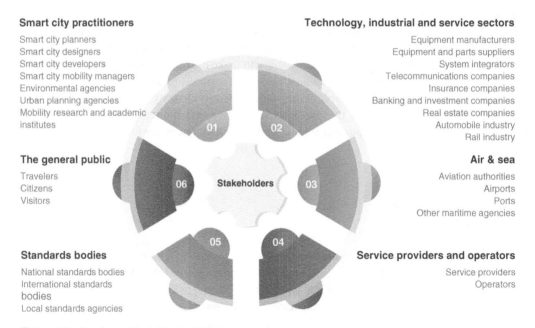

Smart city practitioners

Smart city planners
Smart city designers
Smart city developers
Smart city mobility managers
Environmental agencies
Urban planning agencies
Mobility research and academic
institutes

The general public

Travelers
Citizens
Visitors

Standards bodies

National standards bodies
International standards
bodies
Local standards agencies

Technology, industrial and service sectors

Equipment manufacturers
Equipment and parts suppliers
System integrators
Telecommunications companies
Insurance companies
Banking and investment companies
Real estate companies
Automobile industry
Rail industry

Air & sea

Aviation authorities
Airports
Ports
Other maritime agencies

Service providers and operators

Service providers
Operators

Stakeholders

01 02 03 04 05 06

Figure 6.6 Accelerated insights participants.

Smart city practitioners	Rationale for participation
Smart city planners	Supporting an interface between conventional transportation planning, design and development and smart mobility
Smart city designers	Supporting an interface between conventional transportation planning, design and development and smart mobility
Smart city developers	Supporting an interface between conventional transportation planning, design and development and smart mobility
Smart city mobility managers	Supporting an interface between conventional transportation planning, design and development and smart mobility
Environmental agencies	Representing climate change and environmental impact interests
Urban planning agencies	Supporting an interface between conventional transportation planning, design and development and smart mobility
Freight planners	Representing the interests of freight transportation
Mobility research and academic institutes	Supporting an enabling role as independent neutral advisors
The general public	
Travelers	Providing end user input
Citizens	Providing end user input
Visitors	Providing end user input
Standards bodies	
National standards bodies	Providing input on current and required standards
International standards bodies	Providing input on current and required standards
Local standards agencies	Providing input on current and required standards
Technology, industrial and service sectors	
Equipment manufacturers	Providing information and insight on technology application, solution, and service capabilities

Figure 6.7 Accelerated insights workshop participants and rationale for participation.

Equipment and parts suppliers	Providing information and insight on technology application, solution, and service capabilities
System integrators	Providing information and insight on technology application, solution, and service capabilities
Telecommunications companies	Providing information and insight on technology application, solution, and service capabilities
Insurance companies	Providing information on road user charging and dynamic insurance approaches
Banking and investment companies	Providing information and insight on funding opportunities, identifying investment opportunities to align with public objectives
Real estate companies	Providing input to land use impacts and values
Automobile industry	Representing the interests of the automobile industry in providing input on potential connected and automated vehicle solutions
Rail industry	Representing the interests of heavy rail, light rail, and rapid transit
Air and sea	
Aviation authorities	Representing the needs of air transportation modes
Airports	Representing the needs of air transportation modes
Ports	Representing the needs of C and maritime transportation modes
Maritime agencies	Representing the needs of C and maritime transportation modes
Service providers and operators	
Service providers	Providing information and insight when service possibilities
System operators	Providing information and insight in the needs of system operation
Freight operators	Representing needs of freight operators including intercity and urban logistics

Figure 6.7 (Continued)

6.9.1.7 Involvement of Private Sector Solution Providers and Potential Investors

Traditionally, the private sector is held at arm's length from a smart mobility planning approach. The authors believe that it improves the overall effectiveness of the process if the private sector is involved proactively at the early stages. This is particularly useful in providing input into what is possible, what technology and services can do, and what the practical constraints are at the current time. This also allows an assessment to be made of the capability of the private sector to respond to the public sector's needs for technology-based solutions. While there is a desire to encompass the latest technologies, these have to be proven, robust, and take full account of practical experience. The inclusion of potential investors could also raise a few eyebrows. The authors believe that this is important as there has been a revolution underway on how smart mobility applications are driven, funded, and operated. Many of the current large-scale initiatives are led by the private sector, using investor finance. It is important to develop an effective dialogue between public agencies who have a clear understanding of what is required, and investors who have a clear idea of the business objectives. This also sets the scene for an important partnership between the public sector while stewarding public interests, and the private sector while attempting to develop new industries and services.

It addresses the difficulty of looking into the future by providing input from multiple perspectives. Many heads are better than one, especially in this situation. Being able to incorporate perspective from the public agencies, private solution providers, and investors together into a single agreed vision for the future, is more likely to be successful than the product of a single mind. The cooperative effort in developing this agreed and shared vision is also an important tool in transferring ownership to a wider group of people. This will be essential for the subsequent success of the plan.

6.10 Summary

This chapter has provided the definition of planning and described the objectives of smart mobility planning. It explains the traditional approach to transportation planning and suggests an accelerated insights approach to smart mobility planning. Three essential elements of the smart mobility plan namely a starting point, a roadmap, and an endpoint or vision are explained. The elements of a roadmap are also identified along with the important characteristics of a vision for smart mobility. The importance of incorporating legacy systems and sunk investment in the starting point is also discussed. The most significant part of the chapter is the introduction of a rapidly focused approach to smart mobility planning, known as accelerated insights. This approach is explained in some detail along with the proposed agenda for an accelerated insights workshop. Essential participants and reasons for their involvement in the workshop are also addressed.

The authors hope that the accelerated insights approach will offer readers a fast-track alternative to traditional planning approaches. The intent is not to replace traditional planning but rather introduce a new element into the planning process consisting of rapid plan development and prototyping. This recognizes the progress that has been made in the information technology industry with respect or rapid evolutionary development, fail-fast, and sprint approaches.

The effect of planning is crucial to the success of smart mobility by ensuring effective stewardship of resources and inclusion of all the relevant stakeholders in an agreed plan that can be supported and implemented. After all, the best plan is the one that gets implemented.

References

1 O'Donnell, C.; Koontz, H. Principles of management. McGraw-Hill Book Company, January 1, 1968.
2 Kipling, R. Just so stories: the original 1902 edition with illustrations. Hardcover – August 11, 2019, by Rudyard Kipling (Author, Illustrator), Suzeteo Enterprises.

3 Jay Arthur, L. Rapid evolutionary development, requirements, prototyping and software creation. John Wiley & Sons, Cities, and software engineering practice, 1991.

4 Software Engineering, Evolutionary Model. 30th April 2019. https://www.geeksforgeeks.org/software-engineering-evolutionary-model, retrieved December 31, 2021, at 12:29 PM Central European Time.

5 Wikipedia. Software prototyping. https://en.wikipedia.org/wiki/Software_prototyping, retrieved December 31, 2021, at 12:33 PM Central European Time.

6 UAE Government website. Happiness page. https://u.ae/en/about-the-uae/the-uae-government/government-of-future/happiness, retrieved December 31, 2021, at 2:56 PM Central European Time.

7 Kirkpatrick and Partners. The Kirkpatrick model. https://www.kirkpatrickpartners.com/the-kirkpatrick-model, retrieved January 3, 2022, at 12:53 PM Central European Time.

7

The Essential Elements of Smart Mobility

7.1 Informational Objectives

After you have read this chapter, you should be able to:

- Explain the essential elements of smart mobility
- Describe smart mobility solutions
- Explain some examples of smart mobility business models
- Define the relationship between technologies, solutions, and services
- Describe how smart mobility solutions fit together
- Explain the importance of data and analytics in smart mobility solutions

7.2 Introduction

In this chapter, the essential elements of smart mobility are explored. The objective is to identify and describe all the significant elements that affect the success of a smart mobility implementation. While all the chapters of the book are focused on smart mobility, there are specific chapters that have important relationships with the content of this chapter. These are illustrated in Figure 7.1.

Figure 7.1 also explains why the relationship between the chapters is so important.

7.3 The Essential Elements of Smart Mobility

At first glance, it would seem that smart mobility is all about technology. However, a deeper investigation of smart mobility implementations reveals that the situation is much more complicated. The successful deployment of smart mobility requires consideration of technology and also a range of factors as illustrated in Figure 7.2. Each of these factors is discussed in turn in the following sections.

Smart Mobility: Using Technology to Improve Transportation in Smart Cities, First Edition. Bob McQueen, Ammar Safi, and Shafia Alkheyaili.
© 2024 The Institute of Electrical and Electronics Engineers, Inc. Published 2024 by John Wiley & Sons, Inc.

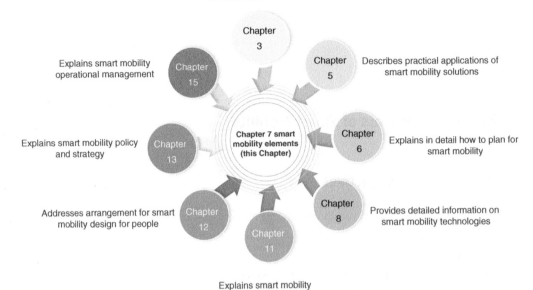

Figure 7.1 Important relationships to other chapters.

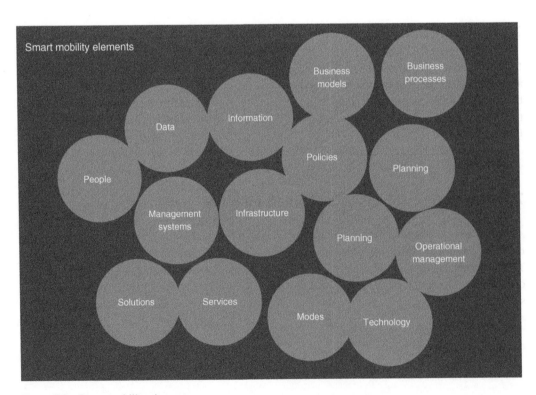

Figure 7.2 Smart mobility elements.

7.3.1 People

This is a critical element in smart mobility. Neglecting, or providing insufficient treatment, to people aspects can result in substandard solutions that are not adopted and do not provide service to the full range of end users. Smart mobility delivers service to the user while the user can also provide feedback on the effectiveness of the smart mobility initiatives. The subject is addressed in detail in Chapter 12, Mobility for People.

7.3.2 Data

Data is a vital raw material in understanding the context within which smart mobility will operate and expected outcomes. However, data is only the first step in a process from data collection to information processing, to the extraction of insight, to the creation of response in action plans. While data is foundational, it must be subject to this process to be valuable and useful. Many of the smart mobility solutions being deployed can deliver service but can also serve as major sources of valuable data. The subject is addressed in more detail in Section 7.7 in this chapter on the importance of data and analytics.

7.3.3 Management Systems

It is very unlikely that smart mobility solutions are implemented in isolation or in a greenfield situation. That are likely to be existing traffic, transportation, and other management systems that must be considered and in some cases interfaces created. Existing management systems can provide input to smart mobility solutions, while smart mobility can provide new and enhanced capabilities to existing management systems [1].

7.3.4 Services

The importance and services of smart mobility were introduced in Chapter 1, and several practical examples are explained in Chapter 5, Smart Mobility Progress around the World. While technologies are powerful, it is the services that are enabled by technologies that deliver outcomes and real value. Services can also include things that are not part of smart mobility such as existing transportation services and other services that exist within the smart mobility ecosystem. These include information services that may or may not carry mobility information, and payment services. Such information services can inform smart mobility deployments and can also be informed by such deployments [2].

7.3.5 Infrastructure

Urban areas feature multibillion-dollar investments in transportation infrastructure. These include road networks, railway networks, and a wide range of transportation facilities. Infrastructure can also include more basic technologies such as road signs and road markings. To be successful, smart mobility initiatives must take account of existing infrastructure of all forms. Infrastructure also represents a significant prior investment which must be fostered and incorporated into the future. While smart mobility can reduce dependence on infrastructure, it can never replace and consequently must partner with infrastructure. Infrastructure can be a foundation for smart mobility solutions and can also benefit through better management enabled by smart mobility.

7.3.6 Information

Information comes in many forms. Typically, it is trends and patterns that can be discovered within large volumes of data, as a result of advanced analysis. As discussed under the data heading, data processing creates information as part of the value chain that is required to extract the maximum value from data. Information can inform smart mobility and smart mobility can generate considerable new information. More detail is provided in the subject in Section 7.7 in this chapter on the importance of data and analytics in smart mobility.

7.3.7 Business Models

For the purposes of this book, business models are defined as ways to do things, the definition of who does what, and the agreement on who pays and who benefits. Smart mobility offers rich possibilities for cooperation between the public and private sector, these can be most successful when based on an agreed business model. There might also be existing business models currently in operation that must be considered when developing a successful approach to smart mobility. Existing business models can shape smart mobility solutions, and smart mobility solutions can lead to the evolution of business models. More detail on business models is provided in Section 7.6 of this chapter [3].

7.3.8 Policies

Smart mobility policy and strategy are covered in detail in Chapter 13, Smart Mobility Policies and Strategies. Policy will interact with smart mobility in two primary ways. It will be required that smart mobility conforms to existing policy, and smart mobility is likely to generate the need to develop new policies. Micromobility is a case in point, were in some parts of the world, the implementation of micromobility solutions has outpaced the appropriate policy and regulation, leading to challenges associated with vehicle speed and indiscriminate parking.

7.3.9 Planning

In the same manner as policy, planning will interact with smart mobility solutions in two primary ways. Smart mobility must adhere to established planning guidelines and plans, while a growing understanding of the capabilities of smart mobility may also influence future planning and plan development. The subject is covered in detail in Chapter 6, Planning for Smart Mobility.

7.3.10 Modes of Transportation

In the same way that infrastructure interaction is an essential component of smart mobility success, then the various existing modes of transportation must also be accommodated within smart mobility deployments. In some cases, smart mobility deployments can include new modes of transportation such as e-scooters and automated shuttle vehicles. Chapter 3, Smart Mobility: A Problem Statement, provides more detail on the modes of transportation available in a typical city.

7.3.11 Business Processes

At the simplest level, these are the protocols and procedures that are used to turn input into output within an organization. Business processes are introduced in Chapter 1, Introduction. Smart mobility solutions interact with existing business processes and can also create the need for new business processes to be defined. Since business processes act as a foundation for business model definition, more detail on business processes can be found in Section 7.6 of this chapter [4].

7.3.12 Operational Management

Smart mobility can have a major impact on transportation operations by raising the level of situational awareness regarding variations in the demand for transportation and operating conditions. Advanced decision support systems can also enhance operational management effectiveness, while information distribution capabilities can significantly improve communications-related operational management. Operational management can also be viewed as a business process within urban transportation agencies. The subject is addressed in detail in Chapter 15, Smart Mobility Operational Management.

7.3.13 Technology

As stated at the beginning of this chapter, technology plays a very important role in smart mobility. Technologies are the building blocks used to create solutions that deliver service. They can be considered as the raw material that is used in the process of creating smart mobility solutions and services. Smart mobility technologies are addressed in detail in Chapter 8, Smart Mobility Technologies.

7.3.14 Smart Mobility Solutions

One of the more significant elements in smart solutions can be described as solutions or applications. These are commercially available solutions or services that represent a combination of technologies designed to deliver a specific set of outcomes. Based on the global progress review described in Chapter 5 and extensive background knowledge from the authors, a review of smart mobility solutions has been prepared. This is intended to provide additional detail to the solutions introduced in Chapter 5, Smart Mobility Progress around the World.

7.3.14.1 Advanced Traffic Management

Traffic management could be considered as the original smart mobility application. Electric traffic signals were first installed in 1914 in Cleveland Ohio, although London can claim an earlier implementation in 1868. The traffic signals in London were installed outside the Houses of Parliament, gas-powered, and used the semaphore arm adapted from a railway signal. Within a year, the installation exploded and killed the policeman operating the system. There were no more traffic signals in London for 50 years after that. The lesson – do not try an unproven technology if not sure that it will work in a high-profile location.

Since those early beginnings, traffic management has evolved into one of the most well-documented, robust applications of advanced technology in transportation. While initially focused on the management of traffic conflicts at urban intersections, advanced traffic management has expanded to include the management of limited access highways, express lanes, and major arterials within cities. For intersection control, the basic three, light system is still in use although the control systems for vehicle detection and timing have improved dramatically over the years. These days the best traffic signal systems are managed on a networkwide basis and feature adaptability, and coordination. Adaptability means that the traffic signal timings are constantly reviewed and aligned to current traffic flows. Coordination means that the operation of one intersection takes account of the operation of surrounding intersections. The most advanced intersection control systems also have special capabilities for pedestrians and bicyclists. Special sensors can be used to detect the presence of pedestrians and bicyclists, giving them the appropriate time to cross. In many cases, special-purpose bicycle lanes are provided to enable them to get ahead of the traffic. For freeway management a combination of sensors, telecommunications, and back-office command-and-control issues to manage routine and nonroutine conditions. This latter condition is often referred to as incident management and involves the detection, verification, clearance, and management of unusual events on the network.

Current applications also make extensive use of variable message signs to communicate to drivers with messages regarding prevailing operating conditions, emergency alerts, and advisories. Express Lane management involves the same techniques with the addition of variable tolling supported by electronic toll collection, to manage the demand for transportation to specific operational objectives. For example, in most express lanes in the United States, the tolls are varied to achieve an average operating speed of 45 mph on the express lanes. Typically, these facilities have non-tolled general-purpose lanes that can be used as an alternative to the paid lanes.

7.3.14.2 Automated Vehicles

These are vehicles that can drive themselves [5]. A considerable amount of research and financial research is focused on the development of vehicle automation technology that can replace the driver. While this is a difficult task, significant progress has been made in the past few years. A vehicle automation system will feature sensors that can detect the operating environment of the vehicle including infrastructure and other vehicles. It also incorporates processing capability to enable the sense data to be turned into decision support. This in turn is fed into the vehicle control system which uses automated braking, acceleration, and steering, to guide the vehicle according to the sensed data. Sensors will also detect the presence of pedestrians and other obstacles. In a similar manner to the connected vehicles described above, automated vehicles will generate significant data. This is likely to be larger than connected vehicle data due to the presence of more sophisticated sensors on automated vehicles including LIDAR and cameras. It is likely that automated vehicles will also require two-way interaction with infrastructure to enable a complete picture of vehicle operation and automated decision-making. Automated vehicles will initially supplement the role of the driver and will ultimately replace the driver. This will have a significant impact on transportation as it sets the scene for on-demand automated vehicles, and more equitable access to transportation for the young, the old, and those not able to use existing transportation services. Vehicle automation could have a considerable cost reduction impact on freight and transit, where the driver is present purely for vehicle operating purposes and not to get to a particular destination. It is estimated that driver compensation contributes about $0.48 per mile to the cost of operating a truck. Automated vehicles will also have a significant impact on short-distance logistics within urban areas such as fast-food delivery and package delivery.

7.3.14.3 Congestion Pricing [6]

There is worldwide interest in the use of congestion pricing for demand management and for the generation of revenue to fund transportation. The basic economic theory of optimizing road use by pricing can be traced back to the British economist Arthur Pigou in 1920 (https://oll.libertyfund.org/title/pigou-the-economics-of-welfare).

Interest, study, and activity in congestion pricing can be summarized by a number of milestones as shown in Figure 7.3.

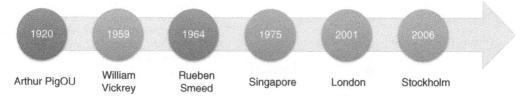

Figure 7.3 Congestion pricing milestones.

Technology options are now available to enable efficiency and effectiveness. Automated payment with minimum intrusion and maximum traffic flow, based on this ability effective, congestion pricing has been enabled on a global basis. Congestion pricing uses electronic toll collection technologies to detect and characterize vehicles passing a particular section of road. Vehicles identifying, using automated number plate recognition or through the use of a radio frequency identification tag attached to the vehicle. Integrated mobility and multimodal transportation management. Data from the roadside sensors used to detect and characterize vehicles is transmitted via a telecommunications network to a special-purpose back-office system. There are various payment strategies in use around the world including pricing by time of day and pricing by vehicle type. Having a date of ID, owner of the vehicle, the congestion charge is calculated, and vehicle owner is informed. The system features significant audit trail capability to enable any challenges or problems associated with the transaction to be investigated. In many of the systems around the world, customer service functions are addressed by a network of partners, including banks, convenience stores, and other public points of access.

The typical roadside infrastructure arrangement for a congestion pricing system is shown in Figure 7.4.

A laser detector (B) senses when a vehicle is passing through the control section of road. The laser then triggers cameras (a) (D), which take photographs of the vehicles front and rear number plates and antenna (C) may also be used. To read data from an RFID tag fitted to the vehicle, the

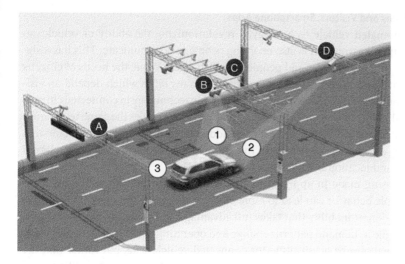

Figure 7.4 Congestion pricing roadside view. *Source:* Prakshaal et al., A unique identity based automated toll collection system using RFID and image processing. © 2018 IEEE. https://medium.com/@lewislehe/a-history-of-downtown-road-pricing-c7fca0ce0c03.

Figure 7.5 Congestion pricing cameras in London. *Source:* Verdict Media Limited / https://www. roadtraffic-technology.com/projects/ congestion / last accessed September 23, 2023.

images from the cameras are analyzed at the roadside using optical character recognition technology to determine the vehicle license plate number. This is then transmitted along with audit images to the back office.

Note that in addition to variations in the technology used and the pricing strategy adopted, there are also different ways to apply the charge. For example, the Stockholm system is known as a cordon crossing approach, where vehicles are detected at the cross-entry or exit lanes to and from the charging zone. Whereas in London, the strategy adopted is known as zonal presence. Vehicles are subject to a charge if they are detected as present anywhere within the charging zone, not just at the entry or exit points. This requires additional sensors inside the zone but leads to greater accuracy by multiplying the number of possible reads of the number plate.

Figure 7.5 shows an example of a camera system used for congestion charging, in London.

7.3.14.4 Connected Citizens and Visitors, Smartphone Apps

While connected and automated vehicle technology is revolutionizing the ability of vehicles to communicate, another parallel revolution is also enabling people to communicate. This has a significant impact on people that live in cities and people who visit cities. While the needs of citizens are an especially important aspect of smart mobility, there are many cities which depend on visitors for commercial and tourism purposes. An important enabling technology in connected citizens and visitors is the availability of position data from smartphones. This enables the same techniques for connected vehicles to be applied to people, with the added benefit of not being specific to a single mode and providing a comprehensive view of the entire trip. Such data is variously referred to as movement analytics and location intelligence.

Significant progress is being made in applying this technology to marketing and other areas where information on people behavior can lead to more efficient processes and operations. This is also true of transportation. Smart mobility that takes full advantage of location intelligence can be much more resilient and agile as demand patterns change and operating conditions vary. It is likely that location intelligence will not be a substitute for connected vehicle data, but rather that the data sources will be combined to provide an even better picture of operating conditions and the demand for transportation. Like connected vehicles, connected citizens, and visitors' services also support two-way communication with the device user, making it possible to provide a range of location-specific services to the traveler.

Figure 7.6 Smart phone shared mobility app. *Source:* vuumly / https://vuumly.com/ / last accessed September 23, 2023.

Here again, practitioners must be ready to develop analytics to take advantage of the new data. It is necessary to equip ourselves to be able to merge data in large volumes from different sources and turn it into information, insight, and action plans as effectively and efficiently as possible. In the past, there was a reliance on estimates and small sample sizes, as this data becomes available, it is now possible to base transportation planning and traffic engineering approaches on highly detailed observed data and reduce dependence on mathematical simulation modeling. This will also enable enhanced user experience and provide both citizens and visitors with a higher level of decision support throughout the entire trip. The support services can also make it easier for travelers to make modal changes and to adapt journeys in the light of changing needs. This will also enable better support for spontaneous travel decisions in addition to those that are preplanned.

Figure 7.6 shows an example of a shared mobility app.

7.3.14.5 Connected Vehicles

Significant progress has been made to apply cellular and dedicated short-range communication technologies to support a two-way link between vehicles and roadside infrastructure. The various communication links involved are described using the following abbreviations:

V2X: a two-way communication link between vehicles and several other elements such as other vehicles, roadside transceivers, or wide-area cellular networks.

V2V: a two-way communication link between vehicles, from one vehicle to another.

V2N: vehicle to network, another way to describe a two-way communication link from vehicles to a communication network such as cellular.

While early work focused on short-range communications, latest implementations have incorporated the use of cellular V2X approaches. It is likely in the future that the three different approaches will be used in combination to create additional capability.

In parallel with the development of wireless communication techniques, there has been significant progress in the use of data inside the vehicle. Engine management systems and carry area networks provide the possibility for a rich stream of data to be received from the vehicle. Now this data is only available to mechanics and technicians and other systems on board the vehicle. By combining this advance with connected vehicle technology, the possibility now exists to externalize vehicle data and use it as the basis for better insight and understanding of traffic and transportation management. At a basic level, the ability to receive data from the vehicle regarding vehicle identification, instantaneous vehicle speed, original origin for the trip, and ultimate destination for the trip will be an important part of the smart mobility revolution. The availability of this data will cause sudden and dramatic changes in approaches and the ability to manage traffic and transportation. This new data can be the basis for more sophisticated transportation planning and more effective traffic engineering.

Vehicle speed data reduces dependence on collecting such data from roadside infrastructure sensors, while also offering a much finer level of detail and a better understanding of the variation of vehicle speed throughout the network. An understanding of the original origin and ultimate destination of the vehicle also enables the placement of existing investments in traffic signal control and advanced traffic management within a wider context. Such techniques now focus on vehicle trip patterns and behavior within the control network and have little information on how the vehicle entered and exited the network. Possessing this additional information will allow a wider, multimodal approach to be taken toward citywide transportation management. Beyond a basic set of data, the following data can also be obtained from connected vehicles:

Road conditions: sensors on board the vehicle detect road geometry changes and condition of the road surface. Sensors also detect average vehicle speed which can indicate traffic conditions being experienced.

Environmental conditions: sensors on board the vehicle measure ambient temperature, detection of windshield wiper use can detect rain.

Operating status of the vehicle: a wide range of data that describes the operating status of the engine, transmission, wheels, and other onboard equipment.

Vehicle usage: additional sensors monitor vehicle speed, location, and average load weight.

Driver behavior: the use of cell phones and Internet-based services can be monitored.

Estimates vary widely, but it is expected that the full set of data that can be obtained from a typical connected vehicle could be as large as 300 TB per day.

Smart mobility practitioners need to get ready to manage large volumes of data from vehicles and efficiently turn it into information. Current analytics must be revised to include the availability of the new data and develop new analytics for new insight and understanding. It is necessary to ensure that data-sharing agreements are in place to enable access to the data from the vehicle. This will include a clear definition of data ownership. There is a multitude of new services that can be supported and new approaches to existing services. For example, the insurance industry can be revolutionized by using data to customize insurance rates to vehicle and driver behavior.

7.3.14.6 Electric Vehicles

Electrically propelled vehicles are not new, but advances in technology over the past few years have made them much more commercially viable. In the past, the use of electric propulsion systems and vehicles was restricted to fleets, where the specialist support required to keep the vehicles operational could be supported and justified. Advances in battery technology and in the manufacturing of electric vehicles have led to the point where electric vehicles are becoming much more prevalent in the market and are likely to dominate the automobile market in the next 5–10 years. Electric vehicles have the important characteristics of being cheaper to operate with zero omissions. Although it could be argued that emissions are not completely avoided but merely shifted from the vehicle to the point at which electricity is generated, i.e. the power station or power plant. This is still desirable, as it is much simpler and cheaper to manage emissions at a central location such as a power station, rather than distributed emission sources represented by millions of vehicles. It is a significant economic aspect of electric vehicles on a geopolitical level. A move toward the use of electric vehicles will reduce dependence on oil.

Electric vehicles must be treated as a single system. While there is still considerable potential for improving the internal combustion engine, it is likely that electric vehicles will play a much more significant role in the future. This means that it is essential to replicate the ubiquity of gasoline filling stations, with appropriate electric vehicle charging infrastructure. In addition to geographic coverage of electric vehicle charging points, each charging point must have the appropriate characteristics for the location. For example, a home charging point can use a lower power rating as it is likely to have at least eight hours to charge the vehicle. Charging points in route will have a need for higher power and faster charging, while charging points at popular destinations such as offices, shopping malls, and schools will require characteristics somewhere in the middle. The expected demand for vehicle charging should also be considered to ensure that unacceptable wait times are not experienced, which within themselves cost congestion. Recent experience with electric taxis in Amsterdam suggests that the distortion in the energy consumption patterns in urban areas is much greater with electric vehicles than expected. This leads to consideration of efficiency and electricity generation and distribution, turning to solar power and microgrid technologies which are often part of the wider smart city picture.

Figure 7.7 shows an example of electric vehicles in use in Norway.

Figure 7.7 Electric vehicles in Norway.

Figure 7.8 Freight and logistics management center.

7.3.14.7 Freight and Logistics Management

In common with transit systems, freight and logistics management involves a significant degree of fleet management. Knowing where vehicles are located at any given time and providing an optimum itinerary a central to the efficiency of freight and logistics. While urban logistics has a different set of needs than intercity freight, there are some commonalities. Like transit, sensors on board the vehicle combined with the telecommunications network are used to transmit data regarding the vehicle back to a central dispatching or command-and-control point. In the back office, sophisticated computer algorithms are used to optimize delivery schedules based on current performance and future customer demands.

Here again, the introduction of automated vehicle technology is expected to lead to significant efficiencies in both urban logistics and intercity freight delivery. The introduction of automated vehicles for intercity freight has already started with road trials of systems that can manage platoons of trucks, either using a single driver or no driver. Automation in urban deliveries has been pioneered by fast food companies such as pizza delivery companies. This whole area is also expected to be impacted by the introduction of urban air mobility which will add a third dimension to urban logistics and offer additional capability for first and last mile connections. Figure 7.8 shows a freight and logistics management center.

7.3.14.8 Integrated Electronic Payment

Smart mobility encompasses a range of options to pay for services without the use of cash. These include electronic toll collection, electronic payment for transit services, and cashless payment for parking. Electronic toll collection involves the use of dedicated short-range communication technologies to support a transaction between the vehicle and suitable roadside infrastructure. Typically, the user has established a prepaid account, usually linked to a credit card account, enabling funds to be deducted when the vehicle is detected crossing a tolling point. Electronic toll collection has substantial value in increasing the capacity of toll facilities and reducing the need

Figure 7.9 Integrated ticketing system. *Source:* https://www.emperortech.com/case-studies/Intelligent-Transportation/Tunion.html

for manual cash collection. Savings result from higher throughput of traffic and avoidance of the cost of handling physical cash including shrinkage or fraud. Initially, electronic payment for transit and parking was supported using smart cards and credit cards, although more recently the use of Host Card Emulation technology has become prevalent. This involves the use of a smartphone as a virtual credit card. Examples include Google Pay, Apple Pay, and Samsung Pay. The user holds the floor next to a receptor device and the wireless communication supports the financial transaction. Integrated payment holds a much larger potential for value and benefits and the integration of the various payment channels to provide what is sometimes known as a universal payment system for a city. This would support the payment of tolls, transit fares, parking fees, and all small-value transactions with a single device and a single account.

In addition to using existing technologies such as smart cards and host card emulation, smart mobility will also feature the use of an emerging technology – blockchain. This technology will enable the type of financial transaction network and distributed ledger required to support citywide electronic payment for all modes. It also has benefits in terms of the reduced cost of operations and a higher level of security.

Figure 7.9 shows an example of an integrated ticketing system from a Chinese solution provider.

7.3.14.9 Intelligent Transportation Systems

This is a range of technologies for transportation management with a particular focus on road traffic management. The term can also be considered as a forerunner to smart mobility with its wider vision of multimodal transportation for all users. A typical intelligent transportation system consists of sensors, telecommunications, a traffic management center for command and control, and information distribution system. A typical layout for an intelligent transportation system is shown in Figure 7.10.

This illustration shows the use of both roadside and on-vehicle systems, combined with a variety of telecommunication techniques to deliver advanced traffic management and traveler information. It shows the use of automated vehicle technology searches intelligent cruise control, aviation, and fleet management is a move toward a multimodal transportation management. Electronic toll collection, in-vehicle navigation systems, and dynamic message signs are all used in support of enhanced traffic and transportation management.

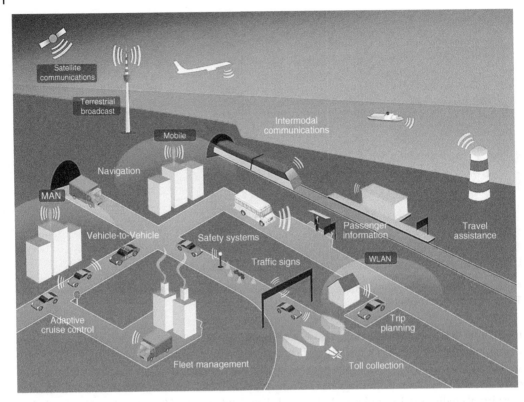

Figure 7.10 A typical intelligent transportation system.

7.3.14.10 Micromobility, Ridesharing, and Smart Mobility Services

This refers to the use of electric scooters, electric bicycles, and other forms of small, lightweight transportation under the auspices of shared use, pay-as-you-go, demand-activated frameworks, designed for personal use. Examples include:

Electric scooters: a standup scooter using an electric motor. Usually designed with a large deck in the center on which the rider stands.

Electric bicycles: a bicycle with an integrated electric motor to assist propulsion.

Golf carts: a small lightweight vehicle originally designed to carry two golfers and golf bags around golf courses. They are also suitable for some types of urban trips. They are electric powered.

Microcars: small cars with an internal combustion engine of less than 700 cc.

Segway: a two-wheeled personal vehicle with the platform between the wheels in which the writer stands while holding onto the handlebars. The vehicle is controlled by the way the rider distributes their weight.

Figure 7.11 illustrates an example of docked electric scooter at a shared mobility site.

Human-powered scooters and bicycles also form part of the original micromobility definition.

Most of the microtransit currently available in cities around the world take advantage of means of electronic payment to enable what is known as the "dockless" strategy. The vehicles are left at strategic locations around cities with no physical or mechanical means of constraint. Users unlock use of the vehicle using a smartphone or a credit card, having previously registered for the use of the service. The vehicles are typically equipped with tracking systems that enable the vehicle speed and location to be determined. To address the issue of one-way journeys which eventually cause a

Figure 7.11 Docked shared mobility electric scooters. *Source:* Deno/Adobe Stock.

displacement from the optimum deployment pattern, staff and vehicles are employed to redistribute the vehicles appropriately. A sophisticated reservation system can also be employed to enable users to make use of the vehicles and make electronic payments. More recently microtransit has been expanded to include the use of any vehicle smaller than a standard transit bus to provide on-demand services. In the future, these will also include the use of automated shuttle vehicles for first mile and last-mile connections to fixed route transit systems. These flexible transit options fit in the gap between the private car and standard transit.

7.3.14.11 Mobility as a Service

Mobility as a service involves the integration of a range of transportation services into a single menu that is available on demand. The transportation services can include public transit, car sharing, bike sharing, taxi, car rental or lease, parking, and combinations of these. Mobility as a Service (MaaS) can be delivered by either public or private sector operators. Essential features of MaaS include the ability to identify and locate transportation service vehicles, a single electronic payment mechanism, a simple user interface that across a wide range of Web-enabled devices including smartphones, and the ability to convert data to information regarding available transportation options and the quality of those options. An important objective is to present the transportation user with the best value proposition for the time of travel.

Location data from mobile devices is one of the critical enablers for mobility as a service. Latest statistics show that more than 80% of the US population uses mobile devices. Mobile devices have become ubiquitous mobile computing devices that can act as the interface to information services and a secure and convenient means of payment. Mobile devices can also play a crucial role in providing the raw data required to manage mobility as a service applications.

Movement analytics data from mobile devices can provide insight into traveler behavior, trip making patterns along with the ebb and flow of transportation demand within the smart city environment. The location data can provide insight into a better way for supply and demand management. The traffic congestion information data from traffic management systems can provide better time estimates of driver arrival and can reduce the time to travel by suggesting the most optimal path in real time.

This data has important implications for understanding supply and demand conditions, prevailing transportation conditions, and the performance of the various services that comprise the mobility as a service application.

The implementation of MaaS also provides a high degree of connectivity between transportation service operators and service users. This can form the basis for the delivery of other services within a smart city such as eGovernment, connected citizen, and visitor services.

First-mile, last-mile issues have always been at the heart of public transportation efficiency challenges. With a few major exceptions, cities around the world have been designed for the private car. This makes regular public transportation services uneconomic and acts as a barrier to investment required to make public transit ubiquitous. By adopting mobility as a service techniques, it is possible to chain together public transportation and other services to create an end-to-end, comprehensive offering to the traveler. This provides the traveler not just with a wider range of choices, but information regarding the choices available at any given time. Mobility as a service changes the traveler's perception about who will provide the services by providing a one-stop shop to gain access to services that can include both public and private sector options. These options can be made available on demand, thus increasing the flexibility available to the traveler. Mobility as a service can also address equity issues associated with transportation service delivery by considering the capabilities and characteristics of the traveler.

Figure 7.12 illustrates an example of mobility as a service operations, indicating have a variety of inputs are used to provide comprehensive information to the traveler.

7.3.14.12   Nonmotorized or Active, Transportation
Nonmotorized transportation includes pedestrian and human-powered transportation modes. These include human-powered bicycles and scooters as well as the basic function of walking from origin to destination. There are many aspects where nonmotorized transportation can be improved using smart mobility technologies. These include customized facilities to enable people to walk across roads, near-miss detection to improve safety at intersections, and sophisticated pedestrian detection and traffic signal timings to create a smoother trip across urban areas for pedestrians.

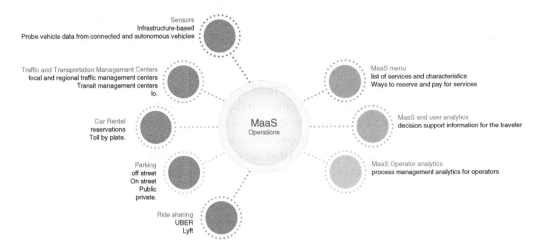

Figure 7.12   An example of MaaS operations.

Figure 7.13 Active transportation support Dublin airport.

Smart mobility really comes into its own as far as pedestrians are concerned when modes are linked together using the connectivity provided by smart mobility. For example, a commuter could walk from home to the nearest micro transit vehicle. The micro transit vehicle could then be used to travel to the nearest bus stop and then deposited at the bus stop. The commuter can then take the bus to the nearest bus stop to work and either walk the rest of the way or pick up another micro transit vehicle. In future, some of these first and last-mile connections could be supported by automated shuttles.

Figure 7.13 provides an example of facilities to support active transportation use at Dublin airport.

7.3.14.13 Public Transportation Management
There are two primary components to transit management – fleet management and passenger information. The use of electronic payment techniques for fear payment can also be considered part of transit management but has been covered as part of integrated electronic payment. Involves the location of all operational vehicles in the transit fleet using automated vehicle location technologies. While this was initially achieved using short-range communication techniques, beacons, these days cellular wireless technology is used to communicate to and from the vehicle. Automated passenger granting technology can also be used to determine the number of passengers on the bus and this data, along with current vehicle location is transmitted to a transit management center or back-office. In the back-office, purpose-built software can be used to optimize transit services in the light of current operating conditions. Changes that can be made include service frequency changes, altered routes, and the early termination of routes to improve system performance.

The automated vehicle location system used for system performance can also be used to provide passenger information. Using either smartphone technology or variable message signs on bus stops, passengers can be made aware of the current performance of buses and how many minutes will elapse until the bus arrives. This can also be extended to home services and mobile services for the information regarding service performances delivered to a home computer or a smartphone device. It is expected that these kinds of passenger information services to be expanded and enhanced to include mobility as a service, where a smartphone device can provide a single point transit availability information, central reservation for multiple modes, single electronic payment, and information on the quality of services.

It is also anticipated that transit service efficiency will be significantly implemented by the introduction of automated vehicle technology, although challenges will have to be overcome. For example, the vehicle operator does more than drive the bus. He or she also represents a customer-facing point of information for travelers and first response in the event of an incident. They also assist customers with wheelchairs. In addition to replacing the driving function, automation will also have to address these additional functions. It is also expected that the definition of transit will be extended to include aviation and maritime services as a multimodal holistic approach to transportation management is enabled by smart mobility approaches.

7.3.14.14 Smart Data Management [7]

There is more data available to transportation professionals today than at any given time in the history of the subject. This is both a challenge and an opportunity. The challenge is that the volume of data is so vast that it would be easy to be overwhelmed by a tidal wave of data. The opportunity lies and efficiently transforming the data to information, extracting insight, and understanding from the information and developing new response strategies and plans based on new insight and understanding. There is the possibility of developing transportation strategies and service delivery models that are much more accurate in matching the needs of the traveler. The challenge with data management from a transportation perspective is essentially a bridge-building problem. The various centers of transportation service provision that currently existed within a city must be networked into a patchwork world of data sources that can be shared across the city and from which analytics can be effectively conducted. Recent advances in data science assist with this immensely. Big data management systems are available to handle blocks of data that in the past would have had to be split into smaller chunks making it manageable. This new ability to handle extremely large volumes of data was initially driven by the needs of the Internet and webpages but now provides new abilities to take an enterprisewide view of data. Because the data is no longer segmented, it is possible to get a clear view across all data. The ability to turn the data into information and apply advanced analytics to the data has also progressed in leaps and bounds. There are new possibilities for understanding trends and patterns and understanding causal mechanisms. This can all contribute to better transportation service delivery across the spectrum from transportation planning to design, to project delivery, to operations, and to maintenance. To achieve all this, a smart approach must be taken to data management, making use of the most advanced technologies, and making organizational improvements required to support data sharing.

Traditionally, data in transportation agencies could be described as stovepiped. Held in individual silos with little connectivity between them and little understanding of what data is available. In some cases, the situation has improved, and data cockpits have emerged. They still do not enable enterprisewide data management but are configured as collections of data and tools to support a specific job or a group of staff. This improves efficiency but does not allow the ultimate efficiency of being able to use data as the utility across all organizational activities. For many years, it has been considered that the objective of database management is to ensure that data goes in once and is used many times. This goal is still not yet attained, and, in many cases, data goes and many times and may not even be used once. The authors have experience with several agencies who have admitted that they do not use all the data at their disposal because it is perceived as too expensive and too inaccessible to use. That is why smart data management is so important. It is also important in paving the way for the application of emerging data science such as Artificial Intelligence and Machine Learning to transportation. Both will need easy, structured access to big data sources.

7.3.14.15 Smart Parking

Smart parking is the application of information and telecommunication technologies to both on-street and off-street parking and urban areas. This is typically for private cars and light trucks. The overall concept is to use sensors of various types to detect if parking spaces are occupied or not, analyze the data, and then provide information to drivers on where the spaces are available. Sensors to detect space occupancy can be radar or cameras. Information can be delivered to the driver via smartphone apps, in-vehicle information systems, or roadside dynamic message signs. The information can also include guidance on how to get from the vehicles current location to this parking spot location some smart parking implementations also include electronic payment arrangements with the driver can pay for the use of the space without cash. Automated sensors are also used for enforcement. In some cases, text messages can be sent to the driver if the time that was paid for is expiring, or if there is another enforcement issue.

Using electronic payment also allows flexible payment structures to manage demand by time of day and duration of parking. Smart parking can also include automated parking structures to enable vehicles to be packed into a tighter space. These systems act like a vending machine, while the driver drops off the vehicle. The system then moves the vehicle to an available parking place. Then the driver can retrieve the vehicle using suitable identification.

A typical smart parking implementation is illustrated in Figure 7.14, along with some of the possible features:

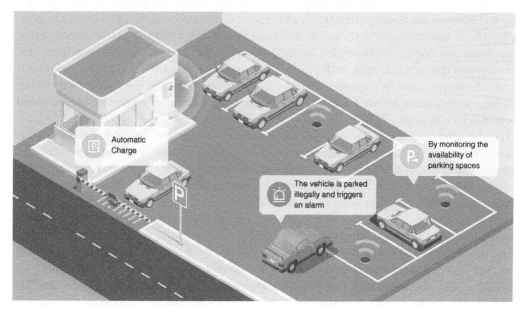

Figure 7.14 Illustration of a smart parking system. *Source:* https://www.mokosmart.com/smart-parking-system-using-iot.

7.3.14.16 Enabling Services

The services that have been described so far share one important characteristic in that they all deliver direct measurable benefit in terms of increased safety, improved efficiency, enhanced user experience, and better environmental management. However, it is impossible to deliver these direct benefit services without the presence of enabling services. As the name suggests the services enable the other services to operate, although they do not deliver a direct measurable benefit. Examples include the operation of telecommunications networks, organizational governance, the management and operation of sensors, and the establishment, and management of public–private partnerships. It can also be argued that smart data management should also be categorized as an enabling service, but the authors consider this to be so important that it has been called out in the previous discussion. Enabling services are in turn fueled by smart mobility technologies such as data collection, information retrieval, visualizations, and sophisticated management techniques.

7.4 How Smart Mobility Solutions Can Fit Together

There is no question that the best outcomes can be achieved for smart mobility solutions when a comprehensive and coordinated approach is adopted. This can be a challenge, as most deployments of smart mobility happened on a piece-by-piece basis over a period of time. There is also a division of responsibility between different public sector agencies and between the public sector and the private sector. Ideally, before smart mobility implementation in the city goes too far, a big picture would be developed that shows how all the pieces are intended to fit together and achieve a coordinated set of objectives. This big picture should take account of existing plans, policies, and an understanding of the direction and motivation of the private sector. Figure 7.15 shows a simplified view of how the smart mobility solutions described in section of this chapter can fit together.

Smart mobility solutions have been divided into back and systems, field systems, and vehicle systems. The backend systems have been subdivided into public and private sectors. That are several important relationships that enable all the solutions to work together. The first is the interaction that is assumed between the public sector and private sector backend systems to ensure that data processing, duplication is avoided, and alignment is achieved with regard to objectives. This assumes that data will also be in both directions. Second is the assumption that telecommunications infrastructure and field systems would be shared between the public and the private sector, avoiding duplicate investment. The third relationship is similar but involves sharing of data to and from in-vehicle systems.

It is also worth noting that it is also assumed that there is integration between the various public sector initiatives. This integration should include aligning objectives, synchronizing plans, and coordinating operations. Ultimately, the attainment of interoperability between the various smart mobility initiatives would act as a foundation for a multimodal approach to the management of transportation service delivery within a metropolitan region.

It is also assumed that smart mobility is not implemented in isolation but is closely integrated with existing conventional transportation solutions and takes full advantage of existing transportation infrastructure. For example, vehicle guidance systems must take full account of the separate guidance provided to drivers using conventional. The processor planning for connectivity in smart mobility is often referred to as architecture development, a component of system engineering [8].

Another important interaction is the alignment and adjustment of the business models adopted by both the public and private sector in developing and delivering the services. It is a white choice of business models available, and summaries are discussed in detail in Section 7.6 of this chapter.

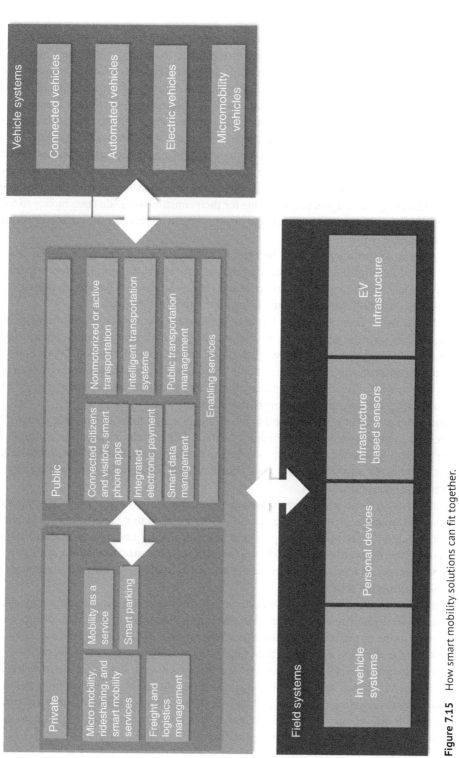

Figure 7.15 How smart mobility solutions can fit together.

7.5   The Importance of Data and Analytics in Smart Mobility Solutions

Recent advances in data availability and science enable probe data from connected and automated vehicles, along with location intelligence from smartphones, to be used to develop a high level of insight into the supply and demand for urban transportation. These are all vital components of an analytic ecosystem capable of integrating, analyzing, and sharing massive streams of real-time data. The ecosystem includes a data warehouse, analytic platform, and oftentimes a data lake. It is what allows a city to optimize resources, improve infrastructures, enhance residents' quality of life, and create sustainable economic development, among other high-impact outcomes.

Many cities are now implementing aspects of smart mobility and realizing the benefits. However, most are settling for only a fraction of the full range of opportunities and returns that smart mobility can offer. This is because they are not integrating and persisting all available data, analyzing it, and then using the insights to drive value for their constituents. Data costs money to collect and store, so making effective arrangements to obtain a return on this investment is crucial. Smart data management is required.

This is often an opportunity lost. The best results from monetizing data are achieved when an entire process is accomplished from collecting data to converting it to information to extracting insights to ultimately defining actions and response strategies. In short, gathering, storing, using, and then acting on data is extremely valuable for smart mobility initiatives.

Data is the lifeblood of smart mobility. It powers apps, which give users basic information. When analyzed, the data provides granular, contextual intelligence for better outcomes. Smart mobility needs smart data management that ingests all available data, shares information, allows app developers to create new products and uses analytics to continuously improve the systems that form the basis of smart mobility. Creating the right conditions for smart data management and exchange entails planning, governing, and executing best practices.

To achieve the best results, smart mobility must integrate data from all available sources. This overcomes the common problem of an inability to share data because of departmental silos. City departments often gather and store their own data to fit their needs. This approach breaks down data silos to bring all data together for rich, robust analytic insights that transcend individual departments or organizations.

Smart data management, when coupled with analytics, can support a wide range of smart mobility services. For example, a city can leverage transportation data that captures where people live and work to help improve public transportation solutions, better understand first-and last-mile needs, and deploy innovative, multimodal transportation solutions. In addition, transportation data supports deeper understanding of driver behaviors and implements improvements made possible through connected vehicles.

Smart data management solutions have become a vital building block and an important starting point as cities recognize the effects of an ever-increasing tidal wave of data and an exponential push for more users. As the smart data management solution grows, it can return revenue or drive down costs for the city. That money can then fund other mobility improvement initiatives.

Mobile phones are powerful tools to connect and involve citizens. In a smart city, a web of connectivity is woven between residents, visitors, and service providers. Crowdsourcing techniques collect data about prevailing conditions, while movement analytics detect trends, patterns, and carefully align transportation supply and demand. This offers an unparalleled ability to understand resident and visitor behavior to develop appropriate response strategies.

A responsive, efficient, and quality transportation system plays a critical role in the overall quality of life for residents, giving them the mobility that matches their lifestyle while providing accessibility to employment, education, healthcare, and other opportunities. Transportation is being transformed within the urban environment with mobility as aservice and automated first-mile-last-mile shuttles as prominent examples within a range of smart mobility solutions. To get the best in the solutions is important to apply insights to develop appropriate responses and strategies. Smart data management enables this by establishing, developing, and connecting data management capabilities across entire urban regions. A smart data management approach will include the following:

- Integration of data from sensors, other automated sources, private sector third-party vendors, and additional city data into a single platform that is manageable and accessible.
- Automated ingestion of multiple data sources into a centralized repository or data lake that allows data to be shared and analyzed.
- The ability to conduct advanced, multi-genre analytics on combined datasets.
- Sharing data under a management regime that controls the level of data access for each user.
- Analytics sharing mechanism that allows all users to share analytics that have been predefined and created while operating the smart data management system.
- The capability to value both raw data and processed information by supporting a data market approach.

Figure 7.16 shows a high-level overview of a smart data management approach for smart city accessibility and mobility.

It is recommended that cities start with a smaller project, test the process, and then scale accordingly. The solution should be specifically designed with a modular and scalable approach in mind that includes both on-premises and in-cloud offerings. This gives the solution the flexibility to answer any question at any time rather than a limited approach that uses predefined, rigid questions. This also fits with the test and iterate approach taken toward demonstration projects by many cities around. The process should begin with a small, highly focused pilot project and then build a business justification for further investment. Figure 7.17 provides an example of a robust methodology that supports incremental approach to implementation.

Approach starts with the development of a roadmap that defines the starting point, the desired endpoint, and milestones along the way. Objectives are then aligned on a metropolitan regional

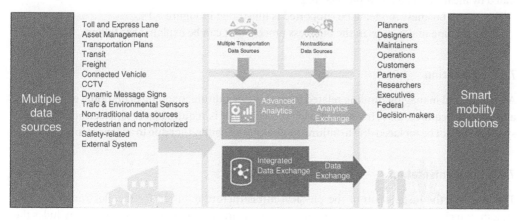

Figure 7.16 Smart data management approach.

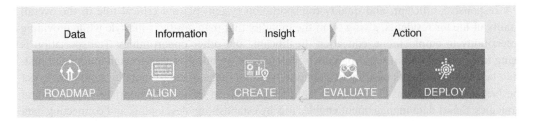

Figure 7.17 Methodology.

level to ensure that the region is on the same page. Then the pilot project, or a series of early-winner implementations can be conducted that possess two criteria:

- Data is currently available to support the project.
- The project will deliver clear and immediate value to the city.

Ultimately, smart data management will support all elements of smart mobility. The ability to migrate from a small-scale pilot to full-scale deployment is based on inherent scalability and a superior implementation methodology.

Experience from around the world has shown that the effective use and sharing of data is a vital component and successful application of smart mobility initiatives. This leads to better outcomes and also enables the vast amount of data generated by smart mobility solutions to be harnessed for further improvement.

7.6   Smart Mobility Business Models

In essence, business models define who does what in delivering a service, who invests the resources, and who achieves the reward. A good way to define a business model is to explain the business processes that will be supported in delivering the service. To provide a practical illustration of business models, a focus has been placed on electronic payment for public transportation. Based on that definition of the standard business process for electronic payment for public transportation, leading US transportation agencies have been studied, and the business models adopted are illustrated by means of a business process map.

The overall business model to be supported is illustrated in Figure 7.18.

There are nine major steps in the business process that can be explained as follows.

7.6.1   Financing

Arranging and managing funding related to capital expenditure, ongoing operational expenditure, and finance required for refreshing the technology. This is required as most electronic payment systems will not be replaced due to failure, or weaknesses, but rather due to advances in technology.

7.6.2   Implementation

This involves the installation of the physical infrastructure required for the electronic payment system. It includes the provision of fare media, smartcard, or cell phone based. It also includes the provision and installation of the readers required to interact with the payment devices, appropriate telecommunications infrastructure, and the backend system, or back office.

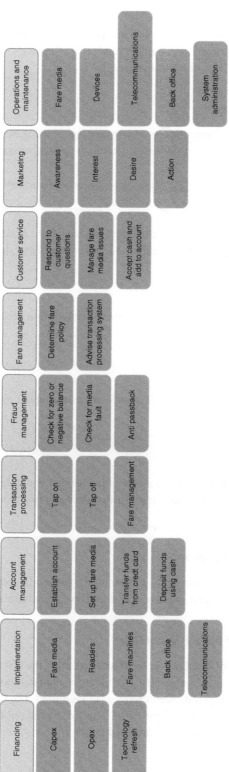

Figure 7.18 Electronic payment for public transportation business process.

7.6.3 Account Management

This covers all the activities required to establish an account for the user, set up the fare media to be used, support a financial transaction to transfer funds from the user's credit card to the backend system and enable the user to deposit funds using cash.

7.6.4 Transaction Processing

This supports the process of the user tapping on our tapping off using the payment media and the reading devices on board transportation vehicles. This also supports interaction with the fare management processes.

7.6.5 Fraud Management

These enforcement processes are required to check if the user has a balance available on the payment mechanism, determining fare media faults, and preventing illegal sharing of payment media.

7.6.6 Fare Management

These processes determine the fare policy and advise the transaction processing system and interact with the transaction processing processes.

7.6.7 Customer Service

These processes support response to customer questions, managing fair media issues, and enabling the user to add cash to an account.

7.6.8 Marketing

These processes are designed to make users aware of the service available, enable them to develop an interest in using the service, move that interest to a desire to use the service, and then finally take action to make use of the service.

7.6.9 Operations and Maintenance

These are the processes required to operate and maintain fair media, payment devices, telecommunications infrastructure, backend systems, and overall system administration.

7.6.10 Possible Public and Private Sector Roles

Based on a study of two leading US public transportation agencies, the Chicago Transit Authority [9] and the Dallas Area Rapid Transit Agency [10], two variations of business models involving different combinations of public and private sector roles were identified. The variation implemented by Chicago Transit Authority is shown in Figure 7.19. The role of the public agency is indicated in blue. While the role of the private sector is indicated in gray.

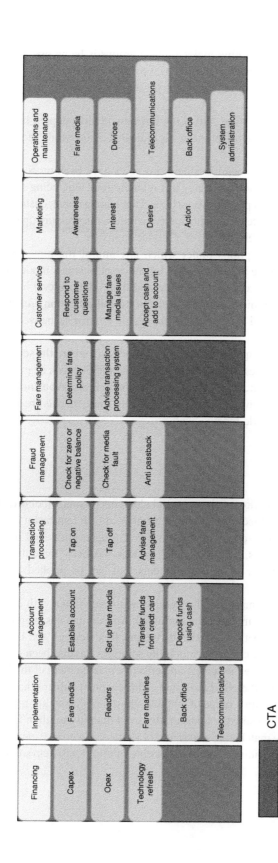

Figure 7.19 Chicago transit authority business model.

This show shows only one of the nine major process groups is being operated by the public sector, with the other eight groups being operated by a private service provider under contract to the public agency. Retaining the fare management process management enables the agency to keep direct control over policy.

The second variation shown in Figure 7.20 is the business model adopted by the Dallas Area Rapid Transit agency. The role of the private sector is indicated in tan, while the role of the public agency is indicated in gray.

This shows that five of the nine major business processes have been outsourced to a private sector service provider, while the public agency has retained four. Financing was outsourced to enable wider access to funding on a pay-as-you-go basis. Fare management, customer service, marketing, operations, and maintenance are all outsourced to relieve the pressure on the public agency resources while supporting a payment system as a service approach.

There are multiple alternative options under which any one of the nine, or a combination thereof, could be outsourced to the private sector. The choice of business model is dependent on the policy of the agency, the resources of the agency, and their comfort level in outsourcing. For example, many agencies in electronic ticketing and electronic toll collection prefer to have a direct relationship with users and avoid having a private sector enterprise between them and the customer.

The same principles can also be applied to mobility as a service, where the service delivery process can be broken down into process steps and then a structured plan can be put together to support the preferred public and private sector roles. Not that there is also a business model which is entirely public sector, or entirely private sector. Unfortunately, these models are prevalent in smart mobility initiatives around the world today.

7.7 Summary

This chapter provides an explanation of the essential elements of smart mobility. It is explained that smart mobility goes beyond technology into a number of different factors. It has described smart mobility solutions, building on the global review of progress provided in Chapter 5. The relationship between technologies, solutions, and services has been reiterated along with an explanation of the different types of business models that can be applied to smart mobility services.

The importance of integrating smart mobility services with each other, with conventional solutions and with existing infrastructure has been discussed. An example has been provided on how this integration can be achieved between backend systems, field devices, and in-vehicle systems. The importance of data and analytics in smart mobility solutions has also been explained in terms of the raw material required to drive the service and effective management of the large volumes of data that can be generated by smart mobility solutions. This chapter confirms that the smart mobility revolution is well underway with a range of tangible smart mobility solutions available that are both robust and field tested. Choosing the appropriate smart mobility solution for the specific circumstances of the city, closely matched to needs and objectives, is a critical element of success. It is hoped that this chapter by providing a menu of possible solutions will help smart mobility practitioners to make wise choices.

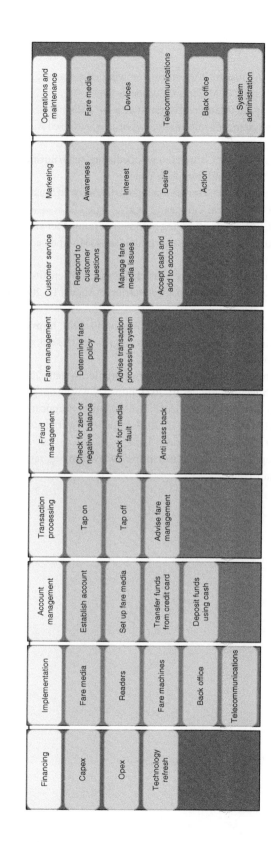

Figure 7.20 Dallas area rapid transit business model.

References

1 Wikipedia. Management systems, August 10, 2023. https://en.wikipedia.org/wiki/Management_system, retrieved August 10, 2023, at 7:34 PM Central European Time.

2 Techopedia. Information systems or information services, August 10, 2023. https://www.techopedia.com/definition/1027/information-systems-or-information-services-is, retrieved August 10, 2023, at 7:37 PM Central European Time.

3 Wikipedia. Business model, August 10, 2023. https://en.wikipedia.org/wiki/Business_model, retrieved August 10, 2023, at 7:39 PM Central European Time.

4 Wikipedia. Business process, August 10, 2023. https://en.wikipedia.org/wiki/Business_process, retrieved August 10, 2023, at 7:41 PM Central European Time.

5 Society of Automotive Engineers. SAE levels of driving automation refined for clarity and international audience, May 3, 2021. https://www.sae.org/blog/sae-j3016-update, retrieved August 10, 2023, at 7:42 PM Central European Time.

6 Mobility Academy. Road user charging: how it works, the challenges and benefits, August 10, 2023. https://www.mobility-academy.eu/course/view.php?id=9, retrieved August 10, 2023, at 7:44 PM Central European Time.

7 Arthur, D. Little, smart data management unlocks growth, August 10, 2023. https://www.adlittle.com/en/insights/viewpoints/smart-data-management-unlocks-growth%C2%A0, retrieved August 10, 2023, at 7:48 PM Central European Time.

8 United States Department of Transportation. Intelligent transportation systems joint program office, national ITS reference architecture, August 10, 2023. https://www.its.dot.gov/research_archives/arch/index.htm, retrieved August 10, 2023, at 7:51 PM Central European Time.

9 The Chicago Transit Authority. Homepage, August 10, 2023. https://www.transitchicago.com, retrieved August 10, 2023, at 7:54 PM Central European Time.

10 Dallas Area Rapid Transit. Homepage, August 10, 2023. https://www.dart.org, retrieved August 10, 2023, at 7:56 PM Central European Time.

8

Smart Mobility Technology Applications

8.1 Informational Objectives

After reading this chapter, you should be able to:

- Provide an overview of technology applications typically used in smart mobility
- Have sufficient knowledge and awareness of technology capabilities and characteristics to make a selection for further detailed evaluation and design
- Define the consequences of the accelerating pace of technology emergence
- Describe the relationship between technology applications, products, services, and outcomes
- Explain how technology applications are identified for smart mobility
- Define challenges and opportunities associated with technology
- Describe sensor technology applications
- Explain telecommunication technology applications
- Define information delivery technology applications
- Describe data management technology applications
- Describe in-vehicle technology applications
- Describe what is on the horizon for smart mobility technology applications
- Provide some advice on how to keep pace with technology change

8.2 Introduction

To play an effective role in the selection of technologies for deployment in support of smart mobility in urban areas, it is necessary to have at least a high-level understanding of the technologies that are available, and the ones that are emerging. This chapter provides an overview of the range of technologies that can be applied within the smart mobility sphere. This includes mature technologies that have been available for a number of years, new technologies that have just entered the market and have been made available quite recently, and emerging technologies that are either still coming along the pipeline, or in initial use. In addition to the characteristics of the technology, it is important to understand the technological capabilities and maturity of the technology as this has an effect on the risks associated with implementation.

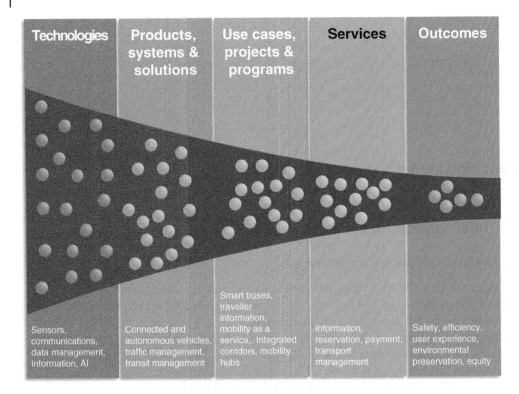

Figure 8.1 From technologies to outcomes.

It is also important to bear in mind the relationship between technologies products, applications, and solutions and the services that are enabled by the technologies. Figure 8.1 illustrates the relationship between these elements:

The range of technologies to be considered for smart mobility deployment is typically not designed specifically for mobility or transportation. Usually, they are designed for larger markets and are adapted for the smart mobility market. For example, the short-range communication application used to support electronic toll collection in the United States, originated as an animal tracking application. It was subsequently amended for high-speed open road use and adapted as an electronic toll collection system. Another point to note is that the fundamental technologies will usually not be addressed in this chapter. A detailed exposition on the physics of fiber-optic for example will not be provided; however, the application of fiber-optic technologies for smart mobility will be addressed. The term "technologies" may be misleading however the smart mobility and intelligent transportation system industry have adopted the term to mean applied technologies rather than the fundamental underlying technology. This is why the title of this chapter is smart mobility technology applications.

Technologies can be categorized under the headings illustrated in Figure 8.2.

These categories have been adopted for the purposes of this book but are not the only categories that could be adopted. As the purpose of the categories is to provide a structured view of technology and not a rigid demarcation, this works for the book. The intention is to provide a broad overview of technologies as typically found in smart mobility applications, explore the nature of these technologies, and provide references for further detail.

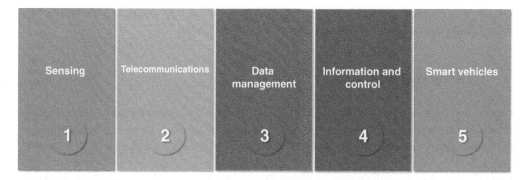

Figure 8.2 Technology categories.

It is interesting to consider how technologies are identified and selected for use within smart city projects and programs. Based on the author's experience in working on such projects on both the public and private sector sides, it is apparent that multiple factors are considered when conducting this activity. The first, and the most important factor in technology identification and selection is the prior experience that the industry or community has accumulated in deploying the technology on previous projects. Another factor is the awareness of the project sponsors and designers of current technology capabilities and constraints. When identifying and selecting appropriate technologies, it is also necessary to establish a clear understanding of the likely performance of the technology compared to the cost of acquiring and implementing it. This cost should be a life cycle cost that includes initial acquisition and long-term operation and maintenance.

The process of technology identification and selection will be addressed in more detail in Chapter 10 when a framework for smart mobility success is defined. Another important factor in technology identification and selection is a clear understanding of the challenges and opportunities associated with the specific technology. Here again, this can be drawn from prior experience or can be as a result of a detailed technology evaluation and assessment exercise. Typical challenges that have been encountered include lack of appropriate expertise in installing and operating the technologies, an initially exaggerated understanding of technology capabilities, and unexpected expense in operations and maintenance.

8.3 What Is Technology?

When researching technology definitions, one of the simplest encountered was as follows:

> Technology refers to methods, systems, and devices which are the result of scientific knowledge being used for practical purposes. [1]

This provides a concise definition of technology and is complete enough to include ways to do things as well as inventions, systems, and devices. Technologies can be considered as the building blocks of smart mobility solutions. New technologies are being developed at an ever-increasing pace, typically for application areas beyond transportation. While this is positive in that it supports technological evolution beyond the resources of transportation, it does present a challenge to those involved in applying technology to transportation due to the need to adapt technologies to suit the needs. In this chapter, the focus has been placed on defining and explaining the technology building blocks that are typically found in smart mobility solutions.

8.4 Technology Overview

As stated above, the range of technologies that can be applied to smart mobility can be categorized under five headings. Technologies drawn from each of these categories can be combined to fully enable smart mobility within a city. The range of technology categories is illustrated in Figure 8.3.

Sensing technologies are the means by which data is collected to provide a basis for understanding current operating conditions and current demand for transportation. Data from sensing technologies can also confirm the effectiveness of implemented strategies.

Telecommunications technologies enable the transmission of data from the location of the sensor to the appropriate data management facility. They also enable information and control commands to be transmitted from the data management or transportation management facility to locations on the transportation network and to smart vehicles.

Data management technologies represent the wide range of hardware and software that can be applied to receive data, verify, and validate it, combine it into a usable data source, and convert data into information. These technologies also manage the storage and accessibility of the data and the ensuing information These technologies also support the conduct of advanced analytics to the data, including the application of artificial intelligence and machine learning techniques to extract insight from the information and enable effective strategies to be implemented.

Information and control technologies enable information to be delivered to the traveler and also support actions such as actuator activation and other remote automated mechanism activation. For example, a remote-controlled barrier could be used because of freeway ramp or divert traffic. Traffic signals are quite a common application of information and control technologies.

Smart vehicles can incorporate the four technology categories described above, but a separate category has been defined due to the importance and the influence that these technologies will

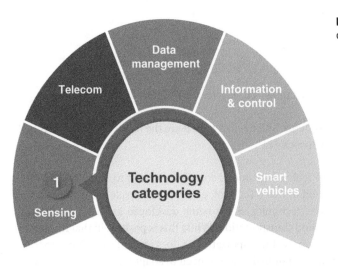

Figure 8.3 Illustration of technology categories.

have in future urban transportation. A smart vehicle can be both connected and automated. Connections between vehicle and roadside infrastructure, vehicle to vehicle, and vehicle to any other transportation system element (V2X) are supported by the application of telecommunications technologies, usually wireless. Telecommunication technologies are used inside the vehicle to transmit data along a car area network. Multiple sensor technologies can be used on board the vehicle to detect the current status of the vehicle, the driver, and the driving environment. Information control technologies are used to deliver information to the driver and operate automated driving systems such as brakes, steering, and accelerator. Smart vehicles include connected vehicles that can support a two-way wireless communication between the vehicle and roadside or back-office infrastructure and other vehicles. Automated vehicles are also included in this category. These are vehicles that have some degree of ability to drive without a driver or provide advanced driver decision support.

In the following pages, this range of categories will be reviewed to provide a high-level understanding of what the technologies are, what they can do, and an initial assessment of the value that they can deliver when combined in support of smart mobility services. Note that this is not intended to be an exhaustive catalog of every technology that is available, but an overview of the most important technologies that have been and are being deployed in smart mobility projects to date.

To simplify the extensive range of information provided in the technology reviews, a consistent structure has been adopted for each technology application category.

8.5 Technology Application Category Overview

At the beginning of each category description, an overview will be provided indicating the technology applications that are included in the category. The percentage that summarizes this will also provide an indication of how widely used the technology application is in current smart mobility implementations. This takes the form of a ubiquity index.

8.5.1 Description of the Technology Application

An overall description and the principles of operation of the technology application.

8.5.2 Operating Characteristics of the Technology Application

The essential operating characteristics of the technology application. This will include things like data that can be collected by sensors, bandwidths, and speed for telecommunication technology applications, data management and processing capabilities for data management, the mechanisms for information delivery for information control technology applications, and the essential characteristics of smart vehicles.

8.5.3 Interesting Aspects and Points to Note Regarding the Technology Application

Where appropriate, additional information is provided including interesting aspects of the technology application and significant points of information.

8.5.4   Sensor Technology Applications

Figure 8.4 summarizes, in alphabetical order, the 14 categories of sensor technologies that will be considered in this section. Note that many of the sensor applications make use of telecommunications technologies such as fiber-optic, wireless, and infrared. A detailed exposition of telecare indication technology applications will be provided in Section 8.5, Telecommunication Technologies.

Figure 8.4 also features an approximate indicator around the circumference. This shows how widespread, or important the technology application is in global smart mobility implementations. The outer circle of each button represents the index as a percentage. The index is intended to provide an overall impression of how widely deployed the technology application is in smart mobility at the current time.

8.5.4.1   Crowdsource

Smartphones and devices have become ubiquitous over the past few years. They represent a two-way communication channel between people, the cloud, and back-office infrastructure. This provides an interesting noninvasive approach to collecting data for smart mobility purposes. Crowdsourcing refers to the willing divulgence of information from many smartphone users for the purposes of planning and research. It is important to safeguard the privacy rights of the individual when doing this. This can be achieved by ensuring that the user has the opportunity to agree to have GPS coordinates transmitted to the cloud or to a back-office. Personal information can also be protected by data aggregation. This particular approach is described under this section and location intelligence in this chapter. The authors differentiate between crowdsourcing and location intelligence as crowdsourcing also involves anecdotal data in addition to the automated collection of GPS data. Crowdsourcing can include video sharing, texting, and information sharing from a large number of people on specific subjects.

8.5.4.1.1   Description of Crowdsourcing

The application is enabled by smartphone technology which is a combination of cellular wireless, sensors embedded in the smartphone, and GPS positioning from a smartphone. It can also include the use of the integrated video in the smartphone as central processing platform is used to receive

Figure 8.4   Sensing technology applications.

and interpret the crowdsource data and this in turn makes use of data management, information, and control technologies as described in the appropriate section.

8.5.4.1.2 Operating Characteristics of Crowdsourcing

Like many of the noninvasive, probe-type approaches to data collection this also involves a sample size rather than a comprehensive set of data. Here again, small sample sizes are not a limitation as data is usually provided that was not previously available. The availability of small sample sizes is better than no data at all, and in many cases, a statistically significant assessment can still be made on a small sample size.

8.5.4.1.3 Interesting Aspects and Points to Note Regarding Crowdsourcing

One of the interesting aspects of crowdsourcing lies in its suitability for public–private cooperation. The smartphone device is connected to a privately operated communication network. Therefore, all data to and from the device is managed by a private company – the service provider. Crowdsourcing data will also be acquired and aggregated by a private-sector company. This requires the public sector to consider ways to effectively engage the private sector to gain access to crowdsourcing data. This also requires a clear understanding of public sector objectives to assist when engaging the private sector.

8.5.4.2 Distributed Fiber-optic Sensing

8.5.4.2.1 Description of Distributed Fiber-optic Sensing

Fiber optic is a telecommunication technology application used for network communications in smart mobility and beyond. As described in the section on fiber-optic communications, it makes use of very thin fibers of glass to transmit data using pulses of light. There are extensive deployments of fiber-optic telecommunication systems around the globe, and it is a major component in the operation of the Internet. Many public agencies involved in transportation service delivery have installed dedicated fiber-optic networks to transmit smart mobility data. One of the major drivers of the need for data bandwidth in smart mobility is use of closed-circuit TV cameras for traffic surveillance. Consequently, that is significant existing deployment of fiber-optic communication cables in close proximity to major highways and other roads. This makes the use of distributed fiber-optic sensing for traffic monitoring, a possibility. Using some sophisticated hardware, it is possible to convert fiber optics running in close proximity to roads into traffic speed sensors. This approach has the advantage of high-resolution speed data capture every 100 m along the highway. The data can be delivered with a high frequency and to a high degree of accuracy. In contrast to some of the sensor techniques described in other sections, the technique delivers close to 100% sample size.

8.5.4.2.2 Operating Characteristics of Distributed Fiber-optic Sensing

The technique makes use of a so-called "dark" fiber from the many already installed and roadside implementations of fiber-optic network communications. A specially developed electronics package is connected to the fiber, either at the roadside or a convenient telecommunications hub for the network. The electronics package sends specially configured pulses of light along the fiber and measures the characteristics of the reflected pulse. The technique is similar to radar and LIDAR but making use of our fiber-optic conduit instead of light or radio waves. A single electronics package installation can monitor up to 80 km of roadway. Analysis of the reflected light pulse converts the fiber into virtual traffic speed sensors every 100 m along the road. The traffic speeds measured represent an average of all lanes across the highway in a given direction. Early tests have shown

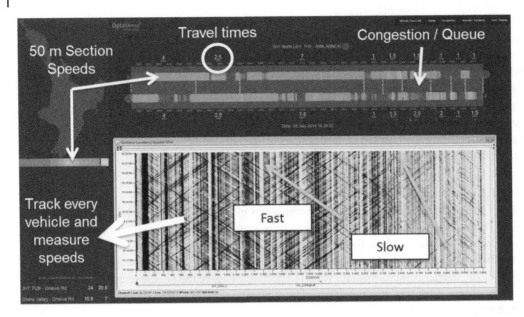

Figure 8.5 Distributed acoustic sensing dashboard.

that as many as eight lanes of traffic can be reliably addressed by the technique. Figure 8.5 provides an illustration of the overall operation of the system.

The light pulse analysis results in a series of traces or trajectories that measure the progress of vehicles along the road. The slope of the trajectory indicates the speed of the vehicle with faster vehicles exhibiting a flatter trajectory and slower vehicles a steeper trajectory as noted in the figure. The trajectory data can be ingested into an analysis system with a graphical user interface which can represent traffic speeds and slowdowns as a traffic radar map, as shown at the top of the figure. This provides a visualization of average traffic speed for each 50 m section of road and a calculated value of travel time over that same section of road. This represents the space mean speed of traffic over the road section.

8.5.4.3 Interesting Aspects and Points to Note Regarding Distributed Fiber-optic Sensing

This is an interesting application of sensing technology that enables prior investment in fiber-optic telecommunications to be leveraged. It can also be considered as a nonintrusive approach to roadside sensing as no onboard installation work is required. This supports cost-effectiveness and promotes safety during the installation of the sensing equipment. The ability to measure traffic speeds every 100 m along the highway also opens up new possibilities for the detection of end of queue and a rapid detection of incident-type conditions. This opens up the possibility of advanced analytics to detect and characterize rapid changes in traffic conditions.

8.5.4.4 Inductive Loops

8.5.4.4.1 Description of Inductive Loops

This could be considered as one of the original technology applications and transportation with respect to sensing. The technology takes advantage of physics. When a voltage is applied to a coil of copper wire, a current is induced in the wire. When a metallic object cuts through the magnetic field generated by the current, it distorts the voltage in the copper wire. The voltage variation can be measured and calibrated to enable the measurement of a number of traffic parameters including vehicle speed, traffic volume, vehicle classification, and vehicle presence over time [2].

8.5.4.4.2 Operating Characteristics of Inductive Loops

The implementation of the technology consists of the installation of multiple loops of copper wire in a slot in the road surface and then sealed with the bituminous sealer. Roadside electronics within the traffic controller cabinet generator voltage which is applied to the copper wire the voltage induces a magnetic field above the roadway. When the magnetic field is cut by a large metallic object such as a vehicle, changes in the voltage can be detected by the roadside electronics. Typically, inductive loops are installed separately for each lane of the road, providing transverse coverage for the roadway. Inductive lip installations may also be implemented at different locations longitudinally along the road to enable accurate speed measurement for the traffic. Since the sensor is buried in the pavement surface it is subject to considerable wear and tear and disruption by various objects that are thrown on the road surface, such as ring pulls from soda cans. Inductive loops have a long history of implementation; consequently, there is a wide and deep base of expertise in the technology that can be applied to calibrate the implementations.

8.5.4.5 Interesting Aspects and Points to Note Regarding Inductive Loops

While there is no available documented data on the subject, commonly accepted wisdom in the industry is that at any given time approximately 25% of inductive loops are not operational due to wear tear and maintenance issues. While the technology is highly accurate, it is also important to take availability into account when evaluating the overall performance of the technology. A performance index representing a combination of accuracy and availability may be an appropriate key performance indicator in this case.

8.5.4.6 Infrared

8.5.4.6.1 Description of Infrared

That are two types of infrared technology applications in use for transportation. The first type is known as active infrared. Infrared light is invisible to the human eye and also safe to the human eye. Beams of infrared light are transmitted across the roadway, as vehicles cross the beam, they reflect a proportion of the light generated and this can be measured and analyzed. The second type, passive infrared does not transmit energy but detects energy coming from two sources. Energy emitted from vehicles, road surfaces, and other objects, energy emitted by the atmosphere and reflected by vehicles, road surfaces, or other objects.

The infrared range is usually divided into three regions: near infrared (nearest the visible spectrum), with wavelengths 0.78 to about 2.5 μm (a micrometers, or micron, is 10–6 m); middle infrared, with wavelengths 2.5 to about 50 μm; and far infrared, with wavelengths 50–1000 μm.

8.5.4.6.2 Operating Characteristics of Infrared

The use of infrared light makes infrared susceptible to visibility and weather conditions. In operation, the technology is deployed as a pair of devices with the first device generating the infrared and the second device collecting the reflected infrared light from the vehicle. One particular application makes use of a rotating prism to create multiple infrared beams enabling time slices of the vehicle to be analyzed for the purposes of classifying the vehicle as well as counting volumes and measuring speeds. In common with video-based devices, the technology will require regular maintenance in the form of lens cleaning to ensure that the performance of the sensors is not degraded.

8.5.4.6.3 *Interesting Aspects and Points to Note Regarding Infrared*

The technology seems capable of reasonably high levels of accuracy for vehicle classification and counting. Minnesota Department of Transportation have started an evaluation of technology capabilities [3].

While the Hawaii Department of Transportation conducted a study in 2007 [4]. The technology has been implemented on a widescale basis to state agencies in Australia and New Zealand, as well in the United States, India, Pakistan, Singapore, South Africa, Turkey, and Europe [5].

8.5.4.7 Internet of Things (IoT)

8.5.4.7.1 *Description of Internet of Things (IoT)*

The Internet of Things (IoT) refers to the expansion of current Internet deployment to include a wider range of devices. The current Internet links computers used by people together into a single network. The Internet of things expands that to include a range of devices. These can include domestic appliances, low-cost sensors for temperature, vibration, sound, and roadside devices for transportation monitoring. In addition to the connection of low-cost devices to a network, IoT techniques can also support local processing of data under the auspices of a technique known as "edge processing." This enables data to be summarized at the edge of the network, reducing the cost of transmitting data to a central location. Transportation examples of IoT include roadside sensors for traffic management, support for a two-way exchange of information between connected vehicles and roadside infrastructure. This technology application is also enabled by the emergence of a number of different technologies. These include low-cost, low-power sensors, Internet communication protocols, cloud computing, machine learning and analytics, and artificial intelligence. The Internet of things refers to the expansion and enhancement of the current Internet to connect things together as well as people.

Currently, the Internet connects computers and laptops and smartphones together enabling people to communicate with each other. The Internet of things adds devices to this list. For example, the refrigerator at home could be connected over the Internet of things to a system that ensures a continuous stock of any particular food item. This could also be extended to reporting to the doctor on the nature of items that are stored in the fridge. The Internet of things could also be used to use a smartwatch to check blood pressure and other vital signs and report that back to the doctor too. Although this sounds a bit too much like life management. Like many other applications of technology, it is important that the user opts and is available so that people can choose or not to use the capabilities of the technology for smart mobility, both vehicles and roadside infrastructure sensors can be considered to be part of the Internet of things. High-speed Internet connections both wireline and wireless can be used to link these just-like devices to central locations. Internet of things is also used to refer to devices that have a degree of intelligence to enable preprocessing of the data inside the device. For example, an Internet of things roadside sensor could collect data regarding vehicle identifications and vehicle speeds and then process the data into assembly before transmission over the telecommunications network to a transportation management center. There is a question regarding the value of big data techniques that will be discussed later. They work better when raw data is sent back to a central location. It is therefore important to balance the motor processing done at the edge of the network, so-called edge processing, with the volume of data sent back to the central location. The advantage of the data being stored in a central repository is that a complete view can then be established across the entire dataset, with more detail supporting more analytics. This will be addressed in more detail in the section on data science, big data, and analytics.

The Internet of things is defined as a network of physical objects, or devices, which are connected together using the Internet. These devices can have sensors and software embedded in them to collect and exchange data over the network. It is estimated that there are more than 7 billion IoT-connected devices in the world today and this is expected to grow to 22 billion by 2025 [6].

IoT devices range from domestic appliances to industrial tools to connected vehicles and roadside sensors for smart mobility.

8.5.4.7.2 Operating Characteristics of Internet of Things (IoT)

The Internet of things is relevant to a range of applications beyond smart mobility. These include smart manufacturing, connected assets, smart power grid, smart cities, connected logistics, and smart digital supply chains. For the purpose of explaining operating characteristics of Internet of Things (IoT), smart mobility applications will be used as the focus. Internet of Things (IoT) uses some form of sensor aligned with local data processing capability (edge processing), and telecommunications to make data collection as efficient and dependable as possible. The bandwidth required for the telecommunications network that links the IoT device to the cloud of the back office is also managed through the application of edge processing. The site operating characteristics of Internet of Things (IoT) will depend on the chosen system architecture and a detailed design of the application. However, the typical operating characteristics can be illustrated with reference to a particular implementation [7]. In this case Internet of things technologies are used to enable sensors attached to the transportation network to collect data regarding the current operating conditions of the transportation network, process this data, and generate transportation Alexa can be provided to travelers using roadside message signs. This has the advantage of avoiding reliance on people or vehicles being equipped with appropriate information delivery devices but represents just one variation in her Internet of things that can be used to revolutionize transportation sensors and information delivery.

8.5.4.7.3 Interesting Aspects and Points to Note Regarding Internet of Things (IoT)

An interesting aspect of the Internet of things is the rapid uptake of Internet of Things (IoT) across multiple industries. Internet of things devices are forming extensive networks that can be used to get a broad understanding of operating conditions and variations in the demand for services across the city. To take full advantage of this, it will be necessary to adopt a partnership approach to the use of these IoT sensors. The traditional procurement model where a public agency will acquire, own, and direct the operation of advanced technology application would be suboptimal in this case. Best use of Internet of things technologies may require a different procurement approach and the adoption of different business models for smart city and smart mobility practitioners.

8.5.4.8 LIDAR

8.5.4.8.1 Description of LIDAR

LIDAR (Light Distance and Ranging) is a sensor technique which operates along similar lines as RADAR (Radio Distance and Ranging). While the latter makes use of radio waves at various frequencies for measurement and range finding, the former makes use of laser or infrared light. LIDAR applied to smart mobility applications is a combination of scanning and 3D imaging. The laser beams used are low-power and safe to the human eye. The laser light bounces off objects within a range of approximately 500 m. Between 8 and 128 individual laser beams can be used to support this operation. The output from a LIDAR device is known as a point cloud and is an extremely detailed digital picture of the entire area within the range of the LIDAR device. While the most prominent application of LIDAR now is in sensing for automated driving systems, LIDAR also has an extremely important role to play at traffic signals, for pedestrian detection and management. It is also likely that LIDAR will continue to play a significant role in automated driving systems both in terms of operations and in terms of data collection required for the introduction of automated driving systems. This will include the development of detailed infrastructure plans for automated driving system support.

8.5.4.8.2 Operating Characteristics of LIDAR

LIDAR has value in both static roadside sensing applications and in-vehicle environments. To illustrate operating characteristics, let us consider LIDAR operation within an automated vehicle. Figure 8.6 shows a typical architecture for an automated vehicle system.

This illustrates the role of LIDAR among a suite of sensors in an automated vehicle architecture.

8.5.4.8.3 Interesting Aspects and Points to Note Regarding LIDAR

One of the significant characteristics of the use of LIDAR is the sheer volume of data produced. The data stream from a LIDAR sensor, depending on its configuration, can be several gigabits per second. This is equivalent to transferring a one-hour high-definition movie every 24 seconds. Managing the volume of data will require the combination of edge processing to compress the data and summarize it, and high-speed communications to transfer the data to the cloud or back-office. LIDAR devices will work better when connected to high-speed fiber-optic communication networks [8].

8.5.4.9 Location Intelligence

8.5.4.9.1 Description of Location Intelligence

Location intelligence, sometimes referred to as movement analytics, is the use of position data from cell phones and smartphones to derive data useful for smart mobility. Location intelligence is in fact a bundle of technologies comprising cellular wireless communications and the global

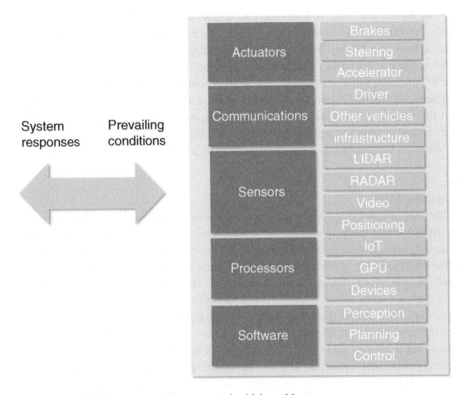

Figure 8.6 LIDAR in context with automated vehicle architecture.

positioning system (GPS) from the wireless communications are used to transmit the location of the device to a central location for analysis. The global positioning system as the name suggests establishes the position of the device on the globe within an accuracy of approximately 5 m [9] depending on the strength of the signal from the GPS, which in turn is influenced by the ability to establish line of sight to the satellites supporting the GPS. GPS and wireless communications will be discussed under separate headings in this chapter.

The term Location Intelligence will be used in this book and provides a definition that includes the collection of a wider range of data while focusing on the data collection activities rather than the analytical ones. It is similar to probe vehicle data from connected and automated vehicles, having the additional advantage that the location is person specific, rather than vehicle-specific. This extends the usefulness of the data to multimodal applications including transit, and nonmotorized transportation. It is very closely related to probe vehicle data, however, instead of the vehicle being the source of data, smart devices such as smartphones or cell phones provide the data. This has the advantage of extending the reach of observed data from vehicles only, to cover other modes of transportation including walking, cycling and micromobility. Location intelligence data involves the collection of latitude, longitude, device identification, and a range of other parameters that describe the accuracy and precision of the data. Location intelligence data, while originally designed for marketing and commercial applications, is extremely valuable to transportation agencies. Both probe vehicles and location intelligence data enable a comprehensive view of an entire agency's region of responsibility. In the case of location intelligence data, this can also involve the characterization of transit, freight, nonmotorized transportation, and micromobility modes, in addition to private cars. It offers the possibility of a complete trip approach to performance management including the evaluation of individual boards and the assessment of all modes working together in a single trip. This will be particularly important when accurately assessing performance from equity, climate, and energy perspectives.

8.5.4.9.2 *Operating Characteristics of Location Intelligence*

Location Intelligence provides a sample of the total dataset available for the location of smart devices within a region. The best estimates suggest that the overall sample size varies between X and Y percent, depending on availability and accuracy of the GPS system in use, and the number of reports provided from the device [10].

A typical dataset supplied by a location intelligence system is shown in Figure 8.7.

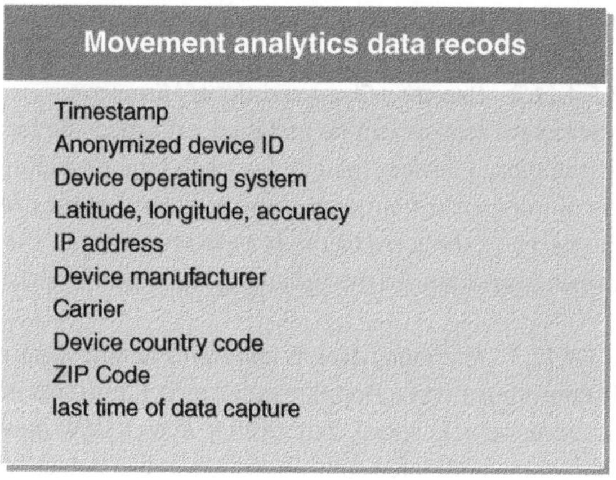

Figure 8.7 Movement analytics data records.

Movement analytics data recods
Timestamp
Anonymized device ID
Device operating system
Latitude, longitude, accuracy
IP address
Device manufacturer
Carrier
Device country code
ZIP Code
last time of data capture

The positional accuracy of location intelligence data is influenced by a number of factors including the strength of the GPS signal, the environment around the user, and personal use of the device. In a recent study [11], the average positional accuracy varied from 21 to 38 m.

8.5.4.9.3 Interesting Aspects and Points to Note Regarding Location Intelligence

Like many technology applications, location intelligence was not originally designed for smart mobility or transportation applications. Most of the location intelligence available today comes from a number of data suppliers whose primary market lies in retail and advertising effects applications. It should also be noted that location intelligence is also used to refer to the process of extracting insight and understanding from the raw data. That are several data platforms available to enable this [12].

8.5.4.10 Magnetic

8.5.4.10.1 Description of Magnetic

Magnetic sensing devices work by detecting local changes in the Earth's magnetic field caused by large metallic objects such as vehicles. This technique is sometimes combined with a variation of inductive looks technology described earlier to increase the data collection capabilities of the device.

8.5.4.10.2 Operating Characteristics of Magnetic

Sensing devices are installed flush to the road surface are several centimeters below the road surface. An onboard battery powers the sensor and the transmission of data to an accompanying roadside unit. This obviates the need for additional slot cutting in the pavement surface to accommodate a wired link between the sensor and the roadside.

Interesting Aspects and Points to Note Regarding Magnetic   A combination of magnetic and micro-inductive lip technology has been promoted as an alternative to the inductive loop. It provides the advantages of less invasive installation, with a small circular hole instead of a long slot. The devices can also support wireless communications to the roadside.

8.5.4.11 Microwave

8.5.4.11.1 Description of the Technology Application

This sensing technique makes use of radar to measure traffic speeds and volumes. Our microwave beam is aimed at vehicles and the reflected energy from the vehicles is analyzed to determine vehicle speed. Different frequencies of microwave energy can be used and often a technique known as Doppler radar is utilized to improve accuracy of speed measurements. This type of technology is also utilized for speed enforcement purposes.

8.5.4.11.2 Operating Characteristics of Microwave

Microwave sensors can be installed, and roadside locations are in overhead gantries. In the latter installation, I devices installed of each lane providing Lane by Lane data collection. While a variety of microwave frequencies and radar techniques are adopted, the principle is the same. A beam of energy in the form of microwaves is aimed at the vehicle stream. As energy is reflected back causing variations in the frequency, this can be measured to ascertain the speed of the vehicle.

8.5.4.11.3 Interesting Aspects and Points to Note Regarding Microwave

If continuous wave Doppler radar techniques are adopted, this has the advantage of providing accurate vehicle speed, but cannot detect slow-moving or stop vehicles. Sensors that transmit frequency-modulated continuous waves can detect vehicle presence as well as vehicle passage and can detect stop vehicles.

8.5.4.12   Passive Acoustic Array

8.5.4.12.1   Description of Passive Acoustic Array

An array of microphones is used to detect sound emanating from vehicles traveling along the roadway. The sideways are subjected to sophisticated processing algorithms. The technology is used to collect traffic volumes and vehicle classification.

8.5.4.12.2   Operating Characteristics of Passive Acoustic Array

Devices utilizing this technology application typically require low-power and have multilane sensor capabilities. They utilize wireless communications to eliminate the need for acquiring and can be installed without lane closure. In addition to vehicle counting and classification, this technology has also been used to identify drivers who are driving the wrong way on and off ramps and freeways.

8.5.4.12.3   Interesting Aspects and Points to Note Regarding Passive Acoustic Array

This technology is particularly robust and adverse weather conditions and low lighting. The technology has also been applied to the detection of trains at railroad grade crossings as a way to support improved safety management. Low power consumption and the ability to use wireless connection techniques make this a useful alternative to wired sensors.

8.5.4.13   Piezo

8.5.4.13.1   Description of Piezo

Piezo sensors are used as an actual detection sensor, installed in the roadway. While most installations are permanent, they can be installed on a temporary basis. Piezo takes advantage of the physical properties of piezo crystals which generate an electrical charge when subjected to pressure, deformation, or vibration.

8.5.4.13.2   Operating Characteristics of Piezo

Piezo technology is most often found in weigh-in-motion applications for heavy trucks. The gross weight of the Actual weight of the passing vehicle can be measured dynamically by piezo sensors installed in a special channel across the road. The voltage generated in the pizza sensor is analyzed by an electronics unit connected to the sensor to provide the required data.

8.5.4.13.3   Interesting Aspects and Points to Note Regarding Piezo

The ability for piezo crystals to generate an electrical charge when deformed is used for example in piezo ignition systems for stoves and gas grills where an electrical charge generated by the deformation of the crystal causes a spark to ignite fuel.

8.5.4.14   Pneumatic

8.5.4.14.1   Description of Pneumatic

Pneumatic could be considered as one of the original sensor applications and transportation. The concept involves stretching the whole rubber tube across the road and then measuring the variation in air pressure caused by vehicles driving over the tube. The technique generates one pulse per vehicle axle and requires algorithms to interpret the relationship between the number of policies and the actual vehicle count.

8.5.4.14.2   Operating Characteristics of Pneumatic

Typically used in temporary current situations, the rubber tubes are stretched across the road that the appropriate location and connected to an interpreter unit. The interpreter monitors the air pressure in the rubber tube and determines traffic volumes from this data.

8.5.4.14.3 Interesting Aspects and Points to Note Regarding Pneumatic

Pneumatic tube technology is typically used in combination with other techniques to create a regionwide picture of traffic flow variations. The technique is also used to establish base counts for before and after studies when a major development affecting traffic volumes is proposed.

8.5.4.15 Probe Vehicle Data

This technology application is usually fine within a package of applications inside the vehicle and referred to as connected vehicle technologies. This technology application has general abilities to deliver information to the driver inside the vehicle and to extract data from the vehicle. This is what is referred to as probe vehicle data, or floating car data. The other aspect of connected vehicles will be discussed under the "information delivery" category.

8.5.4.15.1 Description of Probe Vehicle Data

Many modern vehicles have equipment fitted on the production line that enable support for a two-way communication link between the vehicle, roadside infrastructure, back-office infrastructure, and other vehicles.

Probe vehicle data involves the use of connected vehicle capability to transmit data from the vehicle to other locations. It is also known as floating car data, as the technique involves sampling suitably connected vehicles as the basis for the determination of vehicle speed, traffic volumes, and vehicle origins and destinations. Floating car data refers to a manual data collection technique for assessing traffic speeds using observers in a vehicle being driven in such a manner that the vehicle floats in the traffic stream at the average speed of the stream. This is accomplished by ensuring that the vehicle is overtaken by the same number of vehicles that were overtaken. The technique is a combination of observed data, combined with mathematical techniques to infer total population numbers. Probe vehicle data can be acquired from several private sector providers and offers the potential for a rapid and comprehensive assessment of performance across entire transportation networks.

8.5.4.15.2 Operating Characteristics of Probe Vehicle Data

Probe vehicle data represents our sample of the total data available regarding instantaneous vehicle speed and vehicle location across the network. The sample size varies depending on the source of the data and the total volume of traffic on the network at any given time. Estimates suggest that the sample size varies from 1% to 10%. Sample sizes vary according to the characteristics of the local road network, variations in traffic flow, and the number of vehicles participating in the probe vehicle data suppliers' program.

8.5.4.15.3 Interesting Aspects and Points to Note Regarding Probe Vehicle Data

The small sample sizes involved in probe vehicle data should not be considered as a constraint or limitation. Studies [13] have shown that even small sample sizes can provide statistically significant assessments of vehicle travel speeds on a networkwide basis. Another interesting aspect of probe vehicle data is the collection of the data from moving vehicles over space and time, rather than a fixed location in the road network (infrastructure-based roadside sensors) probe vehicle data provides an estimate of space mean speed, which is the average of vehicle speeds over a given section of Highway. Infrastructure-based roadside sensors provide time mean speed, which is the average of vehicles passing a particular point on the highway. See the following reference for an example of a project that applies probe vehicle data in Scandinavia [14].

8.5.4.16 Ultrasonic

8.5.4.16.1 Description of Ultrasonic

Using the same reflected energy principles as radar and LIDAR, this technique makes use of a focused beam of ultrasonic or acoustic energy. This uses frequencies greater than 20,000 Hz and is typically not audible to humans. Humans can hear sounds from about 20 to 20,000 Hz, so it is possible that some humans can hear ultrasound. Ultrasound is also used by animals like bats and whales to locate prey and to navigate.

8.5.4.16.2 Operating Characteristics of Ultrasonic

Ultrasonic sensors have been used for traffic detection for advanced traffic management in Japan [15]. A focused beam of ultrasound is directed at the surface of the road. Vehicles are detected by comparing the difference between the travel time of the waves reflected from the vehicles and the waves reflected from the road surface. The Japanese advanced traffic management system application also uses other sensing technology including probe vehicle data to provide a complete picture on which decisions are taken regarding traffic signal timing and traffic flow optimization.

8.5.4.16.3 Interesting Aspects and Points to Note Regarding Ultrasonic

Note that ultrasonic sensing is different from passive acoustic sensing. Ultrasound generates an ultrasonic beam directed at the traffic on the road surface. Passive acoustic sensing simply listens and does not generate any acoustic energy. In addition to the application of this technology for traffic detection associated with traffic signals, the technology is widely applied for vehicle applications including distance sensing for automated parking and obstacle detection.

8.5.4.17 Video Image Processing

8.5.4.17.1 Description of Video Image Processing

This technique takes advantage of the progress that has been made in high-speed data processing and closed-circuit TV cameras design. Images from cameras that are focused on the traffic stream are captured and analyzed to provide data on traffic parameters including speed, flow, and turning movements at intersections. Vehicles are identified and tracked within the camera's field of view. In some cases, video image processing is combined with optical character recognition to support automated license plate detection and recognition. The technique can also be applied to the detection of changes in traffic flow on freeways and other limited-access highways, to generate an alert. This can also include one-way driving.

8.5.4.17.2 Operating Characteristics of Video Image Processing

For traffic signal detection applications, special-purpose video cameras are installed at specific locations on the network, usually attached to traffic signals. The zone of interest is specified using on-screen graphics and an onboard processing unit then analyzes the video images from within each zone. The vehicles are identified according to size and shape characteristics, while at nighttime vehicle headlights are used for recognition purposes. For freeway applications, existing pan zoom tilt (PTZ) closed-circuit TV cameras are utilized, usually with a predetermined PTZ setting.

8.5.4.17.3 Interesting Aspects and Points to Note Regarding Video Image Processing

Transport for London adopted image processing technology for the purposes of enforcement related to congestion charging in London. While the automated license plate recognition system at an overall accuracy of about 75%, advantage was taken of cumulative probability. On average, there are 11 possibilities to capture the license plate image from a vehicle making a trip across the congestion zone. The multiple opportunities to collect the license plate image raise the overall probability of detection to the high 90%. It is an important lesson in this and that transport for London carefully matched the capabilities of the technology to the business model to be adopted for the charging. The business model is based on zone presence and not on cordon crossing, enabling advantage to be taken of the cumulative probability technique [16]. Figure 8.8 provides a summary of the advantages and disadvantages of each sensor technology application.

8.5.5 Telecommunication Technology Applications

This is the second technology application area as illustrated in Figure 8.9.

These include a range of wireline and wireless technologies that make use of advanced protocols to ensure the efficient transmission of data from the various subsystems that make up a smart mobility system. From the original copper wire, wireline technologies have evolved to include the use of sophisticated signal processing to transmit data over fiber optics. While the wireless speeds tend to be lower, there is also considerable progress in the speed and bandwidth for these technologies. Wireless technology has an important application in connecting people and vehicles to roadside and cloud-based infrastructure.

In this section of Chapter 8, separate telecommunications technologies will be addressed, as summarized in Figure 8.10.

Again, these are in alphabetical order rather than in order of importance or ubiquity. The telecommunications technology applications described in this chapter are most widely used in smart mobility. The scale around the circumference of the circle provides an assessment of how prevalent the use of the technology is in smart mobility.

Note that all of the wireless telecommunication's applications make use of the same principles of physics. They use electromagnetic waves that travel through the air, generated by an electrified piece of metal which forms waves as it becomes energized. The device that generates the waves is called the transmitter and the device that picks up the waves is called the receiver. The mobile device contains both receiver and transceiver and is often referred to as a transceiver. The data to be carried by the wave can be encoded by a variety of techniques that are either analog or digital. Analog encoding can be conducted by varying the frequency or the amplitude of the wave, this is known as modulation. The different approaches taken to modulation mean that not all devices are compatible. The frequencies available for radio devices fall within the 3 kHz to 300 GHz band of the radio spectrum [17].

Digital modulation involves the use of binary "1"s and "0"s to represent the data. A constellation diagram is used to change the analog carrier wave to discrete binary numbers [18].

Telecommunication technology applications covered in this chapter are those most often found in smart mobility applications. The intention is not to provide an exhaustive catalog of all available telecommunication technology applications but to provide a summary of those that are in common use for smart mobility applications. Figure 8.11 provides a summary of the telecommunication technology applications described in this chapter and typical smart mobility applications for them.

Figure 8.12, published by USDOT, provides a summary of the electromagnetic spectrum and a magnified view of the radio spectrum most used for smart mobility applications.

Technology	Advantages	Disadvantages
Crowdsource	Low cost, wide-area coverage	Variable sample size and data quality, significant resources required for processing
Distributed fiber-optic sensing	Low cost, nonintrusive, leverages existing investment in fiber-optic communications	Does not measure Lane by Lane speeds, accuracy lower for wider road cross-sections
Inductive	High accuracy, well proven and trusted technology	Requires a pavement surface to be cut, subjected to deterioration over time resulting in lower availability statistics
Infrared	Noninvasive, can be fitted at the side of the road and not on the road surface, can measure multiple zones	Operation may be affected by low visibility caused by heavy rain, fog, or snow
Location intelligence	Low cost, capability of wide-area coverage, possibility of rich data stream describing multiple modes of transportation	Low sample size, sample size variability, data quality variability
Lidar	Accuracy, high-resolution, lower cost as volumes increase	Significant processing costs and edge processing techniques required
Magnetic	Less intrusive than inductive loops, low operating cost can be wirelessly connected to roadside infrastructure, short installation times	Battery life can be unpredictable. Cannot detect stationary or slow-moving vehicles (less than 5 miles per hour)
Microwave	Certified installations do not require road closure, impervious to weather conditions, multiple Lane data collection	Cannot detect stopped vehicles or slow-moving vehicles. Traffic in outer lanes may be subject to obscuration by high sided vehicles
Passive acoustic	Not sensitive to precipitation, multiple Lane operation	Low temperatures may affect vehicle count accuracy
Piezo	Simple, low cost, can measure vehicle weight in addition to traffic speeds in traffic volumes	Require slot cutting into the pavement surface and specialist expertise to install
Pneumatic	Simple, low cost	Installation requires highway closure; accuracy can be problematic
Probe vehicle	Low cost, wide-area coverage	Variable sample size and data quality, significant resources required for processing. details regarding preprocessing of data may be obscure
Ultrasonic	Multiple Lane data collection including vehicle height, significant deployment experience in Japan	Extreme temperatures can affect performance
Video image processing	Can monitor multiple lanes and multiple detection zones, easy to modify detection zones, wide array of data available	Lenses must be cleaned at regular intervals, performance affected by visibility levels, reduce performance at nighttime

Figure 8.8 Advantages and disadvantages of each sensor technology application.

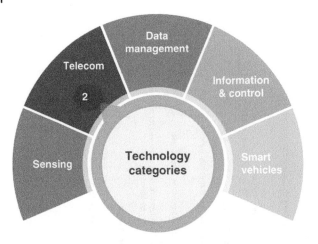

Figure 8.9 Technology application areas.

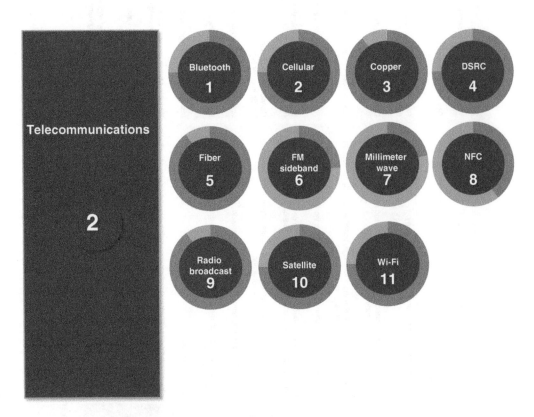

Figure 8.10 Telecommunication technology applications.

8.5.5.1 Bluetooth
8.5.5.1.1 Description of Bluetooth
Bluetooth is an extremely popular wireless communication technique for connecting devices. It enables wireless communication between keyboards and computers, between smartphones and speakers, between tablets in medical devices, or can send messages between thousands of nodes in

	Telecommunications technology application	Common use in smart mobility
1	Bluetooth	Connection between smart devices and vehicles, pro vehicle data
2	Cellular	Connected and automated vehicles, travel information, mobility as a service
3	Copper	Connecting traffic controllers to traffic management centers, local-area networks for transportation management
4	DSRC	Vehicle to roadside communications, electronic toll collection
5	Fiber	Roadside devices to traffic management centers, traffic management centers to traffic management center
6	FM sideband	Driver advisories and travel information
7	Millimeter wave	Vehicle to vehicle communications for connected and automated vehicles, internet of things roadside devices
8	Near Field communications	Electronic payment systems
9	Radio broadcast	Traveler information
10	Satellite	Traveler information
11	Wi-Fi	Travel information, mobility as a service, probe vehicle data

Figure 8.11 Telecommunications technology applications in smart mobility.

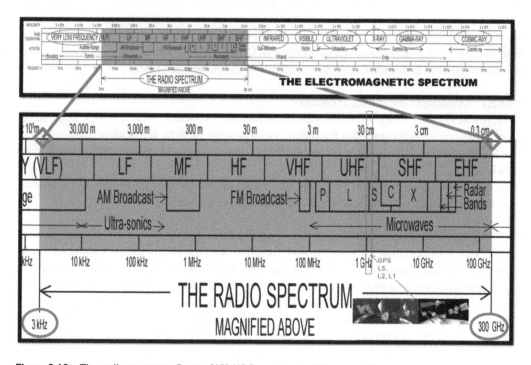

Figure 8.12 The radio spectrum. *Source:* [19], US Department of Transportation.

a building automation solution. Two of the main reasons for its popularity is its versatility and flexibility. It is estimated that 7 billion Bluetooth-enabled devices will ship annually by 2026 [20].

8.5.5.1.2 Operating Characteristics of Bluetooth

Bluetooth utilizes low-power radio to stream data over 79 channels in the 2.4 GHz frequency band. This band is unlicensed and intended for industrial, scientific, and medical usage. It supports point-to-point communications.

In addition to the original Bluetooth classic radio application, there is not a Bluetooth low-energy application. This is designed for very low-power operations and transmits data over 40 channels in the 2.4 GHz frequency band. Bluetooth low energy supports point-to-point network topologies but also broadcast and mesh network topologies.

8.5.5.1.3 Interesting Aspects and Points to Note Regarding Bluetooth

Bluetooth low energy also has the capability to support location services. Bluetooth direction finding enables suitably equipped devices to determine the direction of a Bluetooth low-energy signal. This makes centimeters level location accuracy. This opens up the possibility of Bluetooth direction-finding equipped smartphones to serve as location intelligence or people probe data sources. It is estimated that 560 million Bluetooth location service devices will be shipped by 2026.

8.5.5.2 Cellular
8.5.5.2.1 Description of Cellular

Cellular wireless as the name suggests, divides the coverage area into a number of cells, with at least one wireless antenna at the center of each cell. The antenna has direct communication with each cell phone within the cell. A mobile switching technique is used to seamlessly handoff the cell phone as it moves from one cell to another. This also supports appropriate routing of calls from one cell phone to another or to a landline. Each cell is assigned a range of frequencies that differ from adjacent cells, to avoid crosstalk between cells. Cell phones and smartphone devices typically generate a signal with just enough power to operate in the current cell [21].

8.5.5.2.2 Operating Characteristics of Cellular

Cellular wireless technology was developed throughout the 1950s and 1960s. The FCC approved a major allocation of radio spectrum for mobile radio systems in 1970. It is widely acknowledged that Martin Cooper of Motorola invented the first personal handheld cellular radiotelephone in 1973 [22].

Since the early days of cellular radio, the technology application has undergone a continuous evolution with changing characteristics as summarized in Figure 8.13.

8.5.5.2.3 Interesting Aspects and Points to Note Regarding Cellular

Cellular technology continues to progress with the introduction of 5G and the impending arrival of 6G technology. Recent adaptation for use as a connected vehicle technology offers a prospect of cellular wireless addressing both the short-range communication needs between the vehicle and roadside and the long-distance needs from the roadside back to a transportation management center. 5G is also emerging as a point-to-point wireless alternative to wireline technologies. The combination of new cellular wireless capability with the ubiquity of smartphones provides smart mobility with options for delivery of information to travelers while obtaining data on travel behavior and travel demand.

Date range	Gen.	Common names	Characteristics		
1970s	1g	Advanced Mobile Phone System	Analog modulation	Network of cells with direct wireless communication to mobile phones	Mobile switching center to route the voice calls either to another mobile or to a landline
1990s	2g	Global Standard For Mobile Communications (GSM)	Digital modulation	Digital supports more efficient use of radio frequencies and more phones within each cell	Make use of two variants of the global standard for mobile communications (GSM) -general packet radio service (GPRS) and enhanced data rates for GSM evolution (edge)
Early 2000s	3g	3G	Digital modulation	User equipment authentication and dental and communication security introduced	Enabled by a range of technologies including edge, universal mobile telecommunications system (UMTS) and wideband code division multiple access (w-CDMA)
2009	WIMAX	Broadband cellular networking LTE	Digital modulation		Network wide use of IP-based packet switching, scalable bandwidth, data rates of hundred megabits per second in motion and up to 1 Gbps when static, seamless roaming across networks and geographic territories globally
2019	5G	5g new radio	Digital modulation	Supports telephony and broadband connectivity and infrastructure and industrial applications such as the internet of things, m2m networking and C –V2C	Capable of delivering 10 Gbps download speeds. Also being adapted to replace wireline communication for point-to-point use.
2030	6G	6G	Advanced modulation	The use of multi-gigahertz frequencies to deliver reduced latency and higher capacity	More complex antenna networking

Figure 8.13 Cellular wireless characteristics.

8.5.5.3 Copper

8.5.5.3.1 Description of Copper

Copper wire communications are achieved using an electrical current which is frequency modulated to create a waveform. Typically, direct current is used. A number of different modulation techniques can be applied to deliver the required bandwidth. Copper is used because of its high conductivity of electricity, enabling signals to travel further than other metals. The cable typically consists of two twisted copper wire pairs as the two wires shield each other from external electromagnetic interference [24]. Figure 8.14 illustrates the use of copper wire for ethernet connections at a switch.

8.5.5.3.2 Operating Characteristics of Copper

The copper wire used to transmit the signals are governed by standards developed by the Telecommunications Industry Association (TIA). The standard for copper wiring inside buildings is known as TIA/EIA-568. This group of standards defines the planning and installation of a structured cabling system and a commercial building environment. It defines the physical characteristics of the wire and the connectors. The physical characteristics of the copper wire are split into categories designed to support different functionalities as summarized in Figure 8.15.

Figure 8.14 Copper wires. *Source:* Image from Unsplash [23], Lars Kienle/ Unsplash.com.

Category	Frequency (mHz)	Bandwidth (mbps)	Maximum length (meters)
3	16	10	100
4	No longer in use		
5	No longer in use		
5e	100	100	100
6	250	1,000	100
6a	500	1,000	100
7	600	10,000	100
8	2000	10,000	100

Figure 8.15 Copper wire communications characteristics.

Note that each of the categories is subject to a maximum run length. Exceeding this causes signal reduction over the length of the cable. If the maximum run length is to be exceeded, then switches or repeaters can be inserted into the telecommunications architecture.

8.5.5.4 Asymmetrical Digital Subscriber Line (ADSL)

This is a popular technique for providing Internet access to residential customers. As the name suggests, this communication technique provides asymmetrical communication speeds that are greater toward the customer than away from the customer. In other words, download speeds are higher than upload speeds when using the Internet. The technique can use the same copper wire used for voice calls as frequencies above those used for voice calls are used to transmit data using this technique. The use of the technique is usually limited to within 4 km of the telephone exchange. The latest version of the ADSL standard, ADSL 2+ enables download speeds of up to 24 Mbps and upload speeds of 3 Mbps [25].

8.5.5.5 Ethernet

This is a group of computer networking technologies used in local-area networks, metropolitan area networks, and wide-area networks [26]. Local-area networks interconnect computers and devices within a limited area such as a residence or office building. Metropolitan area networks interconnect computers and devices in a geographic region the size of a metropolitan area [27]. Wide-area networks cover a larger geographical area and can involve leased or rented telecommunication facilities in addition to private dedicated [28]. It is used in combination with copper wire to interconnect computers, appliances, and other personal devices. Ethernet standards define specifications for cables, connectors, and protocols and cover the cables and the hardware (network interface cards) required to support communications (Figure 8.16).

8.5.5.6 Universal Serial Bus (USB)

An industry-standard establishes specifications for cables, connectors, and protocols for connecting peripherals and other computers. There are 14 different connector types of which USB-C is the most recent [30]. The standard is designed to enable data communications between computers and peripherals and also support the transfer of electricity [31] (Figure 8.17).

8.5.5.7 Interesting Aspects and Points to Note Regarding Copper

Due to the speed and capacity of fiber-optic telecommunication networks, the demise of copper wire communications has often been forecast. However, improvements in the ways in which existing copper cables can be utilized continue to emerge. There is also a significant legacy of sunk investment in

Figure 8.16 Ethernet. *Source:* [29], Gavin Allanwood / Unsplash.com.

Figure 8.17 USB. *Source:* Hal Gatewood/Unsplash.com.

copper wire infrastructure. Many telecommunication network architectures feature a combination of fiber and copper, representing a hybrid approach to telecommunication for smart mobility. Effective use of the telecommunication technology applications described requires that a specialist develops a telecommunications network architecture that takes account of the connection needs of the primary components of the system and selects the most appropriate. When it comes to telecommunications technology applications, it is most certainly a question of "horses for courses."

8.5.5.8 Dedicated Short-range Communications (DSRC)

8.5.5.8.1 Description of Dedicated Short-range Communications (DSRC)

This involves the use of wireless telecommunication technology to support a two-way communication link between the vehicle and roadside or centralized infrastructure. It can also support vehicle-to-vehicle communications. The most widely known application of DSRC is in the field of electronic toll collection. As transponder within the vehicle supports a two-way wireless communication link with roadside transceivers enabling the payment of a toll in the form of an electronic transaction electronic toll collection can also be achieved using satellite positioning in communications and automatic number plate recognition with image processing. The two-way communication link between the vehicle and roadside infrastructure can also be utilized for other applications such as collision avoidance and the communication of traffic signal information from the roadside to the vehicle. Electronic toll collection transponders typically operate in the 915 MHz frequency band [32] although the 5.9 GHz band is also made available for dedicated short-range communications in support of vehicle-to-vehicle radio communications. This is currently the subject of some controversy as the FCC wants to reallocate the spectrum for other uses.

8.5.5.8.2 Operating Characteristics of Dedicated Short-range Communications (DSRC)

DSRC can be described as "spot" communications as the two-way link is only supported within the operating range of the roadside transceiver. For 915 MHz applications, the range is a few 100 m, but for 5.9 GHz applications the range can be up to a couple of kilometers.

8.5.5.8.3 Interesting Aspects and Points to Note Regarding Dedicated Short-range Communications (DSRC)

Dedicated short-range communications could be regarded as one of the most ubiquity's telecommunication applications in smart mobility. The technology supports hundreds of miles of toll and express lanes in many parts of the world. Having achieved a legacy of existing infrastructure for dedicated short-range communications, the challenge is to expand the functionality of the infrastructure to include other applications beyond electronic toll collection. These can include connected and automated vehicle applications that make use of two-way communications between the vehicle and roadside or centralized infrastructure to support the transfer of data to and from the vehicle. The future of DSR is in question partly because of HCC consideration of spectrum reallocation and partly because of development in the rival cellular wireless or CV-X application that would obviate the need for dedicated DSRC roadside infrastructure.

8.5.5.9 Fiber

8.5.5.9.1   Description of Fiber

Fiber-optic telecommunication applications revolutionized high-speed, long-distance connections for smart mobility. Most transportation agencies around the world have extensive deployments of fiber-optic communications that deliver high bandwidth and low latency. While copper wire uses electromagnetic waves and electrons passing through the copper, fiber-optic technology makes use of light. Optical fibers are long thin strands of very pure glass about the diameter of a human hair. They are bundled into optical cables and used to transmit light signals over long distances. Data is transmitted as a series of light pulses. The outer layer of the optical fibers is coated with cladding that reflects the light back into the court and enables the light pulse to be reflected over long distances. Some optical fibers can also be made from plastic instead of glass [33].

In a similar manner to copper wire communications, fiber-optic communications make use of modulation techniques to transmit data, this time using light. The two techniques adopted an analog modulation the digital modulation. In analog modulation, the light wave is continuously adjusted to represent the data. In digital modulation, information is encoded into a series of pulses separated by spaces that represent binary data. Digital modulation has the advantage of being less dependent on signal intensity (Figure 8.18).

8.5.5.9.2   Operating Characteristics of Fiber

Laser-generated light is used to provide the pulses of light for fiber optics. The laser source is typically referred to as a transponder. Like multiple transponders can be combined by a wave division multiplexing technique so that their wavelengths are different and can coexist in the same fiber. This technique also allows communications in both directions along the fiber.

Figure 8.18 Fiber optics. *Source:* Denny Müller / Unsplash.com.

Here again, a specialist is required and other to create the telecommunications network architecture for fiber this will take account of the characteristics of the fiber and any redundancy required in the network to ensure that mission-critical applications can always be supported [34].

8.5.5.9.3 Interesting Aspects and Points to Note Regarding Fiber

Recent developments have shown that fiber-optic telecommunications infrastructure can also be repurposed as traffic sensors. This is covered earlier in the section on sensor technology applications. This provides an excellent possibility to leverage prior investment in fiber-optic technologies for telecommunications purposes as continuous traffic sensors along significant length of highway. The most ubiquitous application of fiber-optic and wireline communications lies in the delivery of cable TV to residences worldwide.

8.5.5.10 FM Sideband

8.5.5.10.1 Description of FM Sideband

This application uses wireless technology, existing FM radio broadcasts, and smart techniques to squeeze even more data into the broadcast wireless signal. FM sideband of the portion of the radio signal in either side of the main carrier frequency. They are typically used for data regarding the broadcast such as radio station name and for tuning purposes. Using a low bandwidth approach, it is possible to superimpose traffic messages onto the sidebands. Suitably equipped radios including those in the vehicle can be used to decode the messages, providing an excellent channel for alerts and traveler information over a wide area. Geolocation techniques can also be used to make the travel information and alert specific to a named intersection. The technique is known as Radio Data System – Traffic Message Channel (RDS-TMC) in Europe and FM sideband or Condition Acquisition and Reporting System (CARS) in the United States.

8.5.5.10.2 Operating Characteristics of FM Sideband

The application makes use of two components and RDS TMC receiver which is a special-purpose FM radio tuner and hardware installed at the radio station. It can make use of terrestrial radio broadcast of satellite radio. The radio broadcaster turns the signal on and off when messages are available. Geographic coverage of the message delivery capability is constrained by the number of radio stations adopting the technology. Expansion of the technique to include the use of satellite radio has addressed this; however, in many countries, coverage is not complete.

8.5.5.10.3 Interesting Aspects and Points to Note Regarding FM Sideband

The application can be used in tandem with global positioning system sensor techniques to improve the precision of geolocation. This can be extended to include an integration of the technique within vehicle navigation systems. There is also some concern regarding the security of the data, as it is straightforward to design and develop inexpensive electronics capable of generating false messages. It is also a propensity for radio stations to misuse the message Channel by turning the elect signal on early other than necessary to include commercials.

8.5.5.11 Near Field Communications (NFC)

8.5.5.11.1 Description of Near Field Communications (NFC)

This lies at the other end of the spectrum to wide area networks described earlier. Near-field communications is a set of communication protocols to enable communication between two electronic devices that are not more than 4 cm apart. It could be described as a micro area network. It offers a low-speed connection and operates in a similar manner to a proximity smartcard. This particular useful to support electronic payment for transit and retail applications. It can also support communication between smartphones and other devices such as sensors [35].

8.5.5.11.2 Operating Characteristics of Near Field Communications (NFC)

Near Field Communications (NFC) supports two-way communications use and frequency of 13.56 MHz available in an unlicensed radio frequency band. It can deliver data rates from 106 to 424 kbs.

8.5.5.11.3 Interesting Aspects and Points to Note Regarding Near Field Communications (NFC)

NFC can be considered as an evolution of existing radiofrequency identification techniques, used for electronic article surveillance, and evolved into dedicated short-range communication for electronic toll collection use. NFC offers a low-cost alternative to Bluetooth but with more limited functionality. The extremely limited range is useful for security as it reduces the likelihood of unwanted interception of the signal. NFC offers a viable alternative to the use of smart cards for cash for transit ticketing and is integrated into payment applications from MasterCard, Visa, and Samsung.

8.5.5.12 Radio Broadcast

8.5.5.12.1 Description of Radio Broadcast

Radio broadcasting involves the transmission of audio and data by radio waves to radio receivers for use with the public audience. Radio stations can utilize amplitude modulation or frequency modulation techniques. These are referred to as AM and FM radio. Newer radio stations use digital audio broadcasting including high-definition radio and digital radio mondiale techniques. Television broadcasting uses similar principles to broadcast video signals [36].

8.5.5.12.2 Operating Characteristics of Radio Broadcast

This telecommunications technology application supports one-way communications from a central point, at a network of central points, to a wide audience. It is excellent for the broadcasting of travel information and elects as part of a smart mobility solution. They can also support the FM sideband technique described previously to enable data to be embedded into the signal.

8.5.5.12.3 Interesting Aspects and Points to Note Regarding Radio Broadcast

Radio broadcasting is one of the most mature wireless telecommunications applications. Its origin can be traced back to the early 1900s, with the first commercial introduction in 1916.

8.5.5.13 Satellite Radio

8.5.5.13.1 Description of Satellite Radio

It is a form of broadcast radio, but making use of satellites to broadcast a signal, rather than terrestrial radio masts. Primarily focused on in-vehicle applications and available on a subscription basis.

8.5.5.13.2 Operating Characteristics of Satellite Radio

In the USA, satellite radio uses a 2.3 GHz band for nationwide digital radio broadcasting. Another parts of the world site late radio uses a 1.4 GHz band allocated for digital audio broadcasting. Operating characteristics are well-suited to the delivery of travel messages from a single central location to a wide audience. As an indication of the reach of satellite radio, in 2021 Sirius XM, the largest satellite radio operator in the United States had approximately 34 million subscribers.

8.5.5.13.3 Interesting Aspects and Points to Note Regarding Satellite Radio

An important attribute of this telecommunications technology application is the reach of the broadcast signal. Through the use of satellite technology, messages from a satellite radio system can reach an audience of millions of people at exceptionally low cost. Satellite radio techniques can also be adapted to deliver Internet services using two-way communication, especially in rural areas. As the number of satellites orbiting the earth, or constellations, increases in the coverage of Internet dealt delivered by satellite radio will increase. Satellite radio services are also capable of enhancement to include travel information services, roadside assistance, and stolen vehicle tracking services.

8.5.5.14 Wi-Fi
8.5.5.14.1 Description of Wi-Fi

Wi-Fi uses radio frequencies to transmit information between the device and a router via different radio frequencies. The router can in turn be linked to an Internet service provider by wireline communications or wireless communications. Initially, Wi-Fi was available at the airport coffee shops and hotels. Nowadays it is ubiquitous in most homes and offices. Many cities also provide free Wi-Fi access to enable residents and visitors to access the Internet on the go. The combination of smartphone devices and Wi-Fi provides a valuable channel to communicate with travelers, visitors, and others. This enables information to be delivered to the traveler and information regarding travel behavior and location to be transmitted back to a central location. Detection of Wi-Fi signals from vehicles can also be used as a probe vehicle data technique

8.5.5.14.2 Operating Characteristics of Wi-Fi

Two radio frequencies are used depending on the amount of data being sent. These are either 2.4 GHz or 5 GHz [37].

Wi-Fi enables smartphones and other devices to establish a two-way communication link with routers that can then relay data to wireline or wireless networks. Wi-Fi makes use of 80211 networking standards which have evolved over the years. The current version 80211 AX, also known as Wi-Fi 6, was introduced in 2019. This allows data flow rates up to 9.2 Gb per second and dollars installation of many more antennas on one writer, increasing the number of connections that can be supported at any given time. Some new devices also connect to 0.6 GHz which is about 20% faster than the current 5 GHz frequency [38].

8.5.5.14.3 Interesting Aspects and Points to Note Regarding Wi-Fi

The use of Wi-Fi techniques enables Internet access to be made available on a mobile basis. This can be particularly useful for providing advice and information to travelers where they are in the course of traveling. This information could be regarding the specific mode currently being used by the entire trip making chain. As a consequence, Wi-Fi has an important role to play in mobility as a service technique. Wi-Fi capability continues to be enhanced by the introduction of new cellular wireless services such as 5G and a combination of enhanced Wi-Fi with enhanced capabilities of smartphones. Wi-Fi techniques are now being adapted to use light instead of radio. It is still early days in this technological development, but promises to provide even faster mobile connections, a low line of sight between the device and the router this is often referred to as LiFi.

8.5.6 Data Management Technology Applications

Harnessing the power of data requires a good understanding of the possibilities offered by data science and the value it can bring to transportation management (Figure 8.19).

Data management technology applications include those technologies necessary to support the full spectrum of activities from collecting data, telling into information, extracting the appropriate insight,

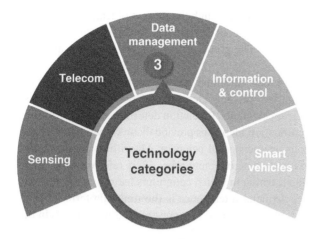

Figure 8.19 Technology application areas.

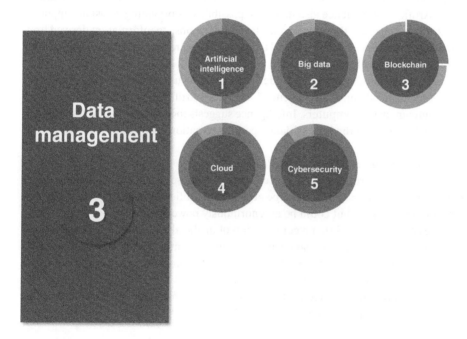

Figure 8.20 Data application areas.

and understanding and supporting relevant response strategies and action plans based on the insight. 5 data management technology applications are showcased in this section of the chapter. Data applications are summarized in Figure 8.20. Each technology application is described in the following sections.

8.5.6.1 Artificial Intelligence

8.5.6.1.1 Description of Artificial Intelligence

Artificial intelligence is a group of technologies that can have considerable impact on the management of data for transportation. Unlike conventional programming where the output is causally related to the input or investment required to program the computer, artificial intelligence holds the possibility of applications that deliver a higher value through the process of learning. Artificial intelligence applications can learn through repeated use, detect patterns, and begin to apply judgments that were formerly the domain of humans. The group of technologies includes machine learning and neural networks.

Machine learning refers to the process by which computer study data to detect patterns for applying rules. This enables them to categorize, predict, identify, and detect. Categorization involves the development of a catalog of different elements. Prediction determines likely outcomes of actions based on identified patterns. Identification identifies previously unknown patterns and relationships. Detection identifies anomalous or unexpected behavior. Algorithms are developed to support these functions. AI also makes use of neural networks, often referred to as Deep Learning, as a means of conducting machine learning in which computers learn to perform some tasks by analyzing training examples. Since artificial intelligence is the umbrella term for both machine learning and neural networks, the term will be used to represent all three subjects in the remainder of this chapter.

As the size of data continues to increase, artificial intelligence applications become an important tool in the management of large volumes of data and the extraction of insight and understanding from big datasets. Artificial intelligence also holds the possibility of providing decision support to human operators and managers as they develop strategies in response to information derived from data. While artificial intelligence has the potential to impact many areas of society, it is particularly interesting to transportation professionals.

The definition of artificial intelligence can be derived from a focus on the two words that comprise the expression "artificial" and "intelligence." The word artificial implies derivation from means other than human, using computers. Intelligence suggests the capacity to develop different responses and apply judgment in the light of data regarding situations and scenarios.

8.5.6.1.2 *Operating Characteristics of Artificial Intelligence*

The current state of the art of artificial intelligence is such that these applications are not ready to completely replace the human. In fact, it could be argued that artificial intelligence may never be a substitute for human judgment but could be an enormously powerful decision-support tool. In the short term at least, it is expected that a combination of artificial and human intelligence will form the most powerful approach to the management of data and information. As is the case with many tools, it is important that the tool is applied in the most appropriate context and within an overall process that maximizes the value of the technology.

The value of applying artificial intelligence to transportation performance management can be characterized under the following headings.

Cost Savings After an initial investment in hardware and software required for artificial intelligence, it is likely that the cost of supporting technical and administrative processes related to traffic data will decrease. The labor component of cost will be reduced, while processing will be achieved in a more cost-effective manner. For example, traffic monitoring, data analysis, labor costs can be avoided by automated analysis, or more efficient decision support.

Efficiency One of the current challenges associated with data management can be the delay between data being collected and turned into valuable information. Many agencies are finding that data is not properly harnessed because of the time required to find and process the data. Application of artificial intelligence should allow processing to be conducted in a shorter period, aligned with the smart data management approach, this should also make answers available within a shorter time frame. For traffic monitoring data, the efficiency is realized by making the results of the analysis available in a timelier fashion and more accessible to a wider range of appropriate users.

Accuracy   Humans are exceptionally good at subjectivity but can be prone to errors when conducting repetitive calculations or tasks. The introduction of artificial intelligence will provide decision support to humans and automate repetitive tasks. This will allow human judgment and subjectivity to be focused on the most appropriate areas, while leaving the repetitive tasks to automation and artificial intelligence. For transportation data, accuracy improvements can be achieved by artificial intelligence taking over the repetitive processing roles of humans including interpolation of data for missing data points and the combination of transportation data from different sources.

Increased Capacity   Because of improvements in the speed of processing and reduced cost, it should be possible to do things that were considered infeasible in the past. These may be things that were cost-prohibitive. Using manual approaches would just simply take too long. By using manual techniques. It should also be possible for a wider range of data to be analyzed to reveal new insight and understanding. For transportation data. It should be possible to analyze a larger number of sensor inputs, allowing the performance management program to be expanded to be more complete.

Figure 8.21 shows a sequence of activities, or process, which converts data to information, extracts insight, and support action based on the new insight. This represents a model of how the conversion process can be supported. Note that this is not a design or a recipe, but a model that can be used to define arrangements for data management. A detailed design would require technology and product choices that are beyond the scope of this document.

At each stage of the process, there are opportunities for the application of artificial intelligence as illustrated in the diagram. For example, artificial intelligence can be applied to data sources and data from sensors to conduct initial verification of the data and to consolidate the data to manage bandwidth requirements for transmission. At this point, duplicate data from the same vehicles passing different roadside transceivers can be detected and consolidated. Use cases such as travel time calculation can be supported by processing within the transceiver device using IoT edge processing techniques. The resulting travel time can then be transmitted back to a central location, rather than the individual vehicle identifications and locations.

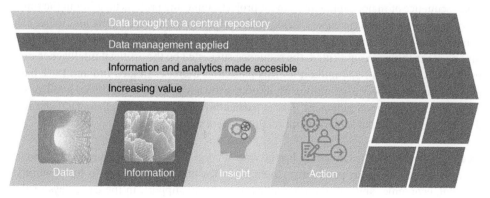

Figure 8.21   Data to outcomes process.

8.5.6.1.3 Interesting Aspects and Points to Note Regarding Artificial Intelligence

Artificial intelligence tools can play a significant role in improving the effectiveness and efficiency of many areas of smart mobility. To bring this to life, four examples are explored – AI for data ingestion and cataloging, AI for project sequencing, AI for sustainability, and AI for decision support in traffic operations.

AI in Data Ingestion and Cataloging   New tools are available that help to automate the process of data ingestion and the creation of data catalogs. One of the challenges of data management lies in bringing data from multiple sources into a single accessible location. While some degree of manual intervention is still required, there are artificial intelligence-based tools that simplify and accelerate the process of bringing the data into a system. These include tools from IBM, Microsoft, and a startup company called Alation. There are also open-source tools available as part of the Hadoop framework including HortonWorks DataFlow, StreamSets Data Collector, Gobblin, Sqoop, and Flume.

Figure 8.22 shows a screenshot from the Alation product that provides a dashboard to assist with the ingestion process.:

AI in Project Sequencing

1) Artificial intelligence and machine learning have been the subject of considerable advances in the past few years. However, a valuable application of artificial intelligence applied to performance evaluation can be found in a specific tool designed to develop optimized project bundles for implementation. The tool is known as Foro DOT and has been applied to enable the best combination of construction projects to be identified. The artificial intelligence system takes account of several factors including project cost, project benefits, location of project compared to other projects, and other specific geometric features. Figure 8.23 shows a dashboard from the tool which explains information about the project bundles, lists parameters for the projects, and calculates the total amount of money that has been saved by bundling them in an optimized manner. The application of artificial intelligence to the development of capital programs and investment plans.

An important aspect of communicating information about artificial intelligence is to be specific about the uses in smart mobility. It also helps if a specific benefit can be defined that results from the application of the technology. One such case is the application of artificial intelligence to the development of capital programs and investment plans. Figure 8.23 shows a screen capture from an application known as Foro, utilized by the Indiana Department of Transportation. This suggests an optimum sequence of major capital investment projects, bundling projects together in the timeline for efficiency. This allows savings in terms of fire travel time for trucks from sites to asphalt and concrete plants and about procurement of maintenance of traffic devices such as barriers and cones, among other things. As indicated, the sequence suggested by the artificial intelligence program resulted in a savings of $34 million for this particular exercise. In fact, Indiana DOT have identified total savings of over $100 million by accepting the advice from the artificial intelligence application, in place of human judgment. This is particularly impressive as Indiana DOT are recognized as among the national leaders in optimizing project sequencing.

AI for Urban Sustainability   Sustainability has become a crucial discussion with respect to smart cities and urban areas. From my perspective, there are at least three dimensions to sustainability: environmental, economic, and continuing public and political support. In the early days of the Intelligent Transportation Systems program, it was quite common to invest a significant amount of

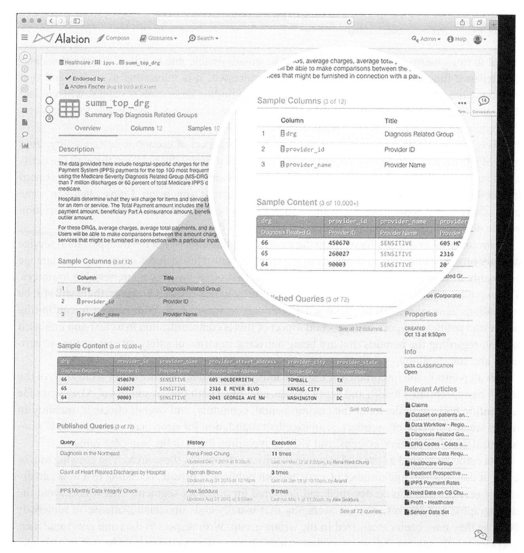

Figure 8.22 Alation dashboard.

Figure 8.23 Project bundling tool.

capital to implement a system only to have it be unsustainable due to a lack of operations and maintenance funding. While this has now been addressed, it is interesting to understand that someone from Mars, would expect the problem to lie in finding large sums of capital, when in fact the issue was a lack of understanding that the application of advanced technology to transportation requires a higher level of sustainable funding.

This is one aspect of economic sustainability. Another aspect of economic sustainability, from more of a private sector perspective, is the ability to build a successful business through the delivery of smart mobility services. As there has been a significant increase in private sector investment in mobility services, then this has also become an important dimension. I have clients in some parts of the world actually asking if there should even be a public sector role in transportation service delivery at all. But that is a discussion for another day.

Environmental sustainability refers to the undesirable side effects that the transportation service delivery process can create with respect to the environment. This includes emissions that damage air quality in urban areas and energy consumption that often involves nonrenewable resources. The migration to electric vehicles will help significantly in this respect; however, there will still be intimations associated with the generation of energy in most cases. With respect to continuing public and political support, an important aspect of this is communication in a clear and unbiased fashion regarding the benefits that are being delivered in terms of safety, efficiency, user experience, and sustainability.

This all points to the importance of the availability of good data and the use of AI to analyze the data. The appropriate breadth and depth of data, combined with the capabilities of AI will provide a scientific basis to understand the environmental, economic, and overall effects of mobility in urban areas. This type of approach is an excellent foundation for strong communications to non-technical decision-makers, citizens, and visitors, regarding the benefits being delivered.

Data could be viewed as the fuel required to power activities in these areas. The growing availability of data from a large number of sources, both public and private, is leading to a new perception of services that can be designed and delivered for a market segment. This would avoid the demographic and market segment averaging that usually ends up causing offense to somebody because they have been categorized in the wrong group. With respect to data and great analytics, averages are the enemy. Advances in data science have moved to the point where it is no longer necessary to aggregate and average data. It is also not necessary to take large datasets and fragment them over multiple computers. There is a two-way opportunity involving a focus on the individual, while aggregating data and managing large datasets, as a single entity. The latter has considerable advantages in enabling an enterprisewide view to be taken over the entire dataset, easily identifying trends and patterns that may not be obvious if the data has been partitioned. These abilities make data even more valuable.

Many people have noted that data can be regarded as the new oil because of its importance. This aligns with the thought that data is the fuel for effective decision-making and implementation. However, the analogy is also useful in that crude oil is dirty, smelly, and nobody wants to touch it when it comes out of the ground. However, when it is subject to a refining process, it results in a range of valuable, highly sought-after products. Data is just like this. As a raw material, it is hard to see the value. However, when it is refined from data to information from which insight, understanding form the basis for action plans, then the value is revealed. Like any raw material, it is what is done with it and how it is done, that influences the value that can be released. Interestingly, the whole concept of sustainability can also be applied to great data management, which in turn can underpin sustainability in urban areas. There are sustainable approaches to data management and others that are unsustainable.

Here's an example of an unsustainable approach. Data is collected from multiple data sources on a speculative basis. There is little understanding of the use to which the data will be put and consequently data collection is focused on delivering the highest accuracy of data across as many collection points as possible. The collected data is then stored in a fragmented and inaccessible manner and in some cases, the fact that the enterprise has even collected the data is lost in the mists of complexity and inaccessibility. This leads to duplication and data collection which further exacerbates the cost of collecting high-accuracy data on an areawide basis. There are several public sector agencies who admit that they do not make use of all the data being collected because it is simply perceived as too expensive to access or too difficult to analyze.

Now let us talk about a sustainable approach to data management. Data is collected in a purposeful manner across multiple data sources with a good understanding of the use to which the data will be put. This understanding required accuracy of data collection and geographical coverage required. The data is then subjected to a fusion process and entered into a data repository with a high level of accessibility. The data repository also provides tools for AI-enabled advanced analytics to turn the data into information, revealing trends, patterns, and insights. Based on this new understanding, better plans and response strategies can be generated to manage the urban environment. Data is also used to monitor the effects of investments and implementations enabling a feedback loop to be created to improve subsequent plans and deployments. This is sustainable. Of course, the data sources can be public, or privately sourced. These days there are considerable offerings in the market for probe vehicle and location intelligence data that can significantly complement public sector data coming from sensors.

How does all this affect sustainability? Well, another aspect of sustainability is the successful identification of threats and opportunities within an urban area. Data plays a crucial role in the successful identification and characterization of this. The availability of high-quality data and subsequent information can serve as the foundation for a scientific approach to project evaluation and benefits definition. In the absence of this scientific approach, then opinion and bias can unfortunately dominate the situation.

It is also important to make the point that this is not just about data collection; it is about the effective use of data to extract insight, understanding, and make better decisions. The effective use of big data and AI goes a long way toward a sustainable approach to sustainability.

AI in Transportation Decision Support   Big data is here representing both a challenge and an opportunity. The challenge lies in what could best be described as a data tsunami. A wave of data that is unprecedented in transportation. This data is coming from traditional and innovative sources including roadside sensors, connected vehicles, automated vehicles, and location intelligence. This latter source extracts position information from smartphones and cell phones, makes it anonymous, and then available as an amazing way of understanding variations in the demand for transportation and operating conditions. While this wave of data will have repercussions across the full spectrum of transportation service delivery activities from planning to design, to implementation, operations, and maintenance, it is expected that transportation operations will be an early pioneer in the application of such data. Due to the stream of rapidly emerging data, now is the time to understand how to harness its power, avoid being overwhelmed, and take all the opportunities that offers. Among the recent advances in data management and science is the emergence of artificial intelligence as a practical approach to managing big data. This article takes a close look at artificial intelligence and potential applications for this powerful technology for transportation operations.

Artificial intelligence, just like big data, is a term that is used widely and describes a platform of technologies from which great things are expected. Expectations are high and it is especially important to manage these expectations with an authentic view of exactly how these technologies can be applied in practice. In some respects, artificial intelligence can be viewed as a shiny object that the transportation industry would like to grasp. Before reaching out and embracing it, however, let us take a closer look at exactly what it is and how it will be relevant to transportation operations.

Before discussing the relevance of artificial intelligence to transportation operations, it is important to have an agreed definition of transportation operations. What is it, what does it do and why is it important? Transportation operations involve a careful balance of two control factors: the available capacity for transportation supply and the current demand for transportation. This is a high-level, snapshot view which serves as an umbrella for a range of activities supported by transportation operations professionals and is summarized in Figure 8.24.

Figure 8.24 identifies opportunities for the application of artificial intelligence to each of these six areas of transportation or smart mobility management. Traffic management involves the management of freeway, arterial, and surface street traffic on a day-to-day basis for both recurring and nonrecurring congestion. Transit management involves two primary elements. The first is fleet management to ensure that the transit fleet is operating effectively. The second is passenger or customer information. This second element also includes electronic fare collection and payment systems. Multimodal management as the name suggests involves planning and operations across all available modes of transportation within a city. This includes the private car, transit, freight, shared mobility, and nonmotorized transportation. Asset management includes the active management of the infrastructure that supports smart mobility service delivery. This includes maintaining an inventory of the network and the assets and ensuring that the appropriate level of maintenance and intervention is provided. Data management involves the collection and preparation of the raw material for artificial intelligence. This also incorporates the activities required to convert data into information, to extract insight from the information, and use analytics to suggest appropriate response strategies. Administration and security include the organizational arrangements required to operate the entire system effectively, ensure the safety of the public, and protect the system from intruders. This also includes the definition and application of data governance principles like previously agreed data sharing and data usage policies.

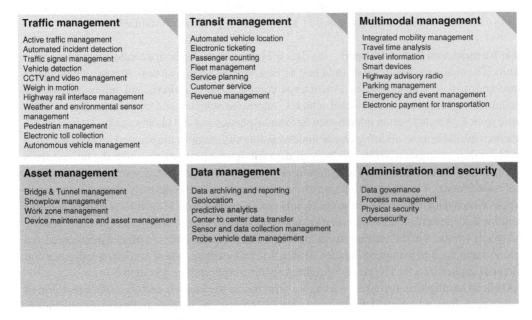

Figure 8.24 Transportation activity summary.

Artificial intelligence is a group of technologies that can have considerable impact on the management of data for transportation. Unlike conventional programming where the output is related to the input or investment required to program the computer, artificial intelligence holds the possibility of applications that deliver a higher value through the process of learning. Artificial intelligence applications can learn through repeated use, detect patterns, and begin to apply judgments that were formerly the domain of humans.

As the size of data continues to increase, artificial intelligence applications become an important tool in the management of large volumes of data and the extraction of insight and understanding from big datasets. Artificial intelligence also holds the possibility of providing decision support to human operators and managers as they develop strategies in response to information derived from data. While artificial intelligence has the potential to impact many areas of society, it is particularly interesting to traffic and transportation professionals.

The definition of artificial intelligence can be derived from a focus on the two words that comprise the expression "artificial" and "intelligence." The word artificial implies derivation from means other than human, using computers. Intelligence suggests the capacity to develop different responses and apply judgment in the light of data regarding situations and scenarios.

The current state of the art of artificial intelligence is such that these applications are not ready to completely replace the human. In fact, it could be argued that artificial intelligence may never be a substitute for human judgment but could be an enormously powerful decision-support tool. In the short term at least, it is expected that a combination of artificial and human intelligence will form the most powerful approach to the management of data and information. As is the case with many tools, it is important that the tool is applied in the most appropriate context and within an overall process that maximizes the value of the technology.

The generic value of applying artificial intelligence to transportation can be characterized in simple terms under the following headings:

Cost   After an initial investment in hardware and software required for artificial intelligence, it is likely that the cost of supporting technical and administrative processes related to transportation will decrease. The labor component of cost will be reduced, while processing will be achieved in a more cost-effective manner. For example, transportation, data analysis, and labor costs can be avoided by automated analysis, or more efficient decision support.

Efficiency   One of the current challenges associated with transportation operations data management can be the delay between data being collected and turned into valuable information. Many agencies are finding that data is not properly harnessed because of the time required to find and process the data. Application of artificial intelligence should allow processing to be conducted in a shorter period, aligned with the smart data management approach, this should also make answers available within a shorter time frame. For transportation operations data, the efficiency is realized by making the results of the analysis available in a timelier fashion and more accessible to a wider range of appropriate users.

Accuracy   Humans are exceptionally good at subjectivity but can be prone to errors and repetitive calculations tasks. The introduction of artificial intelligence will provide decision support to humans and automate repetitive tasks. This will allow human judgment and subjectivity to be focused on the most appropriate areas while leaving the repetitive tasks to automation and artificial intelligence. For transportation data, accuracy improvements can be achieved by artificial intelligence taking over the repetitive processing roles of humans including interpolation of data for missing data points and the combination of transportation data from different sources.

Capacity   Because of improvements in the speed of processing and reduced cost, it should be possible to do things that were considered infeasible in the past. These may be things that are cost-prohibitive or where manual approaches would just simply take too long using manual techniques. It should also be possible for a wider range of data to be analyzed to reveal new insight and understanding. For transportation data, it should be possible to analyze a larger number of sensor inputs, allowing the transportation program to be expanded to be more complete.

AI in Transportation Operations   To bring the subject of artificial intelligence to life with respect to transportation operations, Figure 8.25 was created. This shows some potential applications of artificial intelligence within each of the six categories of transportation operations activities, illustrated in the figure. This is not intended to be an exhaustive catalog as there are many applications of artificial intelligence possible however there should be enough here to interest and inspire consideration of other applications.

All the possible activities have the same elements. Interrogation of an extensive database by a sophisticated data management and analytics system capable of machine learning and adaptation. This will result in more accurate predictions over a longer time span. In addition to predictions, business processes, and operations can be optimized by a detailed analysis of data regarding past performance, leading to new insight and understanding as the basis for better performance management, responses, and strategies.

It is obvious that artificial intelligence is of growing importance in the world of transportation operations. As the industry is faced with challenges and new opportunities associated with big data, the effective application of data science techniques such as artificial intelligence, will be unavoidable if practitioners are to gain the advantages and manage any undesirable side effects. If this information has generated some interest, then the authors suggest that the next step should be to gain a more detailed understanding of what artificial intelligence is, what it can do, and what value it could deliver for the specific job function. Examine the potential application of artificial intelligence to traffic operations activities that have been provided and determine where there are gaps to be filled and enhancements to be made. Consider the type of hardware and software that will be required to effectively conduct artificial intelligence-driven transportation operations.

It is also especially important to consider organizational capability and its readiness to extract insight and understanding from artificial intelligence and the staff's ability and enablement to do something about it. Consider how the internal business processes may have to be realigned to the new opportunity. There is a lot to get ready before the effective application of artificial intelligence. It has been said that data is the new oil, and the authors like this analogy as crude oil is dirty, smelly, and nobody wants to touch it when it comes out of the ground. However, once it has been passed through a refining process, it becomes a wide range of highly sought-after and highly valued products. As data is refined into information and further refined into insight and understanding, it also becomes a range of highly valuable products that can support better decision-making, more effective responses, and a higher level of performance for transportation operations. Data is an extremely valuable raw material that can be applied to many different things related to smart mobility. To provide an illustration, the example of the value of data in supporting sustainability in smart mobility has been explored as follows.

8.5.6.2   Big Data
8.5.6.2.1   Description of Big Data
Big Data may potentially be as important to transportation as the Internet has been in the past. More data leads to more accurate analyses and greater understanding of the underlying mechanisms that affect the performance of transportation services. Using the right approach, the true

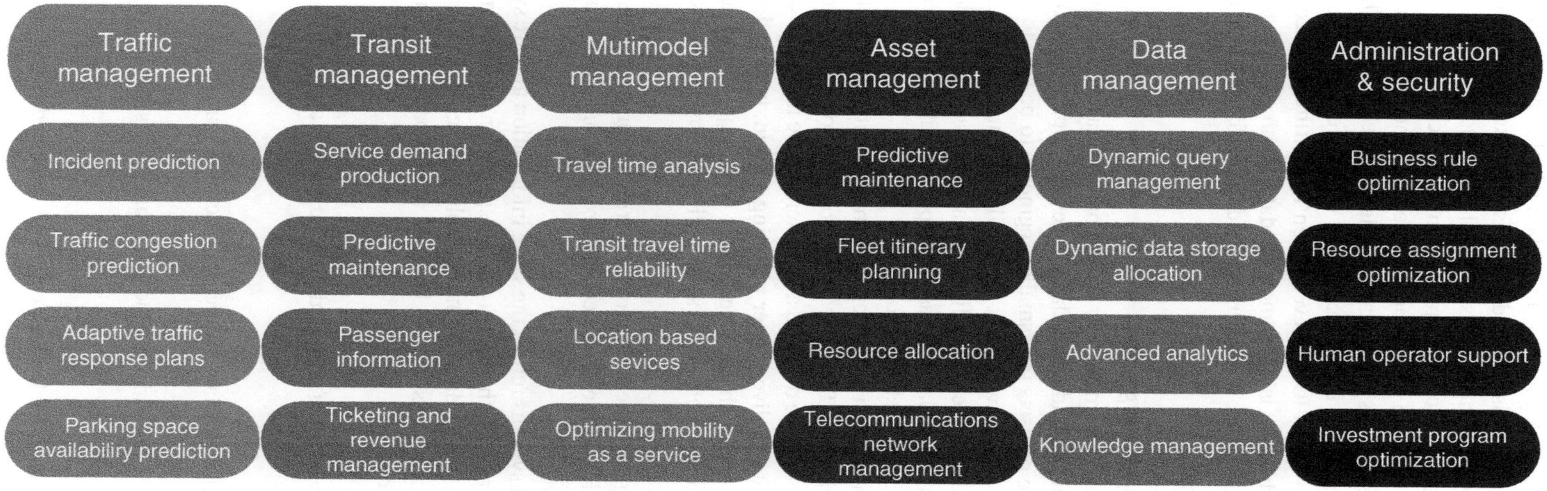

Figure 8.25 Possible applications of AI to smart mobility.

value of Big Data can be realized. A good way to explain it is that data itself is difficult to manage, not that attractive, and its meaning is well hidden. It is a lot like the data that forms the basis for music. The notes on the page and the various musical frequencies come together to form something that is pleasant to the ear because of arrangement and orchestration. The data is brought to life by additional tools and expertise being applied to extract the desired result. Another analogy that has been widely used recently is that data is the new oil, referring to the potential value that can be delivered. This is a useful analogy as crude oil when it emerges from the ground is dirty, smelly, and nobody wants to touch it. However, when it is refined it becomes a series of highly valued and sought-after products. It is the same with data, it can often be perceived as a nuisance and a cost generator with little value, but when distilled into information it can become extremely valuable.

8.5.6.2.2 *Operating Characteristics of Big Data*

Big data is an all-encompassing term for any collection of datasets so large and complex that it becomes difficult to process using traditional data processing applications. In the past practitioners would have had to split large datasets into smaller segments to be able to manage them with the data management hardware and software that was available. Challenges include analysis, capture, correlation, sharing, storage, transfer, visualization, and privacy protection. With advances in hardware and software, the trend to large datasets supports additional information derivable from analysis of a single large dataset of related data.

This is an improvement in the past when separate smaller datasets with the same total amount of data were analyzed individually. This new ability to see across the entire data horizon allows correlations to be found, trends to be detected and patterns to be identified and exploited. The new abilities of big data science will deliver better planning, design, project delivery, operations, and maintenance in transportation and have a significant impact on performance management. Big data, if it is to be successful and effective, has several dimensions as follows.

Completeness   The data must meet expectations in terms of completeness. Data can be complete even if optional data is missing. If the data meets the expectations, then it is complete.

Consistency   Data across all systems reflect the same information and are synchronized with each other across the entire agency. With respect to transportation performance management, this involves ensuring that the same data from multiple sources is compatible and consistent. Techniques such as orthogonal sensing can be used to compare the same data from different sources for quality verification.

Conformity   The data follows a set of standard data definitions and in is in the expected structure and format.

Accuracy   The degree to which the data correctly reflects the real world.

Integrity   All data must be valid and must be traceable from source through all connections to other systems.

Timeliness   Information must be delivered when it is expected, where it is expected and to the right people.

8.5.6.2.3   *Interesting Aspects and Points to Note Regarding Big Data*

Some additional aspects of big data also include advances in data analytics to support greater understanding of the purposes to which data can be put. Advances also include the use of unstructured data and a technique known as Query on Demand. This allows the effort to structure the data in advance to be minimized until such times as the data is needed to support a query and deliver information.

Data science advances also include significant capability in understanding the nature of each query and structuring the data in the most optimized fashion based on the most common queries. This also extends to data storage, where query management separates data into that used most often and least often used. Most often can be kept in more expensive short-term, high-accessibility storage, while least used data can be moved to cheaper, less accessible, longer-term storage.

An interesting development in data management is the emergence of data mesh techniques. Data mesh is a data management and architecture approach designed to create a foundation for extracting value from historical and analytical data at scale. Previous generations of technological approaches to address big data storage and management have focused on the need for a central data repository such as a data lake or a data warehouse. The data mesh approach is designed to address the use of analytical data at scale while applying distributed architecture techniques to allow data to remain at source. This offers the possibility of a federated approach to data storage and handling that takes full advantage of existing investment in data storage and information processing resources.

Taking this approach involves the use of decentralized big analytical data, where the original data owner or source of the data retains ownership of and access to the data. The data owner also takes responsibility for data management and retains ownership of the data. Each owner is also responsible for the steps necessary to ensure that data can be delivered as a product to other groups. The overall data mesh provides user-friendly self-service capabilities to enable each data owner to develop data products and consumer products from others. This background infrastructure enables multiple data on those to share advanced tools for data management and analytics. This approach allows a patchwork of data ownership to be supported and yet still takes advantage of commonality of support tools.

The data mesh approach enables small-scale additional investment to unlock the value of previous investments. It minimizes the disruption of existing systems and provides a way to join them together for the common good while preserving individual management and control. Individual systems are linked together by the mesh. A flow diagram for a data mesh approach is shown in Figure 8.26.

This shows data coming from individual data sources 1 through 5 instead of a central data depository.

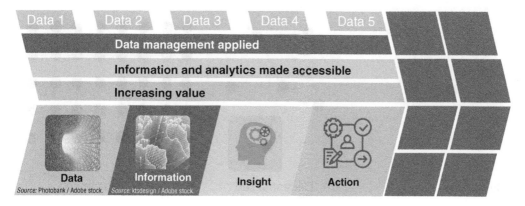

Figure 8.26   Data mesh approach.

8.5.6.3 Blockchain

8.5.6.3.1 Description of Blockchain

Blockchain is one of those technologies where a focus is required on potential benefits rather than the characteristics of Blockchain. The focus on the technology may be stifling the implementation of Blockchain across smart mobility and other applications. The best definition of Blockchain is as follows [39].

"The blockchain is an incorruptible digital ledger of economic transactions that can be programmed to record not just financial transactions but virtually everything of value."

This is also an excellent resource for further information on what Blockchain is and why it is important.

Blockchain is still a relatively new technology and has a lot of hype associated with it. In many cases, there is a misunderstanding about the difference between Blockchain and crypto currencies such as Bitcoin. In fact, Blockchain is the underlying enabling technology behind Bitcoin, but it is not the same. IBM has recently been emphasizing the difference by referring to Blockchain as "Blockchain for Business."

While technology is still emerging, it does hold significant potential for transportation, especially with regard to applications that support an integrated payment system for smart cities. It would seem that the time is right for an investigation into the opportunities and challenges associated with the application of Blockchain to integrated payment systems. This would best be addressed by the design and implementation of a prototype Blockchain-supported integrated payment system within a controlled environment to enable discovery and lessons to be learned.

Although Blockchain is available as an open-source software platform, an investigation of the technology reveals that there will be costs associated with the implementation of Blockchain for integrated payments associated with transportation.

8.5.6.3.2 Operating Characteristics of Blockchain

Blockchain is an approach to digital ledgers for economic transactions that have the following characteristics:

- Distributed – refers to the topology of the network. This is compared to centralized and decentralized network approaches as shown in Figure 8.27.
- Immutable – Blockchain is immutable because of the use of cryptographic hashes. These cannot be reverse engineered, delivering high levels of security and privacy.

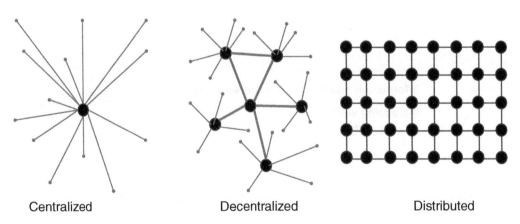

Centralized   Decentralized   Distributed

Figure 8.27 Network topologies.

- Transparent – the complete history of transactions is stored and visible. Data stored in the Blockchain is stored in a way that data integrity is verifiable.
- No requirement for third-party operations – this reduces cost of operations.
- Faster and cheaper than current transaction methods – the avoidance of third-party operational courses listed above, and the use of high-speed processing and telecommunications make Blockchain cheaper and faster than current transaction methods.
- No single point of failure – the distributed network approach and processing avoid the creation of a single point of failure that the entire system would depend on.

As each owner or partner in the Blockchain takes responsibility for their own operations there is no need for an independent third party to manage the operations. This results in operational cost savings. Immutable and tamper evident refer to the high level of difficulty in changing data on the Blockchain, and if it is changed it is obvious that some changes have been made.

8.5.6.3.3 *Interesting Aspects and Points to Note Regarding Blockchain*

Blockchain takes advantage of advanced cryptographic techniques including public key cryptography, zero-knowledge proof (ZKP), and hash functions. Public key involves the use of a pair of public and private key issues for encryption and digital signatures. This is how data owners access their blocks in the Blockchain. The process is illustrated in Figure 8.28.

The sender (1) creates an email message (2), then uses the receivers public key (3) to encrypt the original message, creating an encrypted secure message (4). The encrypted message (4) is sent to servers (5) that store the public keys for both the sender and receiver (6). The servers relay the encrypted message (4) to the receiver (7) who uses their private key (8) to decipher the encrypted message (4), resulting in a readable copy of the original message (9) from the sender (1).

Zero-knowledge proof (ZKP) requires proof of knowledge of a secret without revealing it. This proves that transfer of assets is valid without revealing the owner or the nature of the data ZKP is a set of tools that allow an item of information to be validated without the need to expose the data that demonstrates it. This is possible thanks to a series of cryptographic algorithms through which

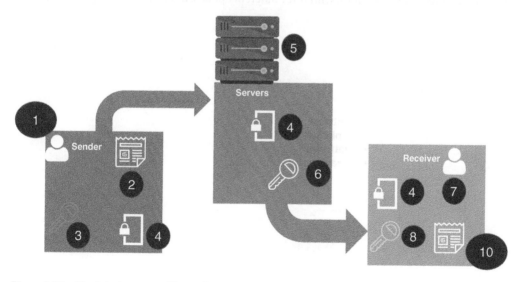

Figure 8.28 Blockchain process illustration.

a "tester" can mathematically demonstrate to a "verifier" that a computational statement is correct without revealing any data. For example, a user could show that they are of the appropriate age to access an item, but not reveal exactly how old they are. Similarly, someone could prove that they earn sufficient income to buy a product, but not share the exact amount of money in their possession.

Hash functions are one-way pseudorandom mathematical functions derived from data blocks and linking blocks together into a Blockchain.

8.5.6.4 Cloud Computing
8.5.6.4.1 Description of Cloud Computing
The overall concept of cloud computing is not new. It is based on network-based computing which dates to the 1960s. The first use of the term is credited to Eric Schmidt who was then Google CEO, who introduced the term at an industry conference in 2006. He explained that there was a new opportunity for data services to be hosted on remote servers. Today, when the term cloud computing is used it refers to a collection of technologies that have been configured to deliver data ingestion, storage, retrieval, processing, and analytics services. This is illustrated in Figure 8.29.

Inside the cloud, there is an ecosystem of technologies that support the following:

- Software applications that support collaboration, communications, monitoring, and content management.
- A platform that supports system operations, manages the database, and controls data govern-ance arrangements.
- Infrastructure including computing power, data storage, and a network that connects it altogether.

8.5.6.4.2 Operating Characteristics of Cloud Computing
There are three different ways to acquire and operate cloud computing services. Services are provided by several large and small companies including Google, Amazon, and Microsoft. The services are offered under three main categories, with three different options, as illustrated in Figure 8.30.

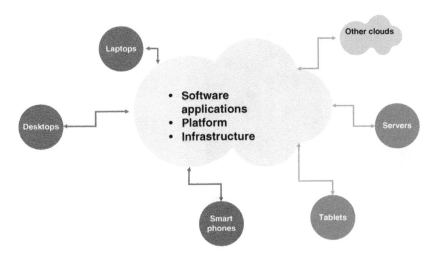

Figure 8.29 Cloud computing.

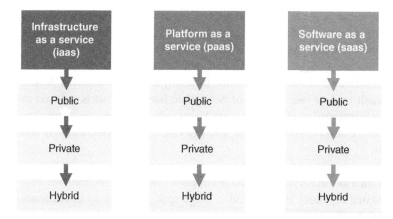

Figure 8.30 Cloud computing options.

Infrastructure as a Service (IaaS) IaaS is an offering of cloud computing where the cloud service provider supplies on-demand access to infrastructure such as networking, storage, and servers. Within the providers' infrastructure, the user runs their own platforms and applications.

Platform as a Service (PaaS) PaaS is an offering of cloud computing where the cloud service provider gives access to a cloud environment in which to develop, manage, and host applications. Access to a range of tools through the platform is available to support testing and development of solutions. The provider is responsible for the underlying infrastructure, security, operating systems, and backups.

Software as a Service (SaaS) SaaS is an offering of cloud computing where the provider gives access to their cloud-based software. Instead of installing the software application on a local device, the user accesses the provider's application using the web or an API.

Private Cloud A private cloud is where the user hosts their on their own data center or intranet. The user owns, manages, updates, and upgrades the cloud ecosystem of server, networking, software, or platform resources. The users own firewall and security solutions are used to protect the data.

Public Cloud A public cloud is where the provider supplies the user with access to their data center infrastructure. The provider is responsible for all management, maintenance, security, and upgrades.

Hybrid Cloud A hybrid cloud is where the user opts to implement a mix of both public and private cloud solutions. The user is responsible for managing how the two services interact, especially the security of data passing between both public and private cloud setups.

8.5.6.4.3 *Interesting Aspects and Points to Note Regarding Cloud Computing*
Approximately 91% of businesses utilize a public cloud while 72% use a private cloud. 69% of businesses use both, hence the overlapping percentages. This is known as a multi-strategy which allows companies to select different cloud services from different providers for different tasks. Some cloud service providers specialize in large data transfers, while others have integrated machine learning capabilities.

The value of cloud computing can be summarized under the following headings.

Flexibility   Cloud infrastructure can be scaled up on demand as workloads vary. There are storage options available depending on security needs and cost considerations. There is a menu of prebuilt tools and features to make solution development as efficient as possible.

Efficiency   Cloud-based applications and data are available from any Internet-connected device. Hardware failures do not result in data loss because of networked backups. Cost is reduced as cloud service providers take advantage of economies of scale. Users only pay for the resources they use. Cloud-based applications are also extremely dependable and can be upgraded effectively and efficiently at one single point.

Strategic Value   Techniques such as virtual private cloud, encryption, and API keys make the data secure. Solutions can be updated regularly and efficiently. Large-scale investments in hardware and software at the client's local office are avoided.

8.5.6.5   Cybersecurity
8.5.6.5.1   *Description of Cybersecurity*
There is a wide range of cybersecurity tools and techniques available. Over the years, cybersecurity has been growing important as awareness of potential threats has grown. This is an important concern for all solutions whether they be on premises or in the cloud.

Cybersecurity arrangements are put in place to protect data and systems from unauthorized users.

8.5.6.5.2   *Operating Characteristics of Cybersecurity*
There is a misconception that in-premise solutions are naturally more secure than the cloud. This is not necessarily the case. Cloud service providers go to great lengths to protect the security of data including the following activities:

- Monthly vulnerability tests.
- Regular security scanning.
- Industrial-grade encryption across the whole transmission path.
- Structured data governance and access arrangements, regularly reviewed.
- Economies of scale make a big difference, the cloud service providers, for example, address millions of users and can afford significant expenditure on data security.

8.5.6.5.3   *Interesting Aspects and Points to Note Regarding Cybersecurity*
When considering cybersecurity, it is also worthwhile considering physical security. While the on-premises solution may be highly secure from a cybersecurity viewpoint, do the physical security measures match up? One simple question for example – when a technician comes to repair a computer or server is he or she required to present credentials? There are known examples of transportation agencies that have lost track of server locations because of lack of physical security and documentation.

In addition to the techniques described above, tools and techniques have been developed within this important subject area. These include tools for access control, anti-malware software, anomaly detection, and application security. Additional tools provide support for data loss prevention, email security, and firewalls.

8.5.7   Information and Control Technologies

Figure 8.31 indicates that Information and Control technologies will be addressed in this section. There are seven information and control technology applications as illustrated in Figure 8.32.

Figure 8.31 Technology application areas.

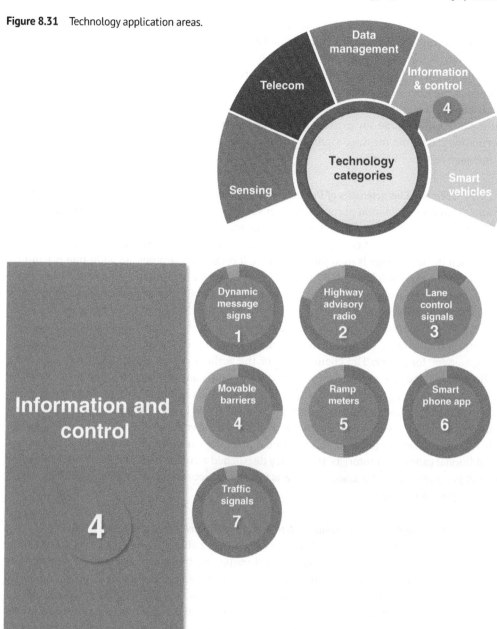

Figure 8.32 Information and control technology applications.

8.5.7.1 Dynamic Message Signs

8.5.7.1.1 Description of Dynamic Message Signs

These are roadside message signs which, unlike static signs, can support changes in the messages displayed. They are typically linked by a telecommunication network to a traffic or transportation management center where the messages originate. An operator in the traffic or transportation management center chooses the appropriate message from a library of messages and indicates to the control system that the particular message should be displayed on a specific sign. A variety of technologies can be employed to facilitate the message change including magnetic flip discs with fluorescent surfaces and more recently fiber optics, one of the ends of the fiber display high-intensity light which can be used to form patterns of dots. In the absence of a significant number of connected vehicles, within vehicle information capabilities, this has become an extremely popular technique for advising drivers with topical information. Dynamic message signs are also known as changeable message signs and variable message signs, or boards.

8.5.7.1.2 *Operating Characteristics of Dynamic Message Signs*

Typically used for information purposes, the signs may or may not have a confirmation feature that enables the operator to ensure that the appropriate message has been selected. This is sometimes achieved by swiveling closed-circuit TV surveillance cameras around to check the message on the same. When the technology is used for electronic toll collection purposes, an audit trail is required and typically the signs will have a built-in camera to be used for confirmation and acknowledgment. It is also important to coordinate the messages on sequences of signs along the highway and this is typically managed by central station software within the Traffic or Transportation Management Center.

Mobile dynamic message signs can also be used for construction zone applications. These typically rely on local programming or wireless communications to change the messages and an electrical generator for power. The client of one of the authors has an interesting experience with mobile dynamic message signs. Heavy rains caused significant flooding in fields adjacent to the freeway. Unfortunately, a cow was stranded on a small hillock surrounded by flood water. After numerous calls from motorists, the agency dispatched a crew to lead the cow to safety. They then used closed-circuit TV surveillance cameras to check on the cow, which had now returned to the same hillock, obviously preferring this location despite the surrounding floodwater. After numerous additional calls from motorists, the agency dispatched a crew with a mobile dynamic message sign and programmed the message "the cow is OK," to reduce motorist's calls. This could be called a "mooving message" sign.

8.5.7.1.3 *Interesting Aspects and Points to Note Regarding Dynamic Message Signs*

The advent of connected and automated vehicles with their ability to support a two-way communication link between the vehicle, roadside, and traffic and transportation management centers means agencies are now reconsidering investment in dynamic message signs at the roadside. While it is unlikely that dynamic message signs will disappear, it is likely that dependence on that information display technology alone will be reduced in the light of a superior ability to provide in-vehicle information to drivers. In vehicle, information delivery has the advantage of longer messages and customized messages for each driver. Dynamic message signs can also be used to change the speed limit on high-speed roads. This requires careful consideration of how enforcement will be achieved and how an audit trail can be developed.

8.5.7.2 Highway Advisory Radio

8.5.7.2.1 *Description of Highway Advisory Radio*

Highway advisory radio involves the use of regional radio broadcasts to inform drivers of specific circumstances and provide travel information [40]. Audio traveler information messages are transmitted via radio to travelers within range of the signal. Travelers and informed of the radio frequency to use and the availability of information by roadside signs that often use flashing beacons to alert travelers that messages are available.

8.5.7.2.2 Operating Characteristics of Highway Advisory Radio

Highway advisory radio has the advantage that it can provide longer message Langston dynamic message signs. However, it often suffers from poor signal quality due to the use of AM radio to transmit and receive the broadcasts. This may not deliver the reliability required for important information because the driver stops listening before the end of the message. The application also lacks a geolocation facility and cannot be tailored to the current location of the vehicle, requiring that the traveler listen to the entire message even if only part of it is relevant.

8.5.7.2.3 Interesting Aspects and Points to Note Regarding Highway Advisory Radio

While this is a low-cost approach, it is likely to be replaced as market penetration of connected vehicle technology increases.

8.5.7.3 Lane Control Signals
8.5.7.3.1 Description of Lane Control Signals

This is a specific application of dynamic message saying technology to enable lane use changes. Figure 8.33 shows an example of lane control signals in operation. The signs are smaller and have a limited number of messages, usually pictograms indicating that the Lane is currently in use for straight ahead, left turn, right turn, or a combination. Lane control signs are typically used when there is a significant difference in traffic patterns between the morning peak, the off-peak, and the evening peak. They are often utilized at traffic signal intersections enabling the transition between lane control sign displays to be managed safely [41].

Figure 8.33 shows an example of Lane control signs on a high-speed highway where the signs are used to control speeds in each lane and if necessary, indicate that traffic should merge into adjacent lanes.

8.5.7.4 Operating Characteristics of Lane Control Signals

Lane control signs are often used as part of an overall strategy of active traffic management. The use is typically complemented by roadside sensors, informational dynamic message signs, and law enforcement control. Arbitration requires a two-way communication link between the signs and a traffic or transportation management center, operating the appropriate management and control software.

Figure 8.33 Example of lane control signs on high-speed road. *Source:* https://ops.fhwa.dot.gov/publications/fhwahop15023/ch7.htm.

8.5.7.5 Interesting Aspects and Points to Note Regarding Lane Control Signals

Lane control signs present a more complicated picture for the traveler. Elaine that is used for one purpose at one time of day can easily be changed to another purpose at a different time, with the propensity to cause confusion. It is important that the information is provided to drivers indicating that changeable lane assignments are in operation and clearly explaining the operational concept.

8.5.7.6 Movable Barriers and Reversible Lanes

8.5.7.6.1 Description of Movable Barriers and Reversible Lanes

As illustrated in Figure 8.34, a specially designed vehicle is used to lift a chain of 1-m-long barrier segments about 6 in. off the ground and replace it at the other side of the lane. This is often referred to as a "zipper" approach. When the vehicle is not in use it requires a special parking space adjacent to the highway. The barriers are typically moved to accommodate tidal flow situations where the EMP direction of traffic is different from the PM and off-peak periods.

8.5.7.6.2 Operating Characteristics of Movable Barriers and Reversible Lanes

It can take several hours for the specially designed machine to move the barrier along significant lengths of Highway. The vehicle operates at approximately 10 miles per hour. This time must be considered when changing lane assignments. The lane being moved is also not open to traffic during the time of operation of the vehicle.

8.5.7.6.3 Interesting Aspects and Points to Note Regarding Movable Barriers and Reversible Lanes

The technique has been in use since 1984 by road authorities and contractors around the globe. The ability to reallocate lanes as a way to avoid major investment and reconstruction of roads and bridges. The barrier segments have a special "T" shaped profile enabling them to be lifted and replaced by the machine and not dragged across the road surface. Machines can transfer the barrier up to 9.1 m in one pass, wider transfers will require multiple passes.

Figure 8.34 Moveable barrier [42].

8.5.7.7 Ramp Meters

8.5.7.7.1 Description of Ramp Meters

Ramp meters, sometimes referred to as ramp signals, are an application of traffic signals technology specifically for freeways. A special traffic signal to aspects red and green is installed at the end of entrance ramp just before the freeway starts. Drivers are only allowed to enter the freeway when the light goes green. The one vehicle is allowed to pass on each green signal. If the entrance ramp is to lanes, often one lane will have a ramp meter and the other lane will be reserved for vehicles of higher occupancy (epically greater than one person in the vehicle). Such vehicles are offered the advantage of bypassing the ramp meter and entering the freeway without stopping. The rate at which green signals are offered to drivers is also according to traffic flow on the mainline freeway. Sensors were placed on the mainland and the data regarding traffic speed and flow is processed by an algorithm that determines the rate of green signal time to be made available to the ramp.

8.5.7.7.2 Operating Characteristics of Ramp Meters

Typically ramp meters work in isolation with no relationship between upstream and downstream ramp from the same freeway. There have been some attempts to coordinate ramp meters [43] that are still in test and evaluation phases.

8.5.7.7.3 Interesting Aspects and Points to Note Regarding Ramp Meters

Ramp meters have been proven successful at increasing traffic flow and speed during peak travel times. This can save the cost of constructing another Lane by achieving better utilization of existing lanes. In some cases, ramp meters have been combined with dynamic tolling and managed lanes. It is interesting to consider the psychology of a driver having just paid a toll to make use of a managed Lane when confronted the red signal before gaining access to the facility.

8.5.7.8 Smart Phone Apps

8.5.7.8.1 Description of Smart Phone Apps

The ubiquity of smartphones makes the use of smartphone applications a compelling way to deliver information to all different types of travelers, citizens, and visitors. A special-purpose smartphone app can be created for traveler information or mobility as a service. Alternatively, an existing app can be expanded for these purposes. For example, an app designed for tourist information could also carry mobility as a service and travel information. The approach has the advantage of being vehicle independent while providing open access to data before, during, and after the trip.

8.5.7.8.2 Operating Characteristics of Smart Phone Apps

This technology application has the advantage that a great many people are completely familiar with the operation of a smartphone. Most smartphones also have geolocation capability in the form of GPS, making it feasible that crowdsourcing applications can be supported to provide information on trip-making behavior and traveler needs. It is important that privacy aspects are considered when implementing this type of crowdsourcing, or probe data collection. Smartphone apps bring the Internet to the mobile device enabling information delivered over static Internet in the home or in the office, to be coordinated and aligned with Internet delivered to the smartphone app.

8.5.7.8.3 Interesting Aspects and Points to Note Regarding Smart Phone Apps

In addition to being an information delivery channel, a smartphone app also has the possibility to support feedback from travelers. It is important that feedback from travelers is obtained under the privacy and legal framework that allows travelers to opt in or opt out of the activity. It is interesting to note that in many cases travelers will trade privacy for an increased level of service, but in any event, it should be voluntary and that should be an ability to opt out of the service.

8.5.7.9 Traffic Signals

8.5.7.9.1 Description of Traffic Signals

This could be described as the original intelligent transportation technology application. Traffic signals were first introduced in the United States in the 1930s and have become one of the most common technologies in use in traffic management today. It is estimated that there are more than 350,000 signalized intersections in the United States alone. Traffic signals enable a timesharing approach to be taken at intersections time where the shared road space is allocated by time to avoid conflicts. In contrast to a space-sharing approach which involves grade separation and significant investment in construction. This latter approach will be inevitable with traffic volumes of two great for traffic signals; however, traffic signals can be a cost-effective approach when volumes do not warrant a grade-separated approach.

8.5.7.9.2 Operating Characteristics of Traffic Signals

There are several different operating modes possible for traffic signals. At a basic level, signalized intersection can be standalone. This means that no account is taken of upstream or downstream intersections and the intersection is programmed to operate in isolation. Signal timings can be according to a fixed time of day plan or varied according to traffic flow derived from sensors in the roadway. This latter approach is known as adaptive traffic signal control as signal timings are adapted to variations in the flow of traffic through the intersection. Traffic signal timings can also be varied to take account of upstream and downstream intersections. This typically requires connectivity between intersections and is known as a coordinated approach to traffic signal timings. Here again, traffic signal timings can be adjusted according to traffic as well as the need to coordinate to ensure smooth progression from one intersection to the other. This is not as adaptive, coordinated traffic signal control. This can be achieved at a local level by coordination between the traffic controllers that control each intersection, or on a networkwide level with a centralized traffic management software system. Such a system uses sensors along the network to collect data and special-purpose algorithms to calculate the optimal signal timings based on current traffic conditions. Traffic signal control systems can also be designed to accommodate emergency vehicles and provide priority to transit vehicles, by sensing such vehicles and providing a green signal to them.

8.5.7.9.3 Interesting Aspects and Points to Note Regarding Traffic Signals

Advances in technology minute traffic signals are continually getting smarter. The availability of lower-cost sensors and the ability for wireless communications means that the cost of advanced traffic signal control is being reduced. Ideally, traffic control on surface streets and arterials will be coordinated with freeway management activities to take a total network approach. In most places around the globe, this has not been achieved yet; however, progress has been made in designated corridor management systems to take account of freeways, arterials, and transit facilities along designated corridors.

8.5.7.10 Smart Vehicle Technology Applications

Smart vehicle technologies are deployed in both connected and automated vehicle systems (Figure 8.35). Many of the technologies deployed in combination with the systems have been described in other sections. Therefore, only those technologies that are unique to connected and automated vehicles will be described in this section. The five selected are illustrated in Figure 8.36.

Figure 8.35 Technology application areas.

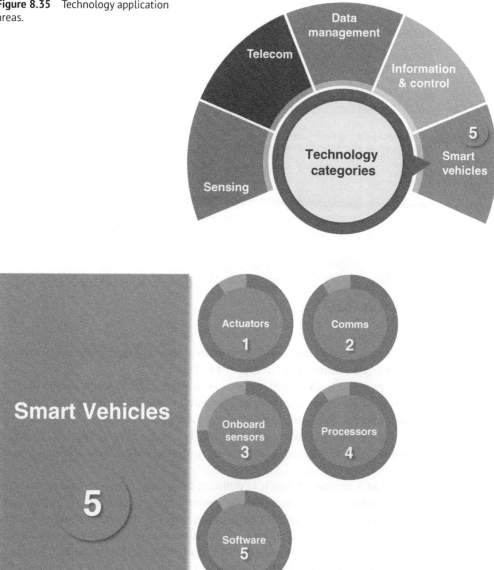

Figure 8.36 Smart vehicle technology applications.

The exception to this will be telecommunications technologies associated with smart vehicles as these are central to the operation of such vehicles. As one of the functions of these telecommunication technology applications is to connect vehicles to roadside and cloud infrastructure and other vehicles, it would be expected that overlap would exist in this area. It is also the reason this section of the chapter contains the fewest technology applications. Smart vehicles can be either connected vehicles or automated vehicles. Connected vehicles as the name suggests feature the use of a two-way communication link between the vehicle and infrastructure. The infrastructure can either be at the roadside or a centralized management or data center.

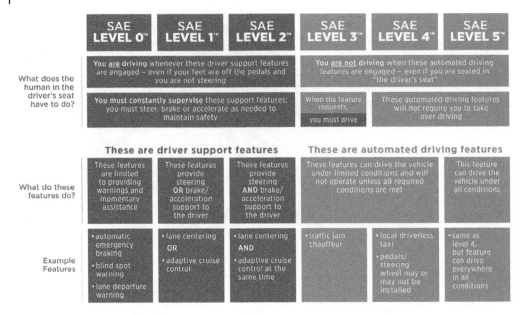

SAE LEVEL 0™	SAE LEVEL 1™	SAE LEVEL 2™	SAE LEVEL 3™	SAE LEVEL 4™	SAE LEVEL 5™

What does the human in the driver's seat have to do?

You are driving whenever these driver support features are engaged – even if your feet are off the pedals and you are not steering			You are not driving when these automated driving features are engaged – even if you are seated in "the driver's seat"		
You must constantly supervise these support features; you must steer, brake or accelerate as needed to maintain safety			When the feature requests, you must drive	These automated driving features will not require you to take over driving	

These are driver support features   These are automated driving features

What do these features do?

These features are limited to providing warnings and momentary assistance	These features provide steering OR brake/acceleration support to the driver	These features provide steering AND brake/acceleration support to the driver	These features can drive the vehicle under limited conditions and will not operate unless all required conditions are met		This feature can drive the vehicle under all conditions

Example Features

• automatic emergency braking • blind spot warning • lane departure warning	• lane centering OR • adaptive cruise control	• lane centering AND • adaptive cruise control at the same time	• traffic jam chauffeur	• local driverless taxi • pedals/ steering wheel may or may not be installed	• same as level 4, but feature can drive everywhere in all conditions

Figure 8.37 SAE J3016 levels of driving automation.

Automated vehicles have varying degrees of automation, best explained using the Society of Automobile Engineers levels of automation as depicted in Figure 8.37.

This shows a steady progress from no automation to full automation through a series of steps that begin with driver support features, move to automation assisted by the driver, and then culminate in full automation. The levels support a gradual introduction of features over time, enabling benefits to be accrued at a steady pace and not as a single lump at the end. This also aligns with automotive industry practice of gradual introduction of new technologies to manage risk.

The authors have used a similar approach to defining the gradual introduction of automation into traffic and transportation management centers, beginning with no automation. The levels progress through more sophisticated decision support, the operator-assisted automation, and then full automation. It makes sense to the authors that a systems approach should be adopted in which both vehicles and back offices are made smart and automated. Connected vehicle technology can be considered as a forerunner and enabler of automated vehicles as the two-way communication link can support advanced driver support features, and enable the automated vehicle to communicate with roadside infrastructure and a centralized command center. For example, Einride in Sweden is developing an automated truck approach in which multiple trucks are controlled from a central command center. It has an SAE level 4 classification [44].

It might well be the case that advanced levels of automation are introduced in trucks and transit ahead of private cars. The current shortage of qualified truck drivers and the fact that in these vehicle types, the driver is not a traveler, but an operator adding operating costs, could be significant factors in this. Figure 8.38 shows the five technology application areas that are addressed under this category along with the ubiquity index around the circumference of each button, as used in the previous technology application sections. These five technology applications are typically implemented in combination as depicted in Figure 8.38.

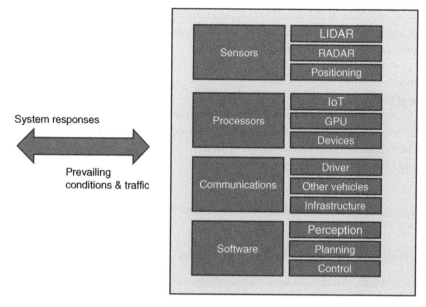

System responses

Prevailing
conditions & traffic

Figure 8.38 Automated vehicle system architecture.

8.5.7.11 Actuators

8.5.7.11.1 Description of Actuators

Actuators are devices to enable automated or remote control of brakes, steering, and accelerator on board automated vehicles. They make use of a combination of communication, hydraulics, and control system to affect this.

8.5.7.11.2 Operating Characteristics of Actuators

Many existing vehicles already feature the use of actuators in so-called "drive by wire" approaches. This obviates the need for a direction connection between the driver, brakes, steering, and accelerator reduces manufacturing costs and vehicle weight. The driver is still in full control of the vehicle but a wired connection between the pedals and brakes, or steering wheel and vehicle wheels enable the driver control. In effect, the driver has "remote control" of the vehicle. Drivers can also be better informed by sensors and communication with external systems such as road condition sensors and traffic signal controllers. Connected vehicle applications enable the delivery of decision-quality information from sources outside of the vehicle. Advanced driver support can also be provided in the form of lane-keeping advisory service and "self-parking" approaches that take some of the burden from the driver. One of the operating challenges with automated vehicles lies in the need to support a significant transition from no automation to full automation. This also includes the market penetration of automated vehicles. For a time, it will be necessary for full, partial, and non-automated vehicles to share road space and coexist safely.

8.5.7.11.3 Interesting Aspects and Points to Note Regarding Actuators

One of the interesting aspects of this technology is the gradual introduction of actuators that is already underway in the car manufacturing sector. A trail has also been blazed by aviation manufacturers who have already migrated to fly-by-wire techniques that make planes lighter and also offer the possibility of a growth path to automation through increasing system intervention.

There are challenges and successes in the aviation sector that could offer practical experience and lessons learned to the automotive sector. The future of aviation and automobiles is converging, especially with the advent of advanced and urban air mobility.

8.5.7.12   Communications

8.5.7.12.1   Description of Communications

Communications are a combination of wireless telecommunications as described in the section on telecommunication technology applications and human communications between the driver and other drivers. The driver must communicate with other drivers regarding intentions and alerts. This is typically achieved through one or more of the following:

- Headlights
- Horn
- Indicators and hazard warning lights
- Hand signals
- Verbal communications, direct or using communication devices

As an interesting comparison, insects also have five means of communication.

- Smell
- Taste
- Sounds
- Light flashes
- Touch

It can be concluded that a driver once inside the vehicle has communication skills equivalent to an insect. The authors will refrain from commenting on how this affects driver behavior. Connected vehicle technology will surely improve driver communication capability.

Modern vehicles also feature a vehicle area network that links all sensors, actuators, and processors for data sharing and coordinated action. A wide range of P codes are transmitted across the vehicle area network for use in vehicle and engine management. The ability to support two-way communication beyond the vehicle will enable this rich stream of data to be externalized. Transportation planning and operations can be revolutionized through access to this data. Even a small subset, comprising vehicle ID, instances vehicle speed, trip origin, trip destination, and current location could be of enormous value.

Vehicle-to-vehicle communications can also deliver dramatic improvements in safety by enabling vehicles to maintain a safe exclusion zone and better coordinate with other vehicles. This can also extend to infrastructure-supported collision avoidance at intersections. There is a significant potential for vehicles and infrastructure to function as a single system through the use of advanced communications.

8.5.7.12.2   Operating Characteristics of Communications

This has been addressed in the preceding section on telecommunications technology applications. With respect to smart vehicles, it can be expected that an aggressive growth curve will be achieved through which dependence on the driver will diminish while the role of automation and communications increases.

8.5.7.12.3   Interesting Aspects and Points to Note Regarding Communications

An interesting aspect of vehicle communications centers on the interface between the driver, the vehicle, and the environment. It is crucial to use technology to provide the driver with better information while maintaining safety levels. It is also interesting to note that better information may mean that drivers are required to make more choices and a larger number of decisions. It may no longer be possible to drive by habit. In fact, there may be an increasing need for additional driver training to equip drivers for a much more dynamic driving environment including route and road geometry changes.

8.5.7.13 Onboard Sensors

8.5.7.13.1 Description of Onboard Sensors

A range of sensor technologies managed by an advanced software management system is used to provide data to the automated system on prevailing conditions. Some manufacturers rely on video sensors and image processing techniques combined with positioning sensors. Others use a combination of RADAR, LIDAR, and video. In the case of smart vehicles, the sensors are installed onboard the vehicles and not at the roadside. As described earlier these are connected by a vehicle area network with data transferred to a central management system and commands transmitted from this system to actuators. Sensors monitor the vehicle, engine, operating environment, and indirectly, driver behavior. This is measured through the use of the vehicle sensors to establish driver behavior through vehicle operating parameters. There are also some smartphone apps emerging that monitor drive behavior directly through computer vision, and AI [45].

8.5.7.13.2 Operating Characteristics of Onboard Sensors

An essential operating characteristic is the ability to provide sufficient data at the right frequency and accuracy to support a clear understanding of the vehicle and operating conditions. For onboard applications, sensors are also required to be durable and not susceptible to vibration and changes in temperature. A degree of flexibility may also be required to fine-tune the data content and format in the light of operational experience.

8.5.7.13.3 Interesting Aspects and Points to Note Regarding Onboard Sensors

Sensors are at the leading edge of a move to enable vehicles and infrastructure. Through the collection of a complete dataset that defines the operating environment and fuses it with data from infrastructure, single system operation become feasible. Vehicles and infrastructure can adapt in alignment and harmony to produce optimum effects in terms of safety, efficiency, and user experience.

8.5.7.13.4 Processors

Description of Processors   In addition to central processing units, approaches to smart vehicles feature extensive use of graphical processing units (GPUs).

GPU (graphical processing units) form part of a wider spectrum of extremely fast and optimized hardware that can be used to support advanced computing applications. GPUs differ from CPUs in that they are specially designed to support an extremely large number of computations, typically associated with high-speed graphical displays and animations. This is ideal for smart vehicle applications that need fast response based on a large volume of data. While the market for GPUs is driven by video gaming, their capabilities are applicable to intense computational needs. This includes the application of advanced analytics on big datasets and in-vehicle applications.

GPUs within the framework of smart vehicles, offer the possibility of handling large volumes of data and converting it to information very efficiently. This is particularly important as the amount of data available to transportation is growing at an accelerated rate. High-speed processing capability is required to distill the data into meaningful information and to support the type of graphics-intensive visualizations that will deliver the desired insight and understanding.

These processors are also utilized to support an Internet of Thing approach to connected vehicles involving a significant amount of edge processing. Computations are conducted at the roadside and onboard the vehicle to optimize the use of the connected and automated vehicle technologies.

8.5.7.13.5 Operating Characteristics of Processors

The primary operating characteristics of processing technologies related to automated and connected vehicles are speed and computational capacity. Like sensors, processes for mobile use must also be robust and not sensitive to vibration of changes in temperature. Form factor and weight are also factors in vehicle manufacturing, along with power consumption.

8.5.7.13.6 Interesting Aspects and Points to Note Regarding Processors

Significant growth in the market for video games is a major driver in the development of high-speed processes capable of managing a large number of computations. The existence of this large-scale market has stimulated the development of the type of processors that connected and automated vehicles will require. While the automotive market is also a large market, the existence of this early gaming market has helped to accelerate the development of the hardware required.

8.5.7.14 Software
8.5.7.14.1 Description of Software

Software in combination with processing hardware provides the ability to take data from multiple sources, convert it into information, develop insight and understanding, and formulate plans of actions, or system responses. The software has three primary tasks within the automated vehicle environment. The first is perception, the software must conduct an analysis of all available data, detect trends and patterns, and to the extent feasible emulate the human perception used in the driving task. The second task is planning. The software must be capable of formulating strategies and plans for the overall control of the vehicle. The final task is vehicle control, the software must supervise the coordinated action of all actuators and sensors onboard the vehicle.

8.5.7.14.2 Operating Characteristics of Software

Since connected and automated vehicles will be made in large-scale production, it is particularly important that any software application is thoroughly evaluated and considered dependable. In the author's opinion, one of the barriers to deployment of advanced technology in automobile manufacturing has been the risk of major recalls because of failures in components or software. When manufacturing millions of units, reliability and risk management are important considerations.

8.5.7.14.3 Interesting Aspects and Points to Note Regarding Software

Once again, the existence of a video gaming market is of considerable advantage to the smart vehicle market. Software developed for one market can successfully be adapted to another. Skills developed in software development and coding in one market can also be applied to another. Lessons learned in one market can also be applied to another. Advances in data science and software engineering have also been of benefit to the smart vehicle market as the software modules or building blocks have become larger and more capable, it is possible to accelerate software development while still maintaining the reliability and risk management required. The use of connected vehicle technology to remotely upgrade software is also a significant factor. There is no need to recall the vehicle to conduct an improvement or an upgrade, simply ensure that the vehicle is connected to the network. It could be said that vehicles are becoming computers and wheels as electronics content of the average vehicle increases and operation of the vehicle becomes increasingly reliant on hardware and software. Another market where lessons can be learned is the one for smartphones where a similar migration has taken place. What was once a device for voice communications, is now essentially a multipurpose mobile computer, connected to the Internet.

8.6 The Accelerating Pace of Technology Change

As stated at the beginning of this chapter, it is not the intention to provide a complete catalog of all possible technology applications within this chapter. The objective is to provide an overview of the typical technologies that can be found in a smart mobility application and provide a high-level description of those technology applications that are directly applicable to smart mobility. This high-level view has been adopted because there is such a large number of technological choices that would need at least a separate complete book to describe this in any level of detail. There is also another challenge. The pace at which new technologies are emerging is accelerating, making it almost impossible to capture the current state of technological progress in a book where evergreen applications and best practices are being described. For this reason, the authors are cooperating with a publishing company in London in the creation of a smart mobility body of knowledge which will provide complementary and more detailed information regarding smart mobility technologies, among other things [46]. This also aligns with the overall objective which is to raise awareness of the characteristics of smart mobility technologies to enable a decision on which ones match the problems. It paves the way for further evaluation of the technologies and the development of detailed designs that incorporate them.

The accelerating pace of change is a concern when it comes to technology application selection and procurement. Some complicated procurements can take so long that technology has advanced before the procurement process is complete. It is essential to take appropriate measures to manage this fact. The first measure involves a continual assessment of the progress being made around the world in the successful use of advanced and emerging technology. An example of such a technology scan is presented in Chapter 5 earlier in this book. It is recommended that any scan is a combination of desk search, literature review, and one-on-one interview techniques. A second measure is to specify the services required, in procurement documents, rather than prescriptive solutions. This enables vendors to offer the best current solution and provides flexibility for incorporating emerging technologies. A third measure is to establish and maintain an effective dialogue with technology and solution providers. This can be facilitated by incorporating a Request for Information at the start of the procurement process, prior to the finalization of any Request for Proposals.

An important challenge in the procurement of rapidly changing technologies is the fact that procurement techniques often lack interactivity. While it is essential to maintain a degree of confidentiality to ensure that the process is fair, integrating opportunities for a managed dialogue with suppliers can be valuable. After all, solution providers are expert in technology capabilities and constraints, while procurement agencies are expert in what they want. An effective dialogue can deliver the best of both worlds.

Finally, take full advantage of curated knowledge sets that have been assembled to make the job of technology scanning more efficient and effective.

8.7 What Is on the Horizon?

In a world of accelerating technological change, it is important to be careful about making predictions. It has been shown that extrapolating the few simple linear sequential basis from past experience, does not work with emerging and disruptive technologies. The view of the future provided here will consist of broad trends and directions rather than detailed technology predictions.

The first major trend is that when it comes to systems and software the technology building blocks are getting larger. As years of experience in developing software for multiple system applications has accumulated, standard building blocks have emerged. These are almost like Lego bricks and as the bricks get larger and more functional it becomes easier and faster to create more complicated and extensive systems software development techniques and adapting to make use of these larger building blocks along with rapid evolutionary employment techniques that feature early prototyping to support interaction with the user group and fail fast techniques to quickly identify things that will not work.

The second major trend is that the time from invention to commercialization is collapsing. When the telephone was first invented, a period of 25 years elapsed from initial invention until full commercialization of the technology. The stairs the time required to collapse from 25 years to address 25 months. Investors have become more sophisticated and startup companies have learned more efficient approaches to commercializing technology.

Regarding the third trend, it is the authors belief that the advent of advanced urban air mobility will revolutionize urban transportation systems by adding a third dimension. The electric vertical takeoff and landing aircraft will be able to make use of urban mobility hubs to offer the convenience of air and travel into an integrated multimodal transportation network for both freight and people.

The fourth trend is that the wider availability of a larger range of technologies is being accompanied by a wider range of best business practices and business models that can fully harness the resources of the private sector were achieving public sector results.

It is a great time to be involved in the application of smart mobility technologies with more solution choices and data to fuel them than at any time in history.

References

1 Collins Online English Dictionary. https://www.collinsdictionary.com/us/dictionary/english/technology, accessed March 8, 2022, at 9:21 CET.

2 FHWA traffic detector handbook: Third Edition—Volume I Publication Number: FHWA-HRT-06-108. Date: May 2006. https://www.fhwa.dot.gov/publications/research/operations/its/06108/01.cfm#ind, accessed May 24, 2022, at 21:47 CET.

3 Evaluation of portable non-intrusive traffic detection system. Minnesota Department of Transportation Research Services Section 395 John Ireland Boulevard. http://www.dot.state.mn.us/research/TS/2005/200537.pdf, accessed May 24, 2022, at 21:47 CET.

4 The Hawaii Department of Transportation. Evaluation of ASIM29. http://www.hawaii-rdtt.com/projects/2007SPRFunded/InvestigationofASIM29.pdf, accessed May 24, 2022, at 21:52 CET.

5 CEOS website. www.ceos.com.au/company/about-us, accessed May 24, 2022, at 21:55 CET.

6 Oracle. What is IoT. https://www.oracle.com/internet-of-things/what-is-iot, accessed May 24, 2022, at 21:56 CET.

7 Research Paper. Development of an IoT based real-time traffic monitoring system for city governance. https://reader.elsevier.com/reader/sd/pii/S2589791820300207?token=C011F2FDC2 4F54A2B05C10179410BAC3C8E5EBD19D644461403422F27D5FF0F1B2746307798E2056B8D456 E7A1D7ED9D&originRegion=eu-west-1&originCreation=20220315124700, accessed May 24, 2022, at 21:58 CET.

8 Texas Instruments. An Introduction to Automotive LIDAR. https://www.ti.com/lit/wp/slyy150a/ slyy150a.pdf?ts=1653361845127, accessed May 24, 2022, at 22:09 CET.

9 http://GPS.gov. https://www.gps.gov/systems/gps/performance/accuracy/#:~:text=For% 20example%2C%20GPS%2Denabled%20smartphonesreceivers%20and%2For%20augmentation% 20systems, accessed May 24, 2022, at 22:10 CET.

10 Maksymova, I.; Steger, C.; Druml, N. MDPI proceedings: review of LIDAR sensor data acquisition and compression for automotive applications, presented at the Eurosensors 2018 conference, Graz, Austria, 9–12 September 2018. Published: 6 December 2018. file:///C:/Users/bob/Downloads/ proceedings-02-00852.pdf, accessed May 24, 2022, at 22:15 CET.

11 MarketingDive. https://www.marketingdive.com/ex/mobilemarketer/cms/news/research/ 22928.html, accessed May 24, 2022, at 22:19 CET.

12 KDnuggets. https://www.kdnuggets.com/2020/01/top-7-location-intelligence-companies-2020.html, accessed May 24, 2022, at 22:20 CET.

13 FHWA Office of Highway Policy Information. Traffic monitoring guide. https://www.fhwa.dot.gov/ policyinformation/tmguide.

14 https://www.nordicway.net/services/probe-vehicle-data, accessed May 24, 2022, at 22:28 CET.

15 UTMS Society of Japan. Traffic control system. https://http://utms.or.jp/english/cont/syusyuu/ index2.html, accessed May 24, 2022, at 22:30 CET.

16 FHWA Traffic Detector Handbook. Third edition. table 1.1. https://www.fhwa.dot.gov/ publications/research/operations/its/06108, accessed May 24, 2022, at 22:32 CET.

17 Commotion Wireless. https://commotionwireless.net/docs/cck/networking/learn-wireless-basics, accessed May 24, 2022, at 22:33 CET.

18 Cisco Meraki. https://documentation.meraki.com/MR/WiFi_Basics_and_Best_Practices/Wireless_ Fundamentals%3A_Modulation, accessed May 24, 2022, at 22:34 CET.

19 FHWA Electromagnetic Spectrum Chart. https:/if/www.transportation.gov/pnt/electromagnetic-spectrum-chart-enlarged, accessed May 25, 2022, at 16:13 CET.

20 Bluetooth Special Interest Group (SIG). https://www.bluetooth.com/learn-about-bluetooth/ tech-overview, accessed May 25, 2022, at 16:14 CET.

21 Qualcomm, Fundamentals of wireless signals and cellular networks. https://developer.qualcomm.com/ blog/fundamentals-wireless-signals-and-cellular-networks#:~:text=A%20cellular%20network% 20consists%20ofradio%20coverage%20for%20its%20area, accessed May 25, 2022, at 16:15 CET.

22 Wikipedia. Mobile radio telephone. https://en.wikipedia.org/wiki/Mobile_radio_telephone#: ~:text=Motorola%2C%20in%20conjunction%20with%20thepublic%20commercial%20mobile% 20phone%20network, accessed May 25, 2022, at 16:15 CET.

23 Unsplash. https://unsplash.com/photos/IlxX7xnbRF8, accessed May 25, 2022, at 16:19 CET.

24 LinkedIn Learning. Electrical systems: communications and data. https://www.linkedin.com/ learning/electrical-systems-communications-and-data/copper-based-communications-systems? resume=false, accessed May 25, 2022, at 16:23 CET.

25 https://en.wikipedia.org/wiki/Asymmetric_digital_subscriber_line, accessed May 25, 2022, at 22:23 CET.

26 Wikipedia. Ethernet. https://en.wikipedia.org/wiki/Ethernet, accessed May 25, 2022, at 16:24 CET.

27 Wikipedia. Metropolitan area network. https://en.wikipedia.org/wiki/Metropolitan_area_network, accessed May 25, 2022, at 16:26 CET.

28 Wikipedia. Local area network. https://http://en.wikipedia.org/wiki/Local_area_network, accessed May 25, 2022, at 16:26 CET.

29 Unsplash. https://unsplash.com/photos/IPApn5olGLk, accessed May 25, 2022, at 16:27 CET.

30 Wikipedia USB. https://en.wikipedia.org/wiki/USB, accessed May 25, 2022, at 16:29 CET.

31 Advanced Global Communications. Copper cabling 101. https://agcworld.com/copper-cabling-101//, accessed May 25, 2022, at 16:31 CET.

32 Wikipedia, EZPass. https://en.wikipedia.org/wiki/E-ZPass#:~:text=Transponders%20may%20be%20put%20inprotocol%20in%20256%E2%80%91bit%20packets, accessed May 25, 2022, at 16:34 CET.

33 How stuff works, how fiber optics works. https://computer.howstuffworks.com/fiber-optic1.htm, accessed May 25, 2022, at 16:36 CET.

34 Tutorialspoint. Wavelength division multiplexing. https://www.tutorialspoint.com/wavelength-division-multiplexing#:~:text=Wavelength%20division%20multiplexing%20(WDM)%20is,directions%20in%20the%20fiber%20cable, accessed May 25, 2022, at 16:37 CET.

35 Wikipedia. Bluetooth comparison. https://en.wikipedia.org/wiki/Near-field_communication#Bluetooth_comparison, accessed May 25, 2022, at 16:39 CET.

36 Wikipedia. Radio broadcasting. https://en.wikipedia.org/wiki/Radio_broadcasting, accessed May 25, 2022, at 16:44 CET.

37 Britannica. How does Wi-Fi work?https://www.britannica.com/story/how-does-wi-fi-work#:~:text=Wi%2DFi%20uses%20radio%20waves2.4%20gigahertz%20and%205%20gigahertz, accessed May 25, 2022, at 16:45 CET.

38 How stuff works, How WiFi works. https://computer.howstuffworks.com/wireless-network.htm, accessed May 25, 2022, at 16:48 CET.

39 Blockchain Revolution. How the technology behind bitcoin exchanging money business underworld, Don tap and Alex Tapscott, 2016. https://dontapscott.com/books/blockchain-revolution/#:~:text=Blockchain%20Revolution%20(2016)&text=Tapscott's%20latest%20book%2C%20entitled%20BLOCKCHAIN,by%20his%20son%20Alex%20Tapscott, accessed May 25, 2022, at 16:49 CET.

40 Washington State DOT. Transportation systems management and operations, highway advisory radio. https://tsmowa.org/category/traveler-information/highway-advisory-radio#:~:text=Highway%20advisory%20radio%20(HAR)%20transmits,alert%20travelers%20of%20advisory%20messages, accessed May 25, 2022, at 16:51 CET.

41 FHWA, Active Traffic Management. https://ops.fhwa.dot.gov/atdm/approaches/atm.htm, accessed May 25, 2022, at 16:52 CET.

42 Roads to the future. Zipper machine on I-95 James River bridge project. http://www.roadstothefuture.com/Zipper_I95_JRB.html, accessed May 25, 2022, at 16:54 CET.

43 Colorado DOT. Smart 25 pilot project: coordinated ramp metering system on I-25 from Ridgegate Parkway to University Boulevard. https://www.codot.gov/projects/smart25pilotproject, accessed May 25, 2022, at 16:56 CET.

44 Einride. https://www.einride.tech/insights/remote-operation-and-the-future-of-trucking, accessed May 25, 2022, at 16:57 CET.

45 Wingdriver. https://www.wingdriver.com, accessed May 25, 2022, at 16:58 CET.

46 COMPASS. https://www.h3bm.com/compass-smart-mobility-body-of-knowledge/welcome, accessed

9

Smart Mobility Opportunities and Challenges

9.1 Informational Objectives

After reading this chapter, you should be able to:

- Describe the relationship between the problem statement, opportunities, and challenges
- Define smart mobility opportunities
- Define smart mobility challenges
- Explain how smart mobility can address opportunities
- Describe how smart mobility can tackle challenges

9.2 Introduction

The opportunities and challenges defined in this chapter have been derived from almost 50 years of collective experience by the authors in working with both public and private sector enterprises in the application of smart mobility technologies. The experience covers a wide range of projects including smart mobility technology implementations, infrastructure implementations, associated business models and technical analysis. The opportunities and challenges are distinct from those described in Chapter 2 when a problem statement for smart mobility was provided. In that chapter societal problems that can be addressed by smart mobility were discussed. In this section specific opportunities and challenges that relate to the planning and implementation of smart mobility will be addressed. It is not expected that the list of opportunities and challenges will be comprehensive or exhaustive, in fact this is not the intent. The objective of this chapter is to provide a checklist and some information based on past experience that will help to create a customized approach for a specific smart mobility implementation.

9.3 Smart Mobility Opportunities

Over the course of many years of experience in both the public and private sectors, the authors have developed a summary of the opportunities that are presented by smart mobility. These are summarized in Figure 9.1 and described in the following sections. In each section, the opportunity category is

Smart Mobility: Using Technology to Improve Transportation in Smart Cities, First Edition. Bob McQueen, Ammar Safi, and Shafia Alkheyaili.
© 2024 The Institute of Electrical and Electronics Engineers, Inc. Published 2024 by John Wiley & Sons, Inc.

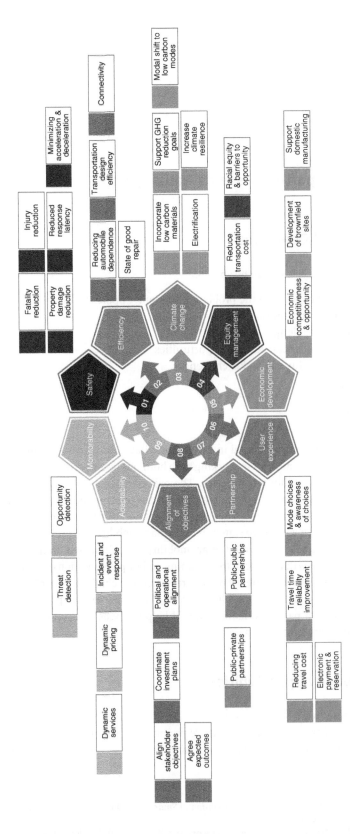

Figure 9.1 Smart mobility opportunities.

explored by describing each subcategory. At the end of each section, a summary table is provided that provides some examples of how smart mobility technology applications can help to address the opportunities for improvement. Chapter 8 on technologies provides a more extensive list of applications that can be used to develop a customized approach for a specific smart mobility implementation.

9.3.1 Safety

There is a growing recognition that fatalities, injuries, and property damage due to road traffic crashes are entirely preventable. Emerging vehicle and infrastructure technologies can be applied to address the situation. Many agencies in the United States for example, have adopted bold goals related to Vision Zero (avoiding crashes completely through a multidisciplinary approach featuring a combination of measures) [3]. This fundamentally different approach to traffic safety is signaled by the use of the term "crash" instead of "accident." It is considered that crashes can be presented but accident cannot. With the advent of advanced transportation technology, smart mobility can play a role in the prevention of collateral damage as a result of the transportation process.

9.3.1.1 Fatality Reduction

This is a very important opportunity as approximately 1.3 million people die every year as a result of road traffic crashes [1]. Such crashes are the leading cause of death for children and young adults aged between five and 29 years [1]. In addition, every year between 20 and 50 million people suffer nonfatal injuries associated with road traffic crashes with many people incurring a disability as a consequence of the crash [1].

9.3.1.2 Property Damage Reduction

Taking account of the property damage associated with noninjury crashes, it is estimated that the total related to road traffic crashes in urban settings is somewhere between 1% and 3% of the gross domestic product of the countries depending on country income [2].

A review of the causal factors related to traffic crashes reveals the following risk factors which can be addressed by smart mobility capabilities. Table 9.1 summarizes the risk factors and suggests some appropriate smart mobility treatments to take advantage of the opportunity.

Table 9.1 summarizes smart mobility opportunities related to safety and offers some suggestions to address them.

9.3.2 Efficiency

9.3.2.1 Reducing Automobile Dependence

An important aspect of transportation efficiency lies in trying to ensure that the most appropriate and most efficient mode is taken by travelers at any given time, for any given trip. The use of the private car can be considered as the "no thought" option, as it is usually the easiest mode of transportation to choose for a journey. People like cars and automobile manufacturers spend a lot of money on marketing and advertising to make sure. Through the use of smart mobility, there is a great opportunity to balance modal use within a city and ensure the most appropriate choice based on origin, destination, time of day, trip purpose, and available choices.

9.3.2.2 Transportation Design Efficiency

This covers all modes of transportation, with a particular emphasis on transportation infrastructure. Through the use of advanced technologies, it is possible to take the opportunity to optimize transportation design efficiency while also addressing transportation service design and operation.

Table 9.1 Smart mobility safety opportunities.

Risk factor	Smart mobility opportunity
Speed	Automated speed enforcement and variable speed limit systems to raise compliance levels and adjust speed limits to travel conditions
Alcohol	Automated vehicles and vehicles equipped with breath alcohol sensors to prevent driving while impaired by alcohol or drugs. Automated vehicles to remove reliance on human control
Equipment	Better equipment monitoring using automated sensors and communication technologies for motorized and nonmotorized transportation. Sensors to detect appropriate use of equipment for passenger vehicles, and for motorcycles. All designed to increase compliance and improve safety
Distraction	In vehicle technology to sense driver behavior such as facial recognition techniques on smartphone apps to alert driver to distraction and human performance issues
Infrastructure	Better road infrastructure condition through the use of asset management techniques to manage asset condition
Vehicles	Electronic stability control, onboard sensors to monitor vehicle and driver capability, automated speed monitoring, and advanced traffic management to minimize acceleration and deceleration. All aimed at improving the safety of vehicle operation
Post-crash care	Automated crash detection, incident management process automation, and resource pre-allocation to improve post-crash response
Law enforcement	Automated enforcement systems combined with staff management systems to improve effectiveness of enforcement

There is an opportunity to combine infrastructure design with smart mobility to increase the flexibility and resilience of transportation designs. With the availability of big data and analytics, it is also possible to ensure that transportation design is optimized for a wide range of travelers.

9.3.2.3 Connectivity

There is a significant opportunity for the application of smart mobility to improve connectivity at multiple levels in urban transportation. Connectivity between modes is an obvious example and should be accompanied by connectivity of transportation services to ensure smooth interchange. Connectivity between objectives and expected outcomes and the capabilities of the solutions being applied should also be provided. Connectivity can also be achieved by coordinating the actions of multiple stakeholders within the city to ensure that objectives are aligned, and services are optimized and coordinated as possible. The design of transportation infrastructure and the application of smart mobility techniques should also ensure that community connectivity is maintained, avoiding the possibility that communities are divided as result of transportation services and infrastructure. Another aspect of connectivity lies in the ease or difficulty in traveling from one part of the city to another. Improvement in this type of connectivity can be measured through the use of an accessibility index.

9.3.2.4 State of Good Repair

The quality of infrastructure and services can be underwritten through the use of smart mobility techniques for asset management. Asset management quality levels can be improved by developing a greater understanding of the needs for asset maintenance and doing a better job managing the asset maintenance process. This can also be achieved as a result of improving the communication

Table 9.2 Efficiency opportunities for smart mobility.

Opportunity	Possible smart mobility solution
Reducing automobile dependence	Mobility as a Service to raise awareness of travel choices
Transportation design efficiency	Combining infrastructure design with smart mobility services, user feedback, use of big data and analytics as design inputs
Connectivity	Service coordination management at the interchange points, stakeholder coordination, big data, and analytics for accessibility evaluation
State of good repair	Advanced asset management and asset condition monitoring

between transportation agencies responsible for service delivery and the end user or traveler. There is an excellent opportunity to engage the traveler through the use of smartphones and crowdsourcing techniques, enabling the traveler to provide data regarding the current state of good repair on roads, bridges, tunnels, transit facilities, and vehicles.

Table 9.2 summarizes smart mobility opportunities related to efficiency and offers some suggestions to address them.

9.3.3 Climate Change

9.3.3.1 Incorporation of Low Carbon Materials
In addition to operational impacts, smart mobility should also incorporate the use of low-carbon materials in transportation system and infrastructure design. This can include the use of techniques to incorporate captured carbon dioxide in the materials to be used for building. It can also involve concrete hardening by incorporating carbon dioxide into the mix. An energy audit can also be completed on the manufacturing process for materials to reduce the carbon footprint. Traditionally high carbon materials can also be substituted with appropriate materials. The carbon footprint assessment should also be extended to the information technology aspects of smart mobility. This can include assessment of energy consumption and emissions for data centers that support data processing and manufacturing processes involved in the various information technology elements of the system. These include sensors, telecommunications, command-and-control, and automated vehicle technologies.

9.3.3.2 Support for Greenhouse Gas Reduction Goals
Support for achieving greenhouse gas reduction goals can be delivered by smart mobility's capabilities with respect to sensing, telecommunications, command-and-control, and information delivery. Progress toward goals can be monitored and measured including the detection of emissions from vehicles and from the overall transportation process. Management strategies can also be implemented that reduce acceleration and deceleration leading to a reduction in greenhouse gas generation by private cars, freight vehicles, and transit vehicles. The management of modal choice and coordination of multimodal services can also support greenhouse gas reduction.

9.3.3.3 Modal Shift to Low Carbon Modes
Smart mobility has a powerful capability to deliver information to travelers at the right time in the right place. This can be used to support a modal shift to low-carbon modes. In addition to providing information on modal choices for any given trip at any given time, information can also be delivered regarding the carbon impact of the trip choice.

Table 9.3 Climate change opportunities for smart mobility.

Opportunity	Possible smart mobility solution
Incorporation of low-carbon materials	Efficient smart mobility solution design
Support for greenhouse gas reduction goals	Monitoring and management of energy consumption and emissions
Modal shift to low carbon modes	Mobility as a Service techniques to inform travelers on choices
Electrification	Charging infrastructure design and location
Increase climate resilience	Micro-level climate resilience planning and information delivery to people and organizations.

9.3.3.4 Electrification

The deployment of electric vehicle charging infrastructure and the use of electric vehicles in place of internal combustion engine vehicles is also an important element of smart mobility. Advanced technology can be used to determine the most appropriate locations for electric vehicle charging infrastructure and the most appropriate charger characteristics. Drivers can be informed of the availability of electric vehicle charging infrastructure using smartphone apps and in-vehicle information technology. The transition from internal combustion to electrically propelled vehicles can be significantly supported by the use of smart mobility techniques for freight, transit, and private cars, in turn supporting a move to zero emissions from vehicles. It could be argued that electric vehicles simply shift the emissions from the roadside to power generation plants. Therefore, it is important to consider the entire supply chain for energy and ensure that clean energy techniques are incorporated. However, even the use of coal-fired generation techniques can be an improvement as emissions are generated from a central location which is more manageable than diffuse emissions from a large number of vehicles.

9.3.3.5 Increase Climate Resilience

As it is impossible to reduce climate risk to zero, while still maintaining economic activity and quality of life, it is necessary to take opportunities to better manage these risks [4]. Smart mobility technologies can be used to understand the risk of climate change and provide support for people and organizations to do their part to improve the situation. Better management and control of transportation processes can also help toward climate resilience, especially if this is combined with suitable land use planning strategies. Smart mobility technology applications can also support microscopic planning for climate resilience as well as microscopic management of transportation modes and services.

Table 9.3 summarizes smart mobility opportunities related to climate change and offers some suggestions to address them.

9.3.4 Equity Management

9.3.4.1 Reduce Transportation Costs

Reducing transportation cost should be part of the goals of smart mobility efficiency techniques. In addition to reducing overall transportation costs, consideration should be given to the cost of transportation as a proportion of household income. Here again, an accessibility index can be used

Table 9.4 Equity management opportunities for smart mobility.

Opportunity	Possible smart mobility solution
Reduce transportation costs	Use of smart mobility techniques to improve operational efficiency of transportation
Racial equity and barriers to opportunity	Big data and analytics to expose inequity and support design of more effective solutions

to measure progress. The accessibility index can include cost of travel for various trips as a proportion of household income.

9.3.4.2 Racial Equity and Barriers to Opportunity

Using big data analytics to discover inequity in the transportation system and define the barriers to opportunity can enable improvement in these areas. There is a lot of discussion about a "market segment of one" approach [5] in marketing enabled by big data and analytics. This same technique can be applied to transportation and mobility to ensure that the opportunity is taken to customize services to the needs of a wide range of travelers, improving equity, and removing barriers to opportunity.

Table 9.4 summarizes smart mobility opportunities related to equity management and offers some suggestions to address them.

9.3.5 Economic Development

9.3.5.1 Economic Competitiveness and Opportunity

There is a growing trend to address economic competitiveness and opportunity-related objectives within smart mobility plans. This can include the use of smart mobility and accessibility to improve access to employment opportunities. It can also include the implementation of a smart mobility system to improve the quality of life in an urban area to the extent that it becomes more attractive for commerce and industry. The opportunity also includes that for workforce development, in which a deliberate attempt can be made to add new skills and new capabilities to the workforce that can support smart mobility in the future. When designing a smart mobility implementation, the opportunity for job creation for different demographics of the population can also be addressed.

9.3.5.2 Development of Brownfield Sites

From a transportation perspective, the brownfield site is property that may have been used in the past for land use that resulted in pollution or contamination of the area. Examples include former gas stations, dry cleaners, or manufacturing facilities that make use of hazardous materials [6]. Incorporating such sites into the implementation of smart mobility could include the development of transportation management centers on the site or the construction of a mobility hub.

9.3.5.3 Support Domestic Manufacturing

Smart mobility can employ emerging and new technologies to achieve its objectives. This can also provide an opportunity to stimulate and foster new businesses that can deliver the new solutions required. When considering a smart mobility implementation, the possibility that small businesses and new industries can be fostered and stimulated by the implementation should be fully considered.

Table 9.5 Economic development opportunities for smart mobility.

Opportunity	Possible smart mobility solution
Economic competitiveness and opportunity	Use of smart mobility techniques to improve attractiveness of city to business and industry, create jobs, and improve accessibility to job opportunities
Development of brownfield sites	Smart mobility management centers and mobility hubs
Support domestic manufacturing	Incorporating emerging technologies into smart mobility implementations

Table 9.5 summarizes smart mobility opportunities related to economic development and offers some suggestions to address them.

9.3.6 User Experience

9.3.6.1 Reducing Travel Cost

The operational efficiency of the transportation delivery process can be improved by the application of smart mobility technologies. This in turn provides an opportunity to reduce travel costs and provide a greater range of travel options. As discussed earlier, it is important to analyze the proportion of household income spent on travel for the different demographics in the urban area.

9.3.6.2 Improving Travel Time Reliability

Research has shown that travelers place a high value on travel time reliability, especially for certain trip purposes [7]. That is an opportunity for smart mobility to improve travel time reliability by improving the connectivity between services and supporting better operational management of transportation service delivery accounting for changes in operating conditions and demand for transportation.

9.3.6.3 Mode Choice and Awareness of Choices

Smart mobility provides an opportunity to offer new modal choices and also increases the awareness of the choices available. New choices such as automated first mile, last mile shuttles, and micromobility services can provide travelers with an alternative service that can be matched to the trip needs. Information services such as Mobility as a Service can also be used to ensure that travelers are fully aware of the choices and make well-informed decisions based on current service availability.

9.3.7 Electronic Payment and Reservation

There is also an additional opportunity to make it convenient for travelers to take advantage of the weight range of services that can be delivered by smart mobility. Electronic payment and reservation services can be used to make it convenient for the traveler to make a cashless payment for a single mode of transportation and also for a complete multimodal trip chain. This can be combined with a central reservation system to make it easier for travelers to take the best decisions for both spontaneous and planned trips.

Table 9.6 summarizes smart mobility opportunities related to user experience and offers some suggestions to address them.

Table 9.6 User experience opportunities for smart mobility.

Opportunity	Possible smart mobility solution
Reducing travel cost	Using smart mobility to increase operational efficiency of transportation service delivery, use of electronic payment systems to reduce the cost of collecting fees
Improving travel time reliability	Improving connectivity between services and improving operational reliability of services
More choices and awareness of choices	First mile, last mile automated shuttles, micro-mobility, and Mobility as a Service
Electronic payment and reservation	Cashless payment and single central reservation system

9.3.8 Partnership

9.3.8.1 Public–Private Partnerships

Smart mobility offers a range of opportunities for public–private partnerships. Emerging technologies and the possibilities of new services attract the private sector to invest in smart mobility. There is a great opportunity for the public sector to understand needs, issues, problems, and objectives and then align these with private sector motivation to develop and deliver new services. Smart mobility techniques can also be employed to measure the effectiveness of such partnerships and the outcomes that the new services are delivering. This also includes a new opportunity to take full advantage of private-sector innovation, to address public-sector transportation objectives, while also building new businesses and new industries.

9.3.8.2 Public–Public Partnerships

In addition to partnerships between the public and private sectors there is an opportunity for smart mobility to foster partnerships between public agencies. This can include data sharing related to big data and analytics and also to the use of information systems to coordinate the activities of different agencies. There is also an opportunity to create a multimodal management system that shares information across the different agencies responsible for different modes of transportation within the city.

Table 9.7 summarizes smart mobility opportunities related to partnership and offers some suggestions to address them.

9.3.9 Alignment of Objectives

9.3.9.1 Align Stakeholder Objectives

The data and information-sharing possibilities afforded by smart mobility can be used to ensure that stakeholder objectives are aligned across the region. There is an opportunity to improve

Table 9.7 Partnership opportunities for smart mobility.

Opportunity	Possible smart mobility solution
Public–private partnerships	Engage the private sector in smart mobility initiative planning to ensure coordination of efforts
Public–public partnerships	Make use of smart mobility information and knowledge transfer capabilities to align public sector and align public sector objectives

Table 9.8 Alignment of objectives opportunities for smart mobility.

Opportunity	Possible smart mobility solution
Align stakeholder objectives	Smart mobility data and information sharing to align stakeholder objectives and support cooperation
Coordinate investment plans	Use of Artificial Intelligence and Machine Learning to optimize investment plans and coordinate across agencies
Political and operational alignment	Use of smart mobility dual-tier decision support systems to align political and operational activities
Agree expected outcomes	Use of big data analytics to define outcomes to support agreement

cooperation and coordination across stakeholders, including the alignment of public and private sector objectives and the coordination of investment plans. This is also an opportunity to communicate objectives across stakeholders as a means of aligning objectives, using information delivery techniques.

9.3.9.2 Coordinate Investment Plans
In addition to the coordination of objectives, smart mobility techniques can be applied to improve the coordination of investment plans. For example, the application of Artificial Intelligence and Machine Learning to project bundling can result in more effective investment plans. These can be coordinated both within an agency and across agencies.

9.3.9.3 Political and Operational Alignment
Smart mobility techniques can also be applied to ensure that nontechnical decision-making is completely coordinated with technical decision-making at the planning and operational level within the city. Smart mobility can provide effective decision support for both levels and ensure that decisions are coordinated between levels.

9.3.9.4 Agree Expected Outcomes
Big data and analytics related to smart mobility can quantify outcomes and form the basis for an agreement on the outcomes expected across the city. This provides an opportunity to define expected outcomes, discuss them, and seek agreement across the various agencies. There is also a valuable opportunity to define expected outcomes and share them with the private sector.

Table 9.8 summarizes smart mobility opportunities related to alignment of objectives and offers some suggestions to address them.

9.3.10 Adaptability

9.3.10.1 Dynamic Services
While it might be difficult or expensive to adjust transportation infrastructure to changes in demand and operating conditions, the services incorporated into smart mobility techniques can support a higher degree of adaptability. The use of smart mobility sensing and telecommunications techniques, combined with big data and analytics can provide the basis for changes in services to match conditions and demand. This can also enable a greater degree of flexibility for operating patterns and services.

Table 9.9   Adaptability opportunities for smart mobility.

Opportunity	Possible smart mobility solution
Dynamic services	Smart mobility data collection and information capabilities used to configure services to conditions and demand
Dynamic pricing	Use of artificial intelligence and machine learning to optimize investment plans and coordinate across agencies
Incident and event response	Improved incident and response activities based on the use of smart mobility data collection and information processing

9.3.10.2   Dynamic Pricing

The use of electronic payment systems can be used to support the application of a dynamic pricing strategy for transportation services in the urban area. There is an opportunity to vary pricing according to demand and operating conditions. This is also an opportunity to support demand management strategies such as congestion avoidance reduction techniques.

9.3.10.3   Incident and Event Response

There is an opportunity to improve the management of incident and event responses through the application of smart mobility techniques. This can result in reduced response latency, raising the overall efficiency of the process and increasing the effectiveness of the operation.

Table 9.9 summarizes smart mobility opportunities related to adaptability and offers some suggestions to address them.

9.3.11   Monitorability

9.3.11.1   Threat Detection

The opportunity for threat detection relates to the collection of data from advanced sensors and the conversion of data into information that provides insight into operational changes that could pose a threat to the transportation system. Threats can include weather conditions, unusually high demand for transportation, service disruptions, or system failures.

9.3.11.2   Opportunity Detection

Similarly, smart mobility techniques can be used to detect opportunities related to the transportation network. This could be an opportunity to provide a new service, or an opportunity to optimize service efficiency, recognizing variations in demand for transportation and operating conditions.

Table 9.10 summarizes smart mobility opportunities related to monitorability and offers some suggestions to address them.

Table 9.10   Monitorability opportunities for smart mobility.

Opportunity	Possible smart mobility solution
Threat detection	Smart mobility data collection and information capabilities used to detect threats and develop responses
Opportunity detection	Use of artificial intelligence and machine learning to identify opportunities and develop plans to take opportunities

9.4 Smart Mobility Challenges

Based on the authors interaction with smart mobility practitioners over the course of many projects, a broad range of challenges are presented by smart mobility implementations. These are summarized in Figure 9.2 and described in the following sections.

9.4.1 Communications

9.4.1.1 Communication to Politicians

Communication to nontechnical decision-makers can become a challenge in a smart mobility implementation. One of the factors is the difference in terminology and language between technical, planning, and operational staff and nontechnical decision-maker politicians. Technical staff tend to make extensive use of key performance indicators to define and measure performance. At the nontechnical, political level that is more focused on effects and outcomes. Therefore, there is a need to bridge the gap by providing a dual-tier decision support process that use technical key performance indicators for technical staff and outcomes for nontechnical decision-makers.

9.4.1.2 Communication to End Users

End-user needs are quite similar to those of nontechnical decision-makers, in that communication should be focused on expected outcomes and a value to the end user. For communication to end users, it is important to paint a picture of what life will be like after the implementation in terms of improved services and travel conditions. Bearing in mind that end users are also typically voters, who are of great interest to the nontechnical decision-makers.

Table 9.11 summarizes smart mobility challenges related to communications and offers some suggestions to address them.

9.4.2 Data Sharing

9.4.2.1 Data Silos and Data Cockpits

Experience suggests that many transportation agencies operate in silos. These can be modal silos or geographical silos. In some cases, certain job functions have created data cockpits in which data and tools have been brought together to support a specific job function or group of job functions. While this is helpful to the specific job function, it does not address the need for more extensive data sharing and accessibility and others to support effective smart mobility implementations.

9.4.2.2 Uncoordinated Data Collection

Uncoordinated data collection can also lead to data gaps and duplication of data collection. In some cases, the data is not effectively converted to information from which insight can be extracted leading to more effective response plans. The lesson has been learned over and over again that data in itself has little value, other than serving as a raw material for information and insight.

9.4.2.3 Ineffective Data Sharing

Ineffective data sharing can occur within an organization and between organizations in a smart mobility implementation. Ineffective data sharing within an organization can led to the situation where data is collected many times and either used once or not at all. Data sharing across organizations can be hampered by the lack of data sharing agreements but also by the lack of data flow even

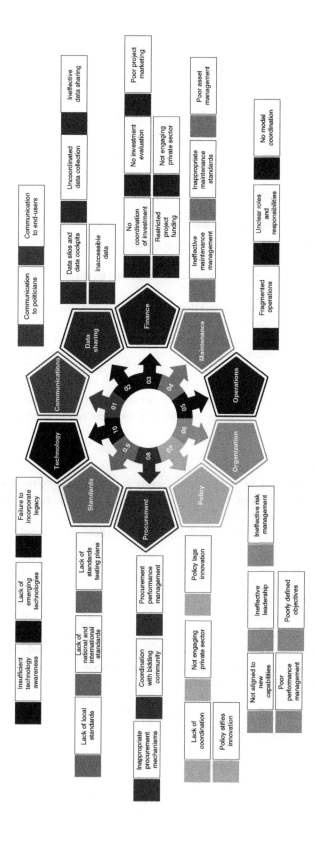

Figure 9.2 Smart mobility challenges.

Table 9.11 Communication challenges for smart mobility.

Challenge	Possible smart mobility solution
Communication to politicians	Advanced decision support featuring a focus on expected outcomes and effects
Communication to end users	Advanced decision support featuring a focus on expected outcomes and specific value to customer groups. Definition of a vision that paints a picture of life after smart mobility implementation for the user

Table 9.12 Data challenges for smart mobility.

Challenge	Possible smart mobility solution
Data silos and data cockpits	Enterprisewide data management system that supports a unified data set, accessible data, and advanced analytics
Uncoordinated data collection	Results-driven data collection planning based on the needs to fill data gaps for analytics
Ineffective data sharing	Data management system that makes data available to all appropriate stakeholders and fully supports data-sharing agreements
Inaccessible data	Data management system that supports full accessibility to data and analytics for all stakeholders

if agreements are in place. Ineffective data sharing can also result due to a lack of recognition of the availability of private sector data sources, especially those related to location intelligence and probe vehicle data.

9.4.2.4 Inaccessible Data

In some cases, agencies collect data that is not accessible due to resource constraints. This is also a symptom of ineffective data sharing. In many cases, a poorly planned technical architecture for data processing results in data inaccessibility.

Table 9.12 summarizes smart mobility challenges related to data and offers some suggestions to address them.

9.4.3 Finance

9.4.3.1 No Coordination of Investment

Integration of silos or cockpits across modes or geographical areas are not restricted to organization and operation but can also affect coordination of investment. While good coordination can be achieved within an agency, often coordination between agencies is less effective, leading to suboptimization of investments. In the worst cases, one investment can at least partially counter the effects of another investment. For example, the use of dynamic pricing to induce a modal shift from private car to public transit along the corridor could lead to a negation of the effects of investments in public transit vehicles due to a large increase in demand.

9.4.3.2 No Effective Investment Evaluation

Often agencies will adopt an index linked approach to budgets, where the previous year's budget is used as the basis for the next year's budget plus an index for cost increases. This avoids having to conduct an effective investment evaluation in terms of return and investment, benefits achieved, and outcomes delivered.

9.4.3.3   Poor Project Marketing

One of the first challenges under this heading is that often projects are the focus for marketing, instead of the services that are supported. It is the services that deliver value and hence provide the best marketing material. Marketing can also be ineffective due to an inability to clearly state the expected outcomes and values for the intended audience. This can involve a lack of definition of the key messages to be projected in the marketing activities and poor choice of marketing information delivery channels.

9.4.3.4   Restricted Project Funding

This can be considered as a symptom of the project focused marketing as discussed above. Inability to clearly define values and benefits for the project can make it difficult for it to compete against other priorities. These other priorities could be related to transportation or could be related to health, social services, or education. To maximize availability of funding, world-class project marketing and service marketing are essential.

9.4.3.5   Not Effectively Engaging the Private Sector

This has been discussed under public-private partnerships above; however, it is worth reiterating that poor engagement of the private sector can have a significant effect on the availability of project and program financing. At worst, private sector investment is duplicated by the public sector, leading to poor stewardship of taxpayer money and ineffective focusing of resources. An example of this is the advanced traveler information system program in the United States during the 1990s. The public sector duplicated some value chain components that could have been delivered by the private sector. This had a dual effect. Funding for effective data collection was restricted, while publicly subsidized competition was created for information delivery channels such as websites. A value chain analysis and a clear understanding of public and private roles could have achieved a much better effect.

Table 9.13 summarizes smart mobility challenges related to finance and offers some suggestions to address them.

Table 9.13   Finance challenges for smart mobility.

Challenge	Possible smart mobility solution
No coordination of investment	Coordinated investment program based on smart mobility planning and architecture
No effective investment evaluation	Use of cost-benefit analysis techniques within a smart mobility framework to evaluate the effectiveness of investments
Poor project marketing	Well-defined evaluation of project values, benefits, and expected outcomes
Restricted project funding	Well-defined evaluation of project values, benefits, and expected outcomes
Not effectively engaging the private sector	Sharing public sector objectives with the private sector in developing a plan for alignment

Table 9.14 Maintenance challenges for smart mobility.

Challenge	Possible smart mobility solution
Ineffective maintenance management	Development and implementation of an effective smart mobility maintenance management system that defines device location and maintenance needs
Inappropriate maintenance standards	Agreement of city or regionwide maintenance standards taking account of the needs of different devices and infrastructure
Poor asset management	Development and implementation of a total asset management system that addresses physical and cyber asset management, including access control

9.4.4 Maintenance

9.4.4.1 Ineffective Maintenance Management

Many public sector transportation agencies are most familiar with the maintenance requirements of asphalt, concrete, and steel. While there has been extensive experience in the application of sensor technologies in roadside infrastructure, there is still a tendency to apply a one-size-fits-all solution to maintenance. It is essential to establish a maintenance management system that takes account of the maintenance needs of individual devices and infrastructure.

9.4.4.2 Inappropriate Maintenance Standards

In a similar manner, there is a tendency to apply one single maintenance standard across all maintenance activities. This is not appropriate when the needs of various devices vary and are different from the maintenance requirements of infrastructure. It is vital that appropriate maintenance standards are defined and applied.

9.4.4.3 Poor Asset Management

Asset management is one of the areas in smart mobility that delivers the lowest value in terms of direct benefits. However, the indirect benefits of asset management are significant. In many cases, agencies have little or no asset management; in some situations, they have multiple asset management systems that are uncoordinated. There are also situations where physical asset management does not match up to the cyber asset management provided by the asset management system. An example of this was experienced with a client who had a very sophisticated cycber asset management system but managed to lose a server when somebody repositioned it without authority. Cyber and physical asset management must both be strong and effective.

Table 9.14 summarizes smart mobility challenges related to maintenance and offers some suggestions to address them.

9.4.5 Operations

9.4.5.1 Fragmented Operations

This could be more kindly described as focused operations in that current transportation in cities tends to be focused on specific modal groups. The exception to this are the large global cities which tend to have a single unitary authority for transportation service delivery. Examples of this include London [8] and Seoul [9]. More typically, in small and medium-sized cities, there is a specialization on specific activities from a planning, design, operations, and maintenance perspective. This can make it challenging to apply smart mobility techniques that are designed for sharing and designed to support application across modes.

Table 9.15 Operations challenges for smart mobility.

Fragmented operations	Use of an advanced decision support system to unify operations with a single unified data set and analytics
Unclear roles and responsibilities	Development of a detailed concept of operations for the smart mobility implementation
Poor modal coordination	Implementation of a smart mobility solution capable of supporting modal cooperation by addressing the needs of each stakeholder

9.4.5.2 Unclear Roles and Responsibilities

Fragmented operations can also make it difficult to define specific roles and responsibilities for a city or areawide smart mobility application. It is important to define roles and responsibilities across the transportation service delivery spectrum including planning, design, project delivery, operations, and maintenance. In many cases, the inability to clearly define roles and responsibilities leads to duplication of activities and activity gaps.

9.4.5.3 No Modal Coordination

This is also related to fragmented operations described above. Agencies can have an extreme focus on the specific mode that they are managing, making it difficult to achieve modal cooperation and coordination across the city or region. This lack of cooperation can vary from splendid isolation to organizational conflict.

Table 9.15 summarizes smart mobility challenges related to operations and offers some suggestions to address them.

9.4.6 Organization

9.4.6.1 Organization Not Aligned to New Capabilities

Many transportation agencies are historically aligned and organized to be effective at managing conventional transportation. The new smart mobility requirements associated with data collection, data management, and information processing can present a challenge to such organizations. In a perfect world, organization would adapt to the optimal technical solution. However, the authors' experience is that in many cases the technical solution is suboptimized to accommodate an existing organization. It is very important to align the organization to achieve the full technical capabilities of smart mobility.

9.4.6.2 Ineffective Leadership

This sounds like a rather harsh description. It is a short way of saying that leadership for smart mobility requires additional skills and capability beyond the conventional transportation approach. Additional skills required include partnership management, influencing, advocacy, and high-level communication skills. This is in addition to the existing technical and managerial requirements for conventional transportation. It is almost always the case that the skill set required cannot be fulfilled by a single individual and may be better performed by a small team. This allows the allocation of technical, business, and organizational aspects to individuals.

9.4.6.3 Ineffective Risk Management

There are some very sophisticated applications of advanced technology in transportation that have little or no risk management elements. When dealing with data, privacy, and advanced technology,

Table 9.16 Organizational challenges for smart mobility.

Challenge	Possible smart mobility solution
Organization not aligned to new capabilities	An organizational assessment conducted as part of the development of the technology aspects of the smart mobility system
Ineffective leadership	Change management and definition of leadership requirements for smart mobility initiative
Ineffective risk management	Development and agreement of a risk management plan
Poor performance management	Making use of the data collection and information processing capabilities of smart mobility to define and implement a performance management process
Poorly defined objectives	Extensive use of prior experience to develop expected outcomes, defined objectives, and agreed as objectives

it is essential to have a thought-out risk management plan in advance of implementation. At a minimum, the risk management plan should address potential risks, the likelihood of them being realized, a risk avoidance strategy, or a risk mitigation strategy, for each defined risk.

9.4.6.4 Poor Performance Management

It is the authors' experience that many agencies focus on performance measurement to the detriment of management. It is important for performance management to include the full spectrum of required activities from data collection, data management, information processing, extraction of insight from information, and the development of performance management plans. This process should also include monitoring and management arrangements to ensure that expected outcomes are achieved. It is expected that a good performance management system will address formative and summative performance management. Formative performance management is near real-time and used to guide operations and implementations. Summative performance management is used to present results to the end user and nontechnical decision-makers regarding prior performance and plans for performance in the next period.

9.4.6.5 Poorly Defined Objectives

This might seem like a trivial issue but there are many situations where an implementation agency is asked why they supported the implementation the answer is somewhat along the lines of "it seemed like a good idea at the time." It is important not just to define objectives but also to agree them across city or regional stakeholders. A simple analogy is that it is important not to press the accelerator in the car until you know where you are going.

Table 9.16 summarizes smart mobility challenges related to organization and offers some suggestions to address them.

9.4.7 Policy

9.4.7.1 Lack of Policy Coordination

Smart mobility covers a wide range of subjects which are often dealt with separately from a policy perspective. Policy as it relates to smart mobility can be fragmented due to the lack of coordination. For example, policy related to advanced air mobility in urban areas also needs to take account of land use planning and the needs of the surface transportation system.

Table 9.17 Policy challenges for smart mobility.

Challenge	Possible smart mobility solution
Lack of policy coordination	Use of the information sharing and effects evaluation possibilities of smart mobility to support better policy cooperation based on an understanding of technology capabilities
Not engaging the private sector	Use of information sharing and effects evaluation to engage the private sector in an effective discussion to align their activity with public sector objectives
Policy lags behind innovation	Careful management of policy definition to ensure that it keeps up with innovation and emerging technologies, balancing the needs of enforcement and compliance with the need to stimulate new businesses and industries
Policy stifles innovation	Use of information sharing and effective valuation of smart mobility to inform policy on technology capabilities and assist in the balance between compliance and fostering industry

9.4.7.2   Not Engaging the Private Sector
In many cases, policy is defined in isolation from a dialogue with the private sector. It can be argued that policy should lead to innovation, but in many cases, innovation has a direct influence on policy. Good policy will take full account of technological capabilities that can often be obtained from the private sector.

9.4.7.3   Policy Lags Behind Innovation
In many cases, policy lags behind innovation. Some good examples of this can be found in the micromobility area where extensive fleets of electric scooters or bicycles have been deployed in advance of policy to regulate such fleets. This has led to undesirable side effects such as excessive speed of micromobility vehicles in inappropriate zones and indiscriminate parking of the vehicles. When policy lags behind innovation, it can also create barriers to new and emerging technology applications.

9.4.7.4   Policy Stifles Innovation
As discussed above, policy lagging behind innovation can cause innovation to be stifled. Unless a balanced approach to policy can be adopted in which the needs of compliance and enforcement are balanced against the need for the encouragement of new businesses, policy can also stall innovation.

Table 9.17 summarizes smart mobility challenges related to policy and offers some suggestions to address them.

9.4.8   Procurement

9.4.8.1   Inappropriate Procurement Mechanisms
Procurement in public transportation agencies can take a long time. Protocols and business process rules have to be obeyed and often procurement processes designed for conventional transportation projects are applied to smart mobility. It is important to build a degree of interaction with the bidding community and to find ways in which to accelerate the procurement process.

Sometimes procurement processes are applied that are inappropriate for emerging and new technologies. Procurement documents are often prescriptive and designed around her specific solution when in fact technology has moved on and new solutions are available that are not

Table 9.18 Procurement challenges for smart mobility.

Challenge	Possible smart mobility solution
Inappropriate procurement mechanisms	Comprehensive scan of prior deployments to ensure that the appropriate procurement mechanisms are chosen and that the speed and flexibility of the procurement mechanism are appropriate to the pace of emerging technologies
Poor coordination with the bidding community	The use of information-sharing techniques to support effective coordination between the procurement agencies in the bidding community. This should also include an assessment of the suitability of the bidding community for the solutions to be procured
Lack of procurement performance management	Definition and application of a comprehensive performance management system for the procurement process. This should include the definition of Key Performance Indicators and the agreement of expected outcomes from procurement

addressed in the procurement documents in many cases this can be avoided by adopting service definitions in procurement documents rather than prescriptive "how-to's."

9.4.8.2 Poor Coordination with the Bidding Community

The public sector agencies are indisputable experts in what is required. Similarly, the private sector solution providers are in the best position to define technology capabilities and possibilities. Unless an effective dialogue is supported between the procurement and the bidding community, these resources and sets of expertise can be missed.

9.4.8.3 Lack of Procurement Performance Management

In the same way that performance management techniques are applied to smart mobility services, it is important to apply performance management to procurement. In many cases, there is not a clearly defined set of key performance indicators or performance management process for procurement. This can lead to delays in procurement and ambiguity in procurement documents. The application of performance management to procurement will also help to ensure that the most appropriate procurement mechanism is chosen for the job in hand.

Table 9.18 summarizes smart mobility challenges related to procurement and offers some suggestions to address them.

9.4.9 Standards

9.4.9.1 Lack of Local Standards

Local standards can be applied to the development of system architecture for smart mobility, data communications, business processes, rules, and practices. While some standards must to be national or international to to achieve economies of scale, there is considerable scope for local standardization as most smart mobility initiatives are highly customized to the locale. Appropriate application of predefined local standards will help to ensure that the smart mobility implementation is relevant, coherent and takes full advantage of opportunities for connectivity and coherence. It also helps to avoid the adoption of proprietary techniques which can cause vendor lock-in and suboptimize the total cost of the project.

Table 9.19 Standards challenges for smart mobility.

Challenge	Possible smart mobility solution
Lack of local standards	Definition of local standards in line with policy and expected outcomes
Lack of national and international standards	Definition of a catalog of national and international standards and incorporation of these into system architecture and plans for smart mobility initiative
Lack of standards testing plans and protocols	Definition of standards testing plans and protocols for those standards applicable to the smart mobility initiative

9.4.9.2 Lack of National and International Standards

At the next level, national and international standards are required and other to ensure that best practices and lessons learned are incorporated into the smart mobility initiative. The application of national and international standards also supports larger market sizes which in turn helps to manage the cost of solutions. National and international standards are particularly relevant to in-vehicle applications where millions of vehicles are manufactured around the globe and delivered to both national and international standards.

9.4.9.3 Lack of Standards Testing Plans and Protocols

The authors have witnessed situations where standards have been developed without the accompanying testing plans and protocols. A standard without these elements makes it difficult to ensure compliance with the standard. These elements also help to shed further light on the nature of the standard and how it should be applied to real conditions.

Table 9.19 summarizes smart mobility challenges related to standards and offers some suggestions to address them.

9.4.10 Technology

9.4.10.1 Insufficient Technology Awareness

As stated earlier, it can be assumed that those in the public and private sector who represent the client for the smart mobility solution are experts in "what" is required. This does not mean that they are also experts in "how" to do it. This expertise is typically found in the private sector among solution providers and prior implementers. These groups have substantial practical experience and lessons learned in the application of smart mobility technologies. The old saying "you don't know what you don't know" is relevant here. It is important that the smart mobility initiative is built on a foundation of knowledge regarding the capabilities and limitations of the technologies being deployed. This can be achieved by technology scans and establishing dialogue with private sector solution providers and prior implementers around the globe. This also extends to a knowledge of the business models that can be applied to get the best from the technology. For example, in the London Congestion Charging Scheme, the decision was made to use nonintrusive license plate recognition techniques for congestion charging, rather than equip every vehicle with a transponder [10]. This avoided significant costs in equipping vehicles with transponders but presented a challenge in terms of the overall accuracy of the license plate recognition system. Even with a large UK license plate, the average accuracy achieved by license plate recognition sensors is of the order of 90% [11]. Transport for London adopted a business model based on recognizing vehicle presence inside the zone and not when the vehicle crossed the cordon to enter or leave the zone. This enables the license plate recognition sensors inside the zone to have

Table 9.20 Technology challenges for smart mobility.

Challenge	Possible smart mobility solution
Insufficient technology awareness	Complete scan of technology capabilities and constraints based on prior global implementation experience
Lack of emerging technology incorporation	Adoption of flexible procurement approaches based on service definition accommodating innovation and emerging technology
Failure to incorporate legacy systems	Thorough assessment of the starting point for the smart mobility initiative including characterization of legacy investments and definition of plans to incorporate them into the smart mobility initiative

multiple opportunities to read the license plate, raising the accuracy through cumulative probability. The business model was chosen for good alignment with technology capabilities and constraints.

9.4.10.2 Lack of Emerging Technology Incorporation

This can happen for a number of reasons. One important reason is a slow or rigid procurement process that fails to keep pace with technological emergence. Another could be the adoption of prescriptive procurement documents rather than the definition of service requirements. This challenge can also emerge when there is a lack of information regarding current technological capabilities and constraints.

9.4.10.3 Failure to Incorporate Legacy Investments

This challenge represents a serious issue. It is highly unlikely that the smart mobility initiative will be implemented in a greenfield site. This is likely to be considerable prior investment in transportation infrastructure and smart mobility. When determining the value and benefits of smart mobility, it is important to recognize the value of prior investment and where possible incorporate legacy investments into the new proposals. Failure to do this sends a negative message that smart mobility initiatives have short-term value and little or no long-term sustainable value.

9.4.10.4 Failure to Incorporate Lessons Learned and Practical Experiences

It is highly unlikely that the technologies being deployed in a smart mobility initiative have not been previously deployed at some location around the world, for transportation or for an allied business sector. Incorporation of lessons learned, and practical experiences is fundamental to ensure the success of smart mobility technology deployment. It is not possible to gain a thorough understanding of technological capabilities and constraints without this.

Table 9.20 summarizes smart mobility challenges related to technology and offers some suggestions to address them.

9.5 Summary

This chapter presents an overview of the opportunities and challenges that have been encountered related to the implementation of smart mobility initiatives. It is expected that this will be used as the basis for customized assessment of smart mobility initiative opportunities and challenges for specific cities and regions. There is considerable variation in the needs and characteristics of individual cities and regions around the globe; therefore, it is not possible to create a customized catalog of opportunities and challenges in this book. However, a great deal of the raw material that would be required to develop such a customized approach has been provided. It is important to bear in mind when identifying and defining opportunities and challenges related to a specific smart mobility initiative that the appropriate range of stakeholders and end users have been consulted to draw on a wide range of experience. The development of such a opportunities and challenges catalog also provides a solid foundation for the development of a risk assessment approach for a smart mobility initiative.

References

1 https://www.who.int/news-room/fact-sheets/detail/road-traffic-injuries, retrieved June 17, 2022 at 6:00 PM Central European Time.

2 García-Altés, A.; Pérez, K. *The economic cost of road traffic crashes in urban setting.* National Library of Medicine, USA, National Center for Biotechnology Information. http://www.ncbi.nlm. nih.gov/pmc/articles/PMC2610566/, retrieved June 17, 2022, at 7:13 PM Central European Time.

3 Vision Zero Network. https://visionzeronetwork.org/about/what-is-vision-zero/, retrieved June 18, 2022, at 12:45 PM CET.

4 World Bank. The adaptation principles: six ways to build resilience to climate change. https:// www.worldbank.org/en/news/feature/2020/11/17/the-adaptation-principles-6-ways-to-build-resilience-to-climate-change, retrieved June 18, 2022, at 2 PM Central European Time.

5 Luxus Worldwide. The segment of one, Bryan Dollery, September 15, 2016. https://www. luxusworldwide.com/blog/the-segment-of-one#:~:text=The%20Segment%20of%20One%20refers individual%20according%20to%20their%20behaviors, retrieved June 18, 2022, at 2:06 PM Central European Time.

6 City of Portland Oregon. Environmental services. What is a brownfield? https://www.portlandoregon. gov/bes/article/316713, retrieved June 18, 2022, at 2:23 PM Central European Time.

7 Carrion, C.; Levinson, D. *Value of travel time reliability: a review of current evidence.* University of Minnesota. file:///C:/Users/bob/Downloads/VORReview%20(1).pdf, retrieved June 18, 2022, at 2:28 PM Central European Time.

8 Transport for London. https://tfl.gov.uk/, retrieved June 18, 2022, at 5:39 PM Central European Time.

9 Seoul Metropolitan Government. https://english.seoul.go.kr/, retrieved June 18, 2022, at 5:40 PM Central European Time.

10 Mayor of London. London assembly. Automated number plate recognition units in London. www.london.gov.uk/questions/2005/0661, retrieved June 19, 2020, at 11:30 AM Central European Time.

11 BBC News. Tuesday, 11 February 2003, 1157, GMT. http://news.bbc.co.uk/1/hi/uk/2748319.stm, retrieved June 19, 2022, at 11:56 AM Central European Time.

10

A Framework for Smart Mobility Success

10.1 Informational Objectives

After reading this chapter, you should be able to:

- Define how to address challenges
- Effectively pursue the values and benefits
- Understand how to use smart mobility planning
- Utilize technology applications
- Take smart mobility opportunities and navigate the challenges
- Define an effective approach to data management
- Understand interactions and how to manage them
- Develop an effective approach to communicating objectives, policies, and outcomes
- Define an effective approach for public and private sector engagement
- Explain how to synchronize initiatives across multiple modes of transportation
- Develop, agreed, and apply strategies and tactics
- Define and agree technical, organizational, and commercial arrangements

10.2 Introduction

Mobility is a complex system consisting of many moving parts and multiple stakeholders. This is a difficult challenge, but it is one that should be addressed, in spite of the difficulties, to improve the chances of success. Several concepts and knowledge sets from the other chapters in the book present at least a high-level view of a potential toolbox that can be applied to maximize the chances of success in smart mobility initiatives. A number of interactions that have a significant impact on smart mobility click success will be defined and identified and the overall interaction between these mechanisms will be explored in the course of this chapter. The essence of this chapter is an explanation of how to harness the knowledge and information delivered in earlier chapters to develop a successful approach to smart mobility. This involves agreement and understanding of the problems to be addressed by the smart mobility implementation. Information on this was covered in detail in Chapter 3, Problem Statement. Since the primary objective is to pursue the values and benefits defined in Chapter 4, these serve as an essential input to the definition of objectives and desired outcomes. In Chapter 6, an accelerator approach was defined for the creation of a

Smart Mobility: Using Technology to Improve Transportation in Smart Cities, First Edition. Bob McQueen, Ammar Safi, and Shafia Alkheyaili.

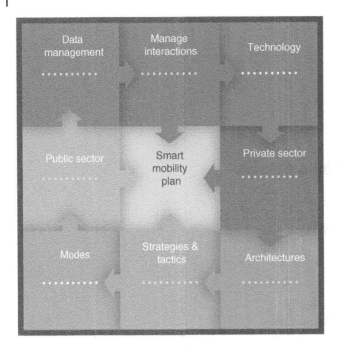

Figure 10.1 The smart mobility success toolbox.

smart mobility plan. This chapter provides complementary information and how to make best use of that plan. A successful approach requires effective application of the technologies described in Chapter 8 and committed pursuit of the opportunities, while navigating the challenges, both described in Chapter 9. Based on the experience of the authors, there are a number of significant strategies that can be applied to raise the probability of success. These have been combined into a toolbox that combines the knowledge and information from these previous chapters and defines the key success factors. The toolbox is a coherent approach to these essential smart mobility success factors. Figure 10.1 illustrates the smart mobility success toolbox. This is not a complete list of all activities that could be supported in developing smart mobility. It represents the most important success factor for smart mobility based on the authors' experience.

10.3 The Smart Mobility Success Toolbox

The toolbox includes effective data management, definition, and understanding of a range of interactions and the definition and application of an effective approach to communications. Full utilization of private sector motivation and resources is also an important feature in the successful approach to smart mobility, running in parallel with the right approach to public sector and stakeholder involvement. Any successful approach to smart mobility will feature effective management of all modes of transportation within a city or region. The toolbox also includes development, agreement, and application of strategies and tactics, complemented by frameworks or architectures that define technical, organizational, and commercial arrangements. The objective of the toolbox is to maximize the probability of success. In this case, success is defined as being able to take advantage of as many of the opportunities presented as possible and either avoid or mitigate the challenges and undesirable side effects.

The following sections provide a discussion on each of the elements of smart mobility success toolbox that explain each element and define its importance.

10.3.1 Data Management

A description of the technology applications that support data management was provided in Chapter 8 on technology applications. It is necessary to take a step beyond the definition and description of technology applications that support data management to discuss how they will be applied for smart mobility success. In the course of several recent projects, the authors have worked with both public and private sector entities on data management approaches. While the objective in most cases is to develop a complete dataset that supports all smart mobility modes across the city or region and is accessible to all relevant stakeholders and users, it is also possible to take a transitional strategy approach to this. A unified dataset can be created that will fully support a targeted number of smart city mobility services mobility and this can be scaled up and developed over time to support all smart city services. Figure 10.2 provides an illustration of a unified dataset. This one has been designed to support the application of artificial intelligence and machine learning to next-generation traffic management. The dataset is used to feed advanced models that can identify traffic phenomena and define appropriate responses. It also enables the effects of the responses of unassociated systems to be identified and evaluated. The decision support system that uses the unified dataset supports a sophisticated approach to adaptive learning that enables the system to deliver immediate benefits "out of the box," while continually improving performance over time. This avoids some of the disadvantages of simulation modeling approaches where an annual recalibration and reconfiguration would be required and consume significant resources. The adaptability approach enables calibration and configuration to happen as the system operates, avoiding these lengthy and expensive exercises.

This shows the incorporation of legacy data sources and new data sources. It is assumed that legacy data sources include existing sensors and roadside infrastructure designed to capture data on prevailing transportation conditions and variations in the demand for transportation. New data sources include the provision of additional privately operated sensors, privately sourced connected

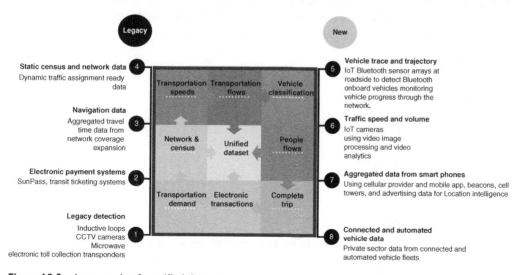

Figure 10.2 An example of a unified dataset.

and automated vehicle data, and location intelligence data from smartphones. These are all defined in detail in Chapter 8. In the case of this example, the unified dataset was designed to support the application of artificial intelligence and machine learning to advanced traffic management and performance management within a city. The same technique could be used to develop a unified dataset for specific services such as mobility as a service, climate change management, or multimodal coordinated management. The unified dataset is a central component in the technical architecture which will be addressed later in this chapter.

Successful data management will efficiently support the process of collecting data, converting it to information, extracting the maximum insight and understanding from the information, and developing appropriate strategies in response plans based on the new insight and understanding. There are many different ways to configure a data management system to address this, but this process could be used to create a checklist to guide the development of the data management system.

10.3.2 Manage Interactions

Interaction is usually defined as a discussion or conversation between two or more people. However, interaction goes a lot deeper than simple exchange of information and involves ownership transfer as well. The ownership transfer can cause influence and actions can influence and be influenced by other actions. When two or more things encounter each other, the product can have a new effect resulting from such an interaction. Interaction can be defined as the particular way in which one thing affects another. Using this definition, then information and knowledge transfer is only one component of interaction.

This seems like a simple thing to do but it is usually the most difficult thing to achieve in the implementation of smart mobility systems. Transportation and mobility systems have grown over time and are extremely efficient because of specialization. Transit specialists focus on transit, freeway management specialists focus on freeway management, and traffic signal specialists focus on traffic signals. In addition to this functional segregation, there is also geographical segregation between cities, counties, states, and national-level transportation agencies. This makes interaction a complex issue. It is exceedingly difficult to identify and define every single type of interaction in smart mobility; however, the most significant have been identified and defined. These are summarized in the following list. The types of interaction are defined and described and how these interactions affect and influence smart mobility success are discussed below.

- People interactions
- Investment interactions
- Technology interactions
- Objectives interactions
- Department interactions
- Modal interactions
- Levels of government interactions
- Public and private sector interactions

10.3.3 People Interactions

This could be the strongest interaction in terms of the influence it has on smart mobility. Smart mobility is all about people and the effects of smart mobility are intended to impact traveler behavior, travel conditions, and improve the quality of life for people in urban areas. In addition to

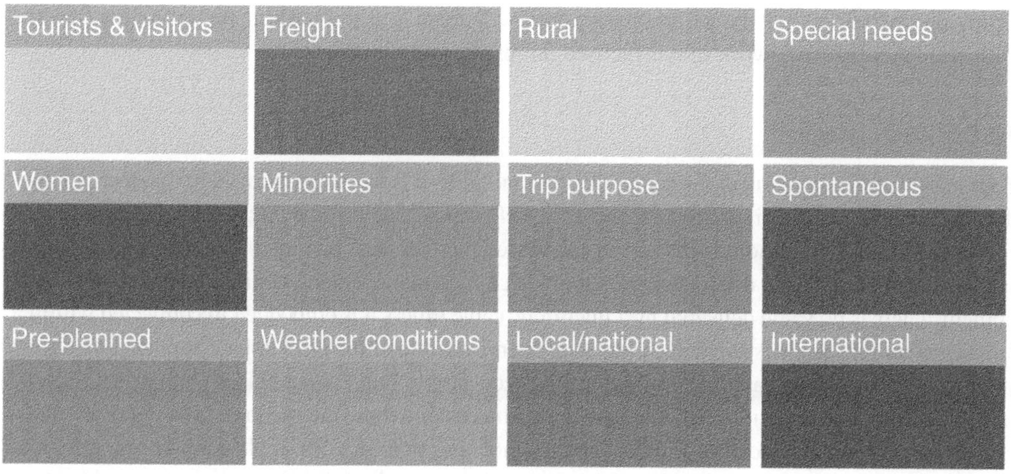

Figure 10.3 Customer analysis framework.

people interactions at the end user level, people interactions are also significant in smart mobility in terms of organizational interaction. After all, organizations are groups of people. In effect, discussing people interactions also means discussing interactions between organizations. In many cases, launching a smart mobility initiative leads to some groups of people being in the same room for the first time. Therefore, it is necessary to support a degree of socialization and understanding each other before moving on to discussing the technical, organizational, and commercial arrangements required for smart mobility. One of the most important objectives in smart mobility is connectivity and this includes connections between systems and people.

It is also important to recognize that the people who use the transportation system are not a single homogeneous group. They have different needs and characteristics, different reasons for travel, and different challenges related to the locale in which to live. The thought of customer understanding analysis needs to be performed and other to ensure that a smart mobility implementation is not designed with an inherent bias toward a particular customer group. Figure 10.3 provides a framework or outline for the customer understanding analysis. This can be used as a starting point for a customized customer analysis based on a specific smart mobility implementation.

10.3.4 Investments

Investments in smart mobility typically manifest themselves as programs and projects, with programs being a collection of projects. It is fairly obvious that one investment in smart mobility such as a mobility as a service initiative, could also have an effect on a separate investment in advanced traffic management. The interaction between investments is an important area for understanding as a detailed understanding of the interaction and consequences of one investment on the other can have a significant impact on improving and optimizing investments and stewarding the money expended.

Considerable effort is applied to attempt to optimize investments in smart mobility. These are typically done manually by applying experience and expertise. Recently initiatives have emerged using artificial intelligence and machine learning to optimize spending programs even more. Whether the optimization is conducted using human and manual methods, or artificial intelligence and computers, it is still essential to understand the interactions between investments to attain optimization. Ideally, prior investments will form the foundation for later investments within a coherent framework that is focused on results and outcomes.

10.3.5 Technology

Technology applications are at the center of smart mobility. It is worth restating here that technologies do not deliver value. As discussed in Chapter 8, technologies are bundled together to form products, services, and solutions, and it is the services delivered by these that provide the outcomes in terms of safety, efficiency, enhance user experience, climate change management, equity, and economic development. There is no value delivered by a fiber-optic cable unless it is embedded in connected within a larger system that delivers services. It has long been understood that technology applications have a significant influence on each other and that it is necessary to build a technology framework to show how the technical applications will fit together and understand these interactions. In many cases, interactions are supported by telecommunications technologies that link one application or one system to another. The technological implication of this interaction is usually defined using a system architecture or a framework explaining how the pieces fit together. The system architecture approach can be even more powerful when, in addition to the technology framework, an organizational and a business model or commercial framework is also added. The business model or commercial framework identifies the underlying business models and commercial arrangements that support the operation of the technology applications. The organizational architecture illustrates the organizational arrangements required to support the technology application. It also shows how stakeholders will interact.

Stakeholders are a specific group of people who must be involved in order for a smart mobility initiative to be successful or those who will be impacted by the effects of the smart mobility implementation. They are known as stakeholders because they are considered to have a stake in the success of the smart mobility initiative stakeholders can come from the public or the private sector, from academia and from the traveler, or the end user group. This group could be viewed as a subset of the people group described above. The difference is that stakeholders, in addition to being end users of the system, are also critical to the successful planning, implementation, and operation of the smart mobility system. Stakeholders are required to understand their roles and responsibilities and assume a degree of ownership in the proposed smart mobility implementation.

Another important aspect of technology lies in what the author's refer to as "cycles of change" – the differing life cycles and pace at which smart mobility elements are replaced.

Effective solution frameworks to raise the probability of success in smart mobility implementations must also take account of cycles of change. These are the natural life cycles for some of the major elements of smart mobility; these include the following list. Each element is described in the following section.

- Smartphones
- In-vehicle equipment
- Bridge infrastructure
- Road infrastructure
- Transit vehicles

10.3.5.1 Smartphones

Almost 98% of the global population has a communication device in the form of a cell phone or a smartphone. Experience assuring that the average lifespan of such devices is approximately 33 months as new models are introduced and new features added [1].

10.3.5.2 In-vehicle Equipment

In-vehicle technologies can be expected to last for the life of the vehicle. However, this could be distorted by retrofitting activities of new technology to all vehicles; however, a conservative estimate can be defined by assuming that in-vehicle technology will last the life of the vehicle. For private cars, the estimated life of the vehicle within the overall global fleet is about 12 years [2]. The average age of a private car in Europe is between 8 and 35 years with a mean of 18 years in Western Europe and 28 years in Eastern Europe, average lifespans of cars vary from 8.0 to 35.1 years, with a mean of 18.1 years in Western and 28.4 years in Eastern European countries [3].

10.3.5.3 Bridges

The average age of a bridge in the United States is approximately 43 years with an intended lifespan of 50 years [4]. A study in Japan reveals an estimated lifespan of concrete bridges to be between 35 and 54 years [5]. It can be assumed that European figures are close to these numbers and that an average lifespan of a bridge around the globe is approximately 50 years. This is accurate enough for the purposes of this section of the book.

10.3.5.4 Roads

In Australia, a study initiated or estimated the average annual lifespan of roads within a specific county to be 39 years [6]. In the United States, the number is between 20 and 30 years [7].

10.3.5.5 Transit Vehicles

One transit agency in the United States, the Massachusetts Bay Transportation Authority (MBTA) in Boston, Massachusetts, adheres to a general standard life cycle of 35 years for rapid transit and light rail vehicles, 25 years for commuter rail locomotives, 25–30 years for commuter rail coaches, and 15 years for buses [8].

Figure 10.4 compares the various life cycles for smart mobility elements and the quantities of each element in use around the world.

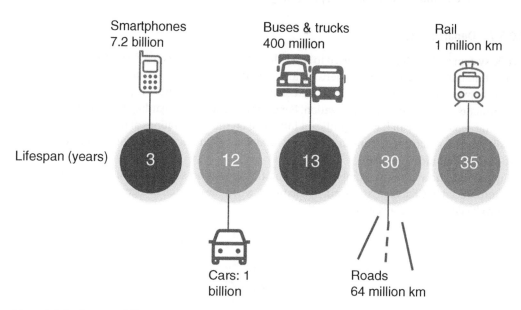

Figure 10.4 Smart mobility elements life cycles and quantities.

There are some interesting points to note. The dominant numbers that smartphones represent with more than 90% of the global population possessing one and a short life cycle of three years mean that this is an excellent channel for delivering information to travelers and non-travelers. Smartphones are particularly interesting as an information delivery channel as the cost of the device itself is borne by the individual and not by the public agency. Smart mobility information can merely be another application that smartphones can address. The long-life cycle times of infrastructure such as roads and rail make it more difficult to integrate emerging technologies. In these situations, it is necessary to adopt a retrofit strategy rather than introduce new technology integrated with new infrastructure. In-vehicle technology also has a challenge of a long-life cycle compared to smartphones at 12 years, making retrofit an important element for this also. Long life cycles for smart mobility infrastructure imply more rigidity and less flexibility than smartphones and in-vehicle applications. In-vehicle applications are also at a disadvantage to smartphone applications because of the shorter life cycle for these devices. Due to concern for human factors in safety, there is a tendency toward in-vehicle integrated systems; however, it must be recognized that smartphones represent the path of least resistance when it comes to introducing new technology and new applications.

10.3.6 Objectives

The objectives that have been defined for the smart mobility implementation are typically not standalone independent objectives but have a relationship or an influence on one another. A smart mobility implementation typically has multiple objectives centered around improving safety, increasing efficiency, and enhancing end-user expedience. Objectives can also include improving equity, managing the impact of climate change, and economic development or job creation. In many cases, these objectives are not mutually exclusive and can either reinforce each other or work against each other. For example, the simplest and most effective way to make a road the safest road in the world to be simply to close it. However, this would not recognize the need to also achieve objective searches efficiency and end-user expedience. A close road would indeed be the safest road in the world because there would be no traffic on it, preventing the achievement of the other objectives.

10.3.7 Departments

While system architecture can be used to show the relationship between different organizations in a smart mobility implementation, there are also departments within each organization. The interaction between the organizations must be considered and so too must the interaction between departments within a specific organization. For example, decisions regarding land use made in the planning department can have a significant influence on traffic operations at the operational level. Ideally, the full spectrum of transportation service delivery activities would be linked together in a two-way communication chain.

Figure 10.5 shows the five components of the transportation service delivery chain. This was first introduced in Chapter 1 and again in Chapter 6 when life cycle challenges were discussed and is reproduced here for convenience.

Figure 10.5 Transportation service delivery activity spectrum.

From	Planning	Design	Delivery	Operations	Maintenance
Planning	Business process improvement that learns lessons from previous projects and feeds and forward into new projects	More effective communication of planning objectives can help to ensure that designs are effective	More effective communication of planning objectives can help to ensure that Project delivery is more effective	More effective communication of planning objectives can help to ensure that operations achieve the original outcomes expected	More effective communication of planning objectives can help to ensure that systems achieve the outcomes expected
Design	Any design difficulties caused by the planning process can be reported to the planning department for future improvement	Business process improvement that identifies lessons from previous projects and incorporates them into new projects	More effective communication of planning objectives can help to ensure that the design process is more effective	Communication of the original design parameters can help to ensure operations are as effective as possible	Communication of the original design parameters can help to ensure that maintenance is as effective as possible
Delivery	Any project delivery difficulties caused by the planning process can be reported to the planning department for future improvement	Suggestions for improving designs the ease project delivery can be communicated to the design department	Business process improvement that identifies lessons from previous projects and incorporates them into new projects	Communication from deliveries operations can ensure a smooth handover and optimize operational management	Information from delivery to maintenance can ensure that maintenance takes full account of the characteristics of the system delivered
Operations	Any operational difficulties caused by the planning process can be reported to the planning department for future improvement	Suggestions for improving the effectiveness of operations can be communicated to the design department	Effective communication between operations and delivery can support a smooth handover of the systems and also ensure that the needs for effective operations are completely addressed	Business process improvement that identifies lessons from previous and current operations and incorporates them into future operations future operations	Communications between operations and maintenance can ensure both operational management and maintenance management effectiveness
Maintenance	Any maintenance difficulties caused by the planning process can be reported to the planning department for future improvement	Suggestions for improving the efficiency and reducing the cost of maintenance can be communicated to the design department	Any characteristics of delivery process that effects maintenance can be communicated and corrected	Maintenance strategies impacts on operational management can be discussed and managed	Business process improvement that identifies lessons from previous and current maintenance activities and incorporates them into future maintenance activities

Figure 10.6 Interactions between departments.

Departments within the transportation service delivery organization can be organized on these functional lines or on geographical lines. Effective support for the interaction between the departments will have a significant influence on the success of the smart mobility implementation. For example, the planning department can benefit from feedback received from operations and maintenance. Similarly, the design department can make practical improvements to infrastructure designs based on feedback and lessons learned from delivery. A full range of examples of how one department can benefit from interaction with another department is summarized in Figure 10.6, which represents a matrix of interactions. Note the interactions within departments are also described in this figure. Processes within departments can also be improved through more effective intra-department interactions.

Business process mapping techniques can be extremely valuable in defining and agreeing on the activities to be supported on the work products that could be exchanged. In many cases, this type of technique can reveal redundancy and parallel activity, offering opportunities for improvement in addition to the identification of opportunities for sharing information [9].

The business process mapping activity can also be a useful tool in bringing different departments together, understanding common ground, identifying workflows, and data requirements. A good approach would be to combine business process mapping with the creation of a data catalog for each department which would provide an excellent basis for the development and implementation of an advanced data management system. Advanced tools are available for business process mapping and for the creation of data catalogs, making use of artificial intelligence to support the process [10].

10.3.8 Levels of Government

Public government agencies tend to be organized in a hierarchical manner. These include national, state, and local government agencies. At the top of the hierarchy of their nontechnical decision-making agencies such as the governor, or the mayor's office. Typically, top-down instructions will

TECHNOLOGY IS EMERGING FASTER THE DISTANCE BETWEEN INVENTION   PRIVATE SECTOR MOBILITY   TECHNOLOGY IS OUTPACING POLICY
                                AND COMMERCIALIZATION HAS   CAPABILITY HAS INCREASED IN LEAPS
                                 CONSIDERABLY SHORTENED   AND BOUNDS

Figure 10.7 Technology driving forces.

come from the nontechnical decision-makers to the technical staff responsible for implementing the top-down instructions or policies. This is one direction of influence within levels of government policymaking and the communication of policy objectives from the top of the government to the operational levels. That is also a flow in the opposite direction that can also be described as interaction. The technical levels of government must show the nontechnical decision-makers that progress is being made and that money is being spent wisely in line with policy objectives. This requires that a link be established between outcomes-based policies and technical key performance indicator (KPI)-driven plans, actions and policy impacts.

Another important influencing factor lies in existing policy and the introduction of new policy. It can be assumed that policy is both the driving force and a primary tool that nontechnical decision-makers use for surrogate investments and actions. There is a need for policy to keep pace with emerging technology. As illustrated in Figure 10.7, technology is emerging faster and the distance between initial invention and commercialization of technology has shortened considerably. Private sector interest in smart mobility and its resources and capabilities has increased in leaps and bounds over the past few years. There is a significant danger that these driving forces will combine to enable technology to outpace policy definition.

Policies are also used to communicate with the public, to the people electing the nontechnical decision-makers, and what the chosen course of action is and what the expected outcomes are. Policies are expected to regulate things to ensure the attainment of desirable outcomes while avoiding or managing undesirable side effects. Policy should also be designed to achieve multiple objectives including safety, efficiency, enhanced user experience, and economic impact such as job creation and industry attractiveness of the city.

One way to approach the complexity of the system within which smart mobility sits is to develop a basic understanding of the decision ecosystem within which decisions are made. Figure 10.8 is an attempt to capture the essential characteristics of such an ecosystem.

Note that this represents a model not a recipe. It is impossible to design an approach that accommodates all context, all situations, all existing legacy investment patterns, and all the objectives. Instead, a model is offered that can be adapted and adjusted to suit specific circumstances and situations. Note also that the model is not intended to represent the typical current situation with respect to organization for decision support for smart mobility. This is a suggested new approach based on the needs and influencing factors that have been described earlier in this chapter.

Describing each of the components in the figure one by one, there are five primary components, the network manager, non-technical decision-makers, Sentinel, public, and service delivery partners. As these are aggregates or bundles, it is necessary to explain the characteristics of each bundle.

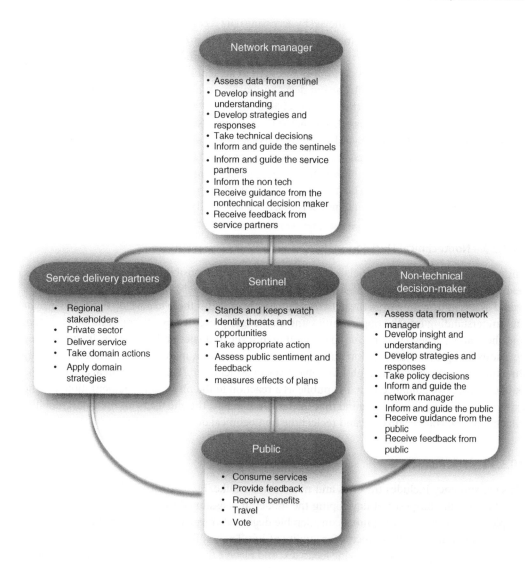

Figure 10.8 Dual level decision-support ecosystem.

10.3.9 Network Manager

This is a role that is not currently being used in any transportation or smart mobility system that has been examined to date. The authors have discussed the concept with many public and private sector participants over the course of several years which has validated the approach, but not led to the implementation of the technique. The role of the network manager is to manage all aspects of smart mobility within a city or even a region. This includes the management of multiple modes, the ingestion of data from multiple data sources and external systems, and the conversion of data to information from which insight and understanding are extracted leading to informed and robust decision-making. The network manager is primarily concerned with the technical performance of the smart mobility system. This is achieved by accessing data received from the centerline using this as the basis for informed decision-making through the application of big data and advanced analytics techniques. The network manager has the role of taking technical decisions and then issuing a range of advice or instructions to a number of different stakeholders. The network manager also has the job of informing the nontechnical decision-makers on how the defined policies are being applied. This also includes how the outcomes-based high-level objectives set by the nontechnical decision-makers are being addressed. The network manager also has the responsibility to engage the Sentinel or sentinels, inform, and guide the service partners as well as provide information to the nontechnical decision-makers. The network manager is also expected to receive feedback from service partners concerning the effects of network management strategies and service partners' objectives and planned actions. The model assumes that the various parties involved will swim like a school of fish. This organizational model is described later in this chapter.

10.3.10 Nontechnical Decision-makers

These are typically elected officials who have political responsibility for gating the network manager through policy and high-level objective statements. Non-technical decision-makers would assess information from the network manager in order for them to develop their own insight and understanding into the opportunities and challenges presented by the smart mobility system and the context. They are expected to develop high-level strategies and responses, include policy decisions and provide information and guidance to the network manager on the intended policy direction to be taken and the high-level objectives to be achieved. This group will also engage the public and receive guidance from the public as a reaction to the planned implementation and the expected outcomes. Bearing in mind that the traveling public, the end users for the smart mobility system are also voters who elect and reelect the nontechnical decision-makers.

10.3.11 The Public

This constituency includes travelers and non-travelers alike. It could also include the young and old allow for the purposes of developing the model; it is assumed that the public will consist of people of voting age who require a considerable degree of travel and depend on the efficiency of the traveling transportation process for quality of life. The public is expected to consume the services generated by the model, provide feedback, and receive the benefits derived from the services. This group is expected to travel or benefit from higher efficiency in travel and transportation. For example, the cost of groceries in the grocery store is related to the cost of transporting the groceries from the source to the store. For example, the largest grocery chain in the United States, Walmart operates 163 distribution centers around the United States and makes use of 7900 drivers to operate a fleet of 8400 trucks and 64,000 trailers. The trucks are driven approximately 700 million miles per year. The scale of the transportation operations makes it extremely sensitive to the efficiency and effectiveness of transportation and mobility having a direct effect on the prices in the stores and hence the quality of life and cost of living for people that do not travel [11, 12].

10.3.12 The Sentinel

A typical traffic or transportation management center that can be found around the world at the moment, the function of the Sentinel is defined in this model and is incorporated along with the network manager into a single traffic and transportation management unit, department, or organizational structure. The network manager typically does not have full responsibility for all modes of transportation and does not have the tools and decision-support necessary to coordinate multiple modes within the city. To make the workload manageable and make a clear differentiation between specific job functions, the network manager has been separated from the data collection and observational needs of smart mobility through the creation of the Sentinel. The primary function of the Sentinel as the name suggests is to keep watch for variations in the demand for transportation and operating conditions being experienced. The Sentinel will also observe the plans and actions for service delivery partners and the travel behavior of the public. The Sentinel is responsible for identifying threats, opportunities, anomalies, and significant changes within the transportation network. Decentralization is responsible for implementing strategies generated by the network manager and measuring the effectiveness of the strategies, reporting this back to the network manager also.

Another reason for separating the Sentinel from the network manager is to recognize that in addition to data collection, the Sentinel has an important role in the management of public–private partnerships and public–public partnerships between the network manager and other agencies responsible for service delivery within the city.

10.3.13 Service Delivery Partners

The service delivery partners group has a number of different elements. The first element of the other public agencies responsible for service delivery of the various transportation modes within the city. This can include transit agencies, cities, counties, and anyone responsible for developing and delivering transportation services. It is vital that these transportation services are coordinated and that the objectives developed and implemented by each agency are aligned to maximize reinforcement and minimize negation. Service delivery partners also include a range of private sector organizations who are responsible for vehicle design, manufacturing sales, micromobility service delivery, electronic payment service delivery, and a wide range of other services that impact effective and efficiency of smart mobility within the city. It is assumed that the network manager's organization has no direct authority over the service delivery partners either the public or the private sector once and that the network manager must feature a higher level of skills and influence, advocacy, communications, and motivation to ensure that objectives are aligned, and that success is achieved. Service delivery partners therefore include regional stakeholders as well as private sector organizations developing and delivering products that impact the quality of smart mobility. These organizations are expected to implement their own policies based on their own objectives while understanding the weather impacts of their actions on other actions being formulated and applied in the city.

The concept of operations for the decision-support ecosystem defined has number of different aspects. The first is the use of formative KPIs.

The network manager is expected to make use of formative KPIs to measure the progress of the strategy implementation. These formative KPIs are provided in near real time to enable incremental course adjustments to be made. The process of collecting data, turning it into information, and measuring the formative KPIs must be fast enough to provide immediate support to the network manager.

The second of the use of summative outcomes measurement. The network manager will also develop summative outcomes measures that explain the progress that has been made toward the achievement of the high-level objectives in the policy direction provided by the non-technical decision-makers. While there will be a direct relationship between the technical KPIs used by the network manager and the outcomes measures, the outcomes measures will be rolled up to provide high-level effects analysis support.

The technical KPIs will be used to support the internal processes within the network manager organization and will consist of a separate set of KPIs for each mode of transportation. It is expected that artificial intelligence and machine learning will be applied to the measurement of these KPIs and to the definition of new KPIs as the behavior of the smart mobility system as a network is understood and as the need for new KPIs is revealed. It is also expected that the KPIs will be used within a framework that supports two levels of learning. The first level of learning is related to the acquisition of knowledge regarding the configuration of the network at any given time, whereas the second level of learning relates developing a detailed understanding of the network response to various strategies and changes in operating conditions at any given time. These are illustrated in Figure 10.9.

The third is the use of performance and operational management.

It may seem that performance management and operational management are the same thing; however, there are important differences. Operational management is all about running the system and managing immediate actions and effects. Performance management relates to assessing the effectiveness of the operations by measuring the outcomes of the operational strategies and changes endued in the network. From an organizational perspective, it may be useful to separate the staffing groups that are performing the operational management from those conducting the performance management to ensure an unbiased and independent assessment of the performance of the operations. It should also be noted that performance assessment is learning driven, while operational management is execution driven. Both functions will have different operational objectives and key performance indicators. However, it could also be argued that both of these jobs should be conducted by a single unit as the data required for operations and performance assessment is similar. It could be argued that effective operations integrate both functions and does not need a third party. Under any operational strategy performance management and operational management should be treated as a binary pair that are very closely related and depend on each other for success. If the choice is made to separate the two functions, then there is a possibility that the performance assessment role could be addressed by a private sector entity under contract to the public sector.

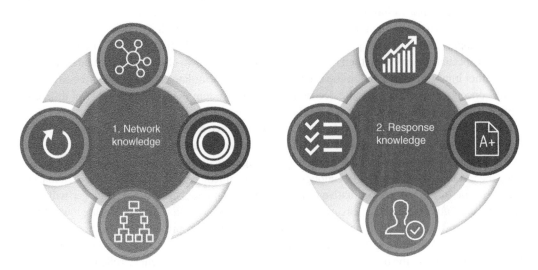

Figure 10.9 Learning cycles.

Effective operational management needs an awareness of tactical key performance indicators and also an awareness of enterprise-wide targets. Enterprise targets could also be described as citywide targets and will include technical performance of the transportation network and progress toward the achievement of political objectives that can also include equity, economic development, and climate change.

The fourth revolves around the learning aspects of performance management. Both performance management and operational management require a complete picture and understanding of the network being managed. The completeness of the picture needs to be assessed and verified. For example, many travelers assume that they know best and should choose the route, time, and even modes of transportation by themselves; however, most drivers have an incomplete picture of the total network operation and the total picture regarding operating conditions and to the demand for transportation at any given time. This complete picture of the network operates in response is central to the success of a smart mobility implementation. Any toolbox approach to smart mobility should have this capability edit center.

The fifth lies in the need for higher-level advice and control. The model assumes that there is no centralized control or authority over all city mobility operations consequently some messages from the network manager will be instructional in nature, while someone will be advisory, and some may even be for consulted purposes only. For example, in the event of an emergency situation, messages from the network manager may be instructional and more like commands than advice or guidance. However, as the service partners have responsibility over their own operations, some messages will be of an advisory nature only. They will simply advise service partners on appropriate action, leaving it to the service partner to decide if action is required. This points to the need for effective communications if the messages from the network manager at to be rigorously evaluated and to be put into action. Another reason for the separation of the network manager in the Sentinel roles is that the network manager must maintain a strategic approach. Sometimes the detailed data regarding network operation and response can constrict the effective strategic approach.

Nontechnical decision-makers and advisors have a crucial need to understand the link between technical actions and political objectives and policies. Here again, effective communications between and from the network manager are vital for this to operate effectively. As discussed earlier, the limited lifetime of a political position also requires effective communications to ensure political stability and continuing support for the smart mobility implementation. As positions change, new politicians are elected and other depart. It is essential to mix short-term projects with short-term results with longer-term projects that require a longer period of time to bear fruit. Multigenerational projects can suffer from instability and political support. As it is unavoidable to implement that progress is being made and success has been achieved. The value and benefits of these projects must be communicated extremely effectively to incumbent politicians, those emerging as political leaders, as well as the general public.

Effective communication also mean that as well as being documented as successful with KPIs and outcome measurement, there should be a perception among nontechnical decision-makers that success is being achieved. It is important not to let a gap develop between the reality of the progress being made and the perceived progress. Being seen to be effective is just as important as being effective in this case.

As noted earlier, it is important to have a direct relationship between key performance indicators or technical level, tactics, and strategies that are driven by political objectives, plans, and policies. This is illustrated in Figure 10.10. This also provides some examples of Key Performance Indicators, strategies, tactics, and outcomes. This represents a bridge between technical performance and political objectives.

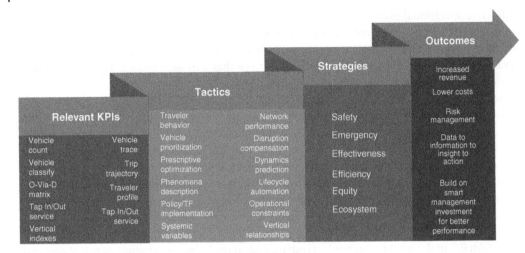

Figure 10.10 The bridge from key performance indicators to outcomes.

10.3.14 Public Sector and Private Sector

The public sector has already been addressed in this section under people and organizational interactions. However, interaction also takes place between the public and the private sector. Historically investment in and direction of smart mobility initiatives has been driven by the public sector. However, over the past 10 or 15 years, the emphasis has switched from public-sector invest-ment to private-sector investment in smart mobility technologies, especially connected and auto-mated vehicles. The more successful approaches to smart mobility would recognize the interaction between the public sector and the private sector in terms of the investments being made, the tech-nology applications and the overall objectives being pursued. These interactions can take the form of policies that have been set by the public sector that the private sector must adhere to our new capabilities that have been generated by the private sector through private sector investment that the public sector could take advantage of. A good example of private sector initiatives and leader-ship lies in micromobility. Around the world private sector companies have emerged that support the planning, management, and operation of micromobility services. These can include electric scooters, electric bicycles, and other electric vehicles that play a significant role in the so-called first mile last mile element of a total transportation system. It is like that the private sector will also play an important role in the mobility as a service application required to provide information to travelers regarding the different modes of travel available and how to create the trip chain using multiple modes. Also, automated vehicle applications and the private sector have taken a consider-able leadership role. For example, in the United States alone, there are at least 1400 field trials where automated vehicles are being evaluated on the open road [13]. These are being driven by and fueled by the private sector.

A simple model of the required interaction between the public and the private sector in estab-lishing a successful partnership is shown in Figure 10.11.

The public sector starts by developing a working understanding of current technological capa-bilities and constraints. This can be based on discussions with private sector solution providers and/or a scan of prior deployments, seeking lessons learned, and practical experience from other public sector agencies. With this knowledge, the public sector would then engage the private sector in discussions about shooting resources and aligning objectives so the public sector objectives may

Figure 10.11 Public–private partnership approach.

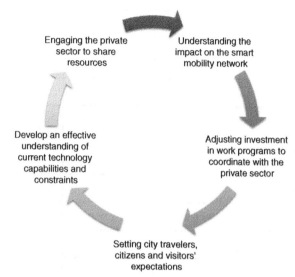

be addressed by private sector resources. This discussion would also include measures that will be taken to build the market for the private sector in return for cooperation.

The public sector together with the private sector must then understand the impact, the outcomes, and the effects of the proposed partnership initiative on the smart mobility network. It will also be necessary for the public sector to adjust currently planned investments and what programs to coordinate with the private sector initiative. This goes hand-in-hand with actions to set city travelers, citizens, and visitors' expectations with regard to the new smart mobility implementation. This would include information on both costs and benefits.

10.3.15 Modes of Transportation

Modes of transportation can also interact and influence each other. A list of smart mobility modes was provided in Chapter 3 when the smart mobility problem statement was addressed. Figure 10.12 is reproduced here for convenience.

An important aspect of smart mobility is the ability to support multimodal management across all modes of transportation in the city. It is important as thorough job is done in identifying the appropriate transportation stakeholders, understanding their role within city and regional transportation, and bringing them to the table to agree on the main planning and design elements of the proposed smart mobility solution. This will also facilitate the incorporation of legacy and sunk investment as each agency is likely to have read considerable prior investment in elements within their management control. It is important to understand how improvements in one board have an effect on another mode as this is a key consideration in effective multimodal management. Figure 10.13 provides a simplified view of a multimodal trip chain from original origin to ultimate destination.

The illustration assumes that a complete trip consists of a "first mile," a "middle mile," and a "last mile." The entire trip from origin to destination could be made using just a single mode of transportation, such as walking if the distances, weather conditions, and trip purpose are appropriate for that mode. The trip could also be made through a multimodal combination of an automated first-mile shuttle from origin to the first hub, commuter rail for the middle mile an eScooter or

Modes of transportation used in cities	Subcategories			
Aerial tramways	Cable cars	Monorails		
Animal powered	Horses	Carts		
Automated vehicles	Automated cars	Automated shuttles	Automated buses	Automated trucks
Bicycle				
Boats				
Buses				
Cable transport	Cable cars			
Cars				
Ferries				
Flying	Urban air mobility	Helicopters		
Freight	Urban logistics	Trucks		
Funicular railways				
Inland waterways				
Micro-mobility	Electric scooters	Electric bicycles		
Multimodal trip chains	Can encompass all the other modes			
Pipelines				
Ride sharing				
Subways				
Trains				
Walking				

Figure 10.12 Transportation modes.

other micromobility vehicle for the last mile. It becomes obvious that a crucial aspect of effective multimodal service is the careful management of interchange between modes at the hubs.

When considering investments and improvements, it is also important to take this complete trip, multimodal perspective for example, smart mobility could increase the attractiveness of public transit, reducing the use of private automobile travel. This is an objective of many smart mobility implementations, as dependence on the private vehicle may well be one of the most important factors in efficiency

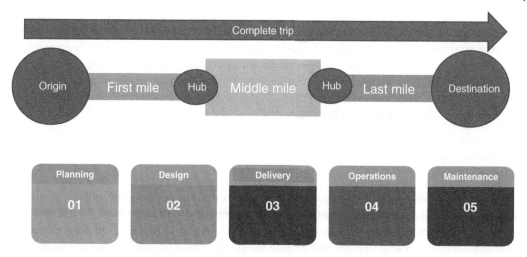

Figure 10.13 Complete trip chain simplified view.

and climate change. While the different modes of transportation tend to have their own smart mobility management system, it is clear that real success in urban smart mobility relies on the capability to have multimodal management approaches. These approaches need to take account of the demand for transportation, the capacity of each mode of transportation within the city, and the temporal variations of demand caused by factors such as weather conditions and special events.

10.3.16 Strategies and Tactics

Strategies are another area where there is interaction that must be accounted for within a successful smart mobility initiative. For example, an operational strategy that attempts to create a modal shift in favor of public transit could be adversely affected by another operational strategy that makes private car use more attractive, such as better signal timings, or express lanes. When deciding on the optimal operational strategy, it is necessary to take a big-picture view and consider how the operational strategies will interact with each other to create either reinforcing order and negating affect. Operational strategies from the private sector should also be considered by the public sector as part of this activity.

If all levels of decision-making, from nontechnical decision-makers to technical staff, are to be effectively addressed it is also important to consider the political life cycle. The generation of political positions varies around the globe and on the position held, but it can be assumed the average life cycle for a politician can be between two and eight years. This is important as projects that require implementation periods greater than the political life cycle may suffer from instability in policy objectives and a lack of sustainability in political support.

Because of the variability of demand for mobility and variations in the operating conditions, it is necessary to integrate high-level resilience and adaptability into any implementation. This is particularly the case as the underlying infrastructure which is often the subject of improvement is very rigid and has a long-term life cycle. The application of smart mobility technologies and services can add a degree of resilience and adaptability to the rigidity of the long-term infrastructure and investments that it represents.

Resilience and adaptability require a high degree of capability in understanding threats, opportunities, changes in the demand for transportation, and changes in operating conditions. In fact, these are essential characteristics of a smart mobility implementation.

Mobility services are also delivered within the wider context of the economy. Changes in the economy which caused the loss or creation of jobs, for example, can have a second impact on the demand for transportation. The recent Covid-19 pandemic has also shown that other factors which affect the economy can also affect the demand for transportation and the operating conditions in which transportation is expected to operate efficiently and effectively. Strategies must be developed that take account of the wider context of the economy and the need for resilience and agility. It is of value to consider strategies in at least three distinct categories:

- Transportation
- Citywide operations
- Informational

Transportation strategies are not only designed to optimize the operation of individual modes such as private car, transit, and freight within the city but also address the need for multimodal cooperation and coordination of the strategies across the total mobility services landscape. Citywide operations take this a stage further and include the consideration of mobility, water, and energy services within a city as an integrated operation. Informational strategies are designed to ensure the quality of the information produced by the smart mobility initiative and ensure that the right information goes to the right user at the right time. Informational strategies also address the needs of system operation and ease of access to data and analytics. Strategies can be further detailed into operational tactics for transportation management center operations. While a higher-level strategy can be linked to political goals and objectives.

10.3.17 Architectures

One of the authors collaborated on a seminal book on the subject of ITS architectures in the early 1990s [14]. As it deals with so-called evergreen principles, it is still a relevant text that provides useful information on how to develop and use a smart mobility architecture. The most well-known example of a national architecture for smart mobility is the US National Intelligent Transportation System architecture [14]. The current version of the architecture is described from the following perspectives.

- Enterprise
- Functional
- Physical
- Communications

The enterprise view defines the roles and responsibilities for each organization within the overall smart mobility initiative and explains the relationships between them. The functional view describes processes and data flows. This is sometimes referred to as logical architecture. The physical view describes real-world objects such as subsystems and devices along with the interfaces that connect physical objects together.

Figure 10.14 provides an illustration of the entire architecture, identifying each of the views and showing how they fit together to deliver services.

The system architecture for the specific smart mobility application, which is a methodology for developing the architecture, the technical tools to be used in support of the architecture development, and the precise nature of the visualizations used to illustrate the enterprise, functional, physical, and communication views. Note that the architecture is much more than a technical tool to show how things can be put together, it is also a vital tool in understanding the business models, the organizational arrangements, and the interplay between stakeholder enterprises required for a successful outcome.

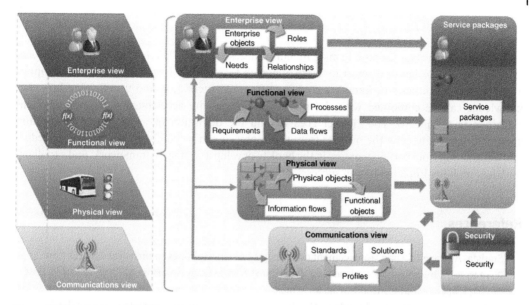

Figure 10.14 US National ITS architecture overview. *Source:* [15] / U.S. Department of Transportation/ Public domain.

10.4 The Application of the Smart Mobility Success Toolbox

It has been noted that this chapter is about the application of the accelerated planning approach discussed in Chapter 6. Several other factors and considerations have been identified in Figure 10.15, which illustrates a methodology for bringing all together and applying the smart mobility success toolbox. It illustrates how knowledge and information can be drawn from Chapters 3, 4, 6, 8, and 9 to provide inputs into the framework that represents the smart mobility success toolbox.

Chapter inputs can be combined with the information provided in the description of the smart mobility success toolbox to create an application approach for the smart mobility plan generated by the accelerated planning process.

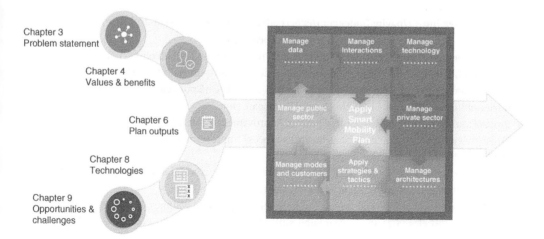

Figure 10.15 Using the smart mobility success toolbox.

10.5   Summary

This chapter presents a proposed smart mobility success toolbox. The intention is to serve as a checklist for the design of a smart mobility approach to a specific implementation. The chapter also provides a synthesis of knowledge sourced from Chapters 3, 4, and 9 indicating how the knowledge can be combined with the plan developed in the accelerated planning approach explained in Chapter 6. The chapter goes beyond the definition of an accelerated planning approach, It explains how the outputs from the approach can be combined with the knowledge and information from earlier chapters in the book to develop a strategy, a framework, and a toolbox for smart mobility implementation success.

References

1 Statista. Average lifespan (replacement cycle length) of smartphones in the United States from 2014 to 2025. https://www.statista.com/statistics/619788/average-smartphone-life/, retrieved June 21, 2022, at 10:50 PM Central European Time.

2 Cascade Collision Repair. Thursday 8 October 2020. cascadecollision.com/blog/what-is-the-average-life-of-a-car/, retrieved June 21, 2022, at 10:53 PM Central European Time.

3 Held, M.; Rosat, N.; Georges, G.; Pengg, H.; Boulouchos, K. Lifespans of passenger cars in Europe: empirical modelling of fleet turnover dynamics. *European Transport Research Review* 2021, 13, 9. https://doi.org/10.1186/s12544-020-00464-0, retrieved June 21, 2022, at 10:56 PM Central European Time.

4 2017 Infrastructure Report Card, American Society of Civil Engineers. chrome-extension://efaidnbmnnnibpcajpcglclefindmkaj/https://www.infrastructurereportcard.org/wp-content/uploads/2017/01/Bridges-Final.pdf, retrieved June 21, 2022, at 10:58 PM Central European Time.

5 Japan Society of Civil Engineers. A statistical study on lifetime of bridges. chrome-extension://efaidnbmnnnibpcajpcglclefindmkaj/viewer.html?pdfurl=https%3A%2F%2Fwww.jstage.jst.go.jp%2Farticle%2Fjscej1984%2F1988%2F392%2F1988_392_73%2F_pdf&clen=847943&pdffilename=1988_73.pdf, retrieved June 21, 2022, at 11 PM Central European Time.

6 Orange City Council. Your say homepage. https://yoursay.orange.nsw.gov.au/roads/news_feed/how-long-do-roads-last?posted_first=true, retrieved June 21, 2022, at 11:03 PM Central European Time.

7 Paving Finder. https://pavingfinder.com/expert-advice/how-long-does-an-asphalt-road-last/, retrieved June 21, 2022, at 11:07 PM Central European Time.

8 Program for Mass Transportation. Priorities for achieving a state of good repair. chrome-extension://efaidnbmnnnibpcajpcglclefindmkaj/viewer.html?pdfurl=https%3A%2F%2Fwww.ctps.org%2Fdata%2Fpdf%2Fstudies%2Ftransit%2Fpmt%2FPMT_Ch5.pdf&clen=1368754&chunk=true, retrieved June 21, 2022, at 11:09 PM Central European Time.

9 Process Maker. How to apply business process improvement. Matt McClintock, June 16, 2020. https://www.processmaker.com/blog/business-process-improvement/, retrieved June 21, 2022, at 11:12 PM Central European Time.

10 Alation. Products. Find, understand, and govern data. https://www.alation.com/product/data-catalog/, retrieved June 21, 2022, at 11:13 PM Central European Time.

11 WMW. 50 years and still rolling strong. Supply chain. https://one.walmart.com/content/walmart-world/en_us/articles/2021/09/50-years-and-still-rolling-strong.html, retrieved June 21, 2022, at 11:16 PM Central European Time.

12 Walmart. A behind-the-scenes look at how Walmart delivers. https://corporate.walmart.com/newsroom/business/20160602/a-behind-the-scenes-look-at-how-walmart-delivers, retrieved June 21, 2022, 11:17 PM Central European Time.

13 SAE. Unsettled topics concerning the field testing of automated driving systems. https://saemobilus.sae.org/content/EPR2019009, retrieved June 21, 2022, at 11:20 PM Central European Time.

14 McQueen, B. and McQueen, J. *Intelligent transportation systems architectures*. Artech House, 1999. https://books.google.fr/books/about/Intelligent_Transportation_Systems_Archi.html?id=qRxPAAAAMAAJ&redir_esc=y, retrieved June 21, 2022, at 11:22 PM Central European Time.

15 United States Department of Transportation. *National ITS reference architecture. Architecture overview*. Reference Architecture. https://www.arc-it.net/html/architecture/architecture.html, retrieved June 21, 2022, at 11:24 PM Central European Time.

11

Smart Mobility Performance Management

11.1   Informational Objectives

After you have read this chapter, you should be able to:

- Define smart mobility performance management
- Describe the dimensions of smart mobility performance management
- Explain an approach to smart mobility performance management
- Discuss the use of a balanced scorecard approach to smart mobility performance management
- Explain the role of performance management within operational management
- Discuss how technology can be applied to performance management for smart mobility

11.2   Introduction

Performance management encompasses the activities and resources required to compare the expected performance of smart mobility with that actually realized. Performance management encompasses the activities and resources required to ensure that the intended effects were actually achieved. The definition of performance management varies according to the subject being considered. For example, from a people perspective, The Society for Human Resource Management (SHRM) defines performance management "as the process of maintaining or improving employee job performance through the use of performance assessment tools, coaching, and counseling, as well as providing continuous feedback." From a transportation planning perspective, The Santa Fe Metropolitan Planning Organization defines transportation performance management as "a strategic approach that uses system information to make investment and policy decisions to achieve national performance goals." Both definitions show that performance management uses information regarding the performance of the system, the person to understand the current level of performance, and then apply tools and resources to improving performance, hence achieving preestablished goals. For the purposes of this book, it is assumed that smart mobility performance management is defined as follows:

> The establishment of goals and objectives for the range of activities required to deliver smart mobility services. The measurement and management of smart mobility applications and service delivery, within the urban environment. The use of appropriate data, and the

Smart Mobility: Using Technology to Improve Transportation in Smart Cities, First Edition. Bob McQueen, Ammar Safi, and Shafia Alkheyaili.
© 2024 The Institute of Electrical and Electronics Engineers, Inc. Published 2024 by John Wiley & Sons, Inc.

application of tools and resources to ensure the attainment of goals and objectives and identify opportunities for performance improvement. Providing support for the actions required to improve performance, and evaluation of the effectiveness of those actions.

Smart mobility performance management is an essential element in the successful application of smart mobility technologies and services. It reflects the satisfaction of the user, enables problems at various stages of service delivery to be diagnosed and solved, and provides a detailed understanding of the demand for transportation and variations in transportation operating conditions. Performance management can also serve as the foundation for a business justification for investment in smart mobility solutions. At the political level, performance management can confirm that political promises are in fact being achieved and considerable progress has been made toward their attainment. Performance management also has a powerful role in ensuring that smart mobility service operations are conducted efficiently and effectively while being fine-tuned to the operating environment.

This chapter explores smart mobility performance management building on the definition of outcomes, values, and benefits discussed in Chapters 3–5, and 7 by taking a deeper look into the performance management aspects of smart mobility. Performance management is addressed from two perspectives: summative and formative. Summative performance management takes a retrospective look at previous performance and provides information required to guide future practice, typically with significant investment. Formative performance management provides rapid feedback to enable incremental course corrections to ongoing deployments. Most types of performance management require the definition of key performance indicators and an assessment of how to measure the data required for the indicators. There is an important management principle that must be applied: "if you can't measure it, you can't manage it." For smart mobility. The principal is extended as follows: ". . . and if you're only measuring it, you are still not managing it." Therefore, the treatment of performance management in this chapter will include a detailed assessment of how to measure performance and how to use the data to manage performance. As stated earlier, it is important to complete the process from data collection to information processing, to the development of insight and understanding, to the definition of action because of the new insight and understanding. The authors believe that this is central to the successful treatment of performance management, recognizing the new possibilities that are now available thanks to modern data collection technologies, information technologies, and data science.

11.3 Smart Mobility Performance Management

Typical mobility performance management approaches around the world have a high degree of stove piping, or silos. In the same way that smart mobility initiatives tend to be implemented on a project-by-project basis with a focus on the project rather than the bigger picture, performance management is also focused on the project level. At the next level up, the department responsible for the smart mobility implementation may also have an internal approach to performance management. In turn, the agency within which the department resides can also have a performance management approach, which may integrate with the department, but not with other agencies. The concept of the mobility modulor was introduced, in Chapter 12, mobility for people, and it was discussed how different smart mobility initiatives could fit together in Chapter 7 on Smart Mobility Initiatives.

This raises a question. Why not have a single system that satisfies the needs of mobility for people, smart mobility solutions, and performance management? In fact, the author subscribed to review that the ultimate system for smart mobility management would in fact be completely integrated and support multimodal management. To our knowledge, no such system exists anywhere in the world at the present time. The authors have been involved in several initiatives to design and implement a completely integrated multimodal management system which resulted in sophisticated design concepts without implementation. Implementation barriers such as funding and organizational issues tend to function as barriers to deployment.

Our experience suggests that a better way is to take an incremental approach toward the ultimate system. Starting with an area that is a particular challenge, or of specific interest to the agency. Adopting lean and agile development techniques. It is possible to build from these initial stepping stones to create the comprehensive system that is required. Therefore, the focus of this chapter is the definition and explanation of a comprehensive smart mobility performance management system that could form the basis for an ultimate mobility management system on a multimodal basis. Some agencies may choose to adopt performance management as a starting point, as it tends to be a particularly crucial element in any city approach to smart mobility. Others may choose electronic payment systems as a starting point. If there is a sunk investment in toll roads, express lanes, or congestion pricing. Many agencies around the world have chosen a traffic management system as the starting point. When this begins is very much a product of the current situation, prior investments, political landscape, and specific needs of the city.

11.4 Smart Mobility Performance Management Approach

Effective and comprehensive performance management for smart mobility only to address a series of stovepipes, silos, or channels relating to the current management of smart mobility.

11.4.1 Across the Service Delivery Process

The service delivery process for smart mobility was introduced in Chapter 10. An important part of a smart mobility approach would be to ensure that is comprehensive by building bridges across the existing silos or stovepipes. There are multiple dimensions to the stovepipes that can be listed as follows.

11.4.1.1 Policy
Non-technical decision-makers can often establish policy with the accompanying political promises, without effective consultation with other departments responsible for the delivery of transportation service if policies not coordinated with the other steps that it is possible for investment decisions to be disconnected from the other steps in the smart mobility service delivery process.

11.4.1.2 Planning
Plans can often be defined without due regard to current design, implementation, operations, and maintenance challenges and experience. Ideally, there will be feedback loop from design, operations, and maintenance back to planning.

11.4.1.3 Design
Designs conducted with a stovepipe can provide the opportunity for a disconnect between design, planning, and policy. There is also an important need to evaluate the effectiveness of prior designs, building lessons learned into future ones.

11.4.1.4 Implementation

Here again, lessons learned during implementation can give feedback on the design, planning, and policy. Strong communications between design and implementation is also necessary.

11.4.1.5 Operations

It could be argued that staff involved in operations must deal with issues and challenges created by those earlier in the process. Lessons learned from operations can be most effective when communicated along the process.

11.4.1.6 Maintenance

Similar to operations, maintenance can provide valuable information to earlier steps in the process, especially with regard to the performance of systems and devices.

Typically, each of the steps in the process is subject to performance management designed specifically for that step.

This is illustrated in Figure 11.1. This figure also lists some of the problems that occur within each of the stovepipes, or channels, which can be addressed by more effective and comprehensive performance management.

There is a need for performance management for each step in the smart mobility service delivery process and across the entire process.

Another important dimension of smart mobility centers around data. Each department and agency involved in the service delivery process for smart mobility will typically have its own data collection and analysis process. This is further fragmented by data sources with data from roadside sensors treated differently from data from vehicles, data from people, and other systems. This can lead to greater expense and data collection and could be possible and lower quality of data. This parallel activity continues with information processing and extraction of insight from the information, leading to lost opportunities for sharing investments, and activities. The opportunities also lost to ensure that the data and information used for performance management are compatible with that used for advanced traffic management. There is a need for performance management for data.

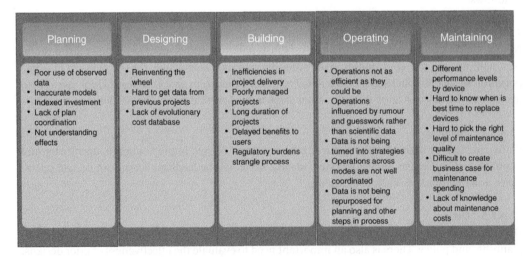

Figure 11.1 Smart mobility service delivery process.

A similar stove piping activity also takes place between the public and the private sector. 20 years ago. It could justifiably have been claimed that the entire smart mobility effort was being driven by the public sector. In terms of investments and activities. Today, this is no longer the case, with the majority of smart mobility initiatives, especially in urban areas, being funded and implemented by the private sector. While the public sector has limited control over private sector activities and investments, it is obvious that the needs to be some form of coordinated performance management.

11.5 A Comprehensive Smart Mobility Performance Management Approach

Given the foregoing challenges, the suggested approach to smart mobility performance management is both comprehensive and results in analytics and key performance indicators are suitable for a wide range of users. The high-level process for performance management for smart mobility is the same one that has been discussed earlier in the book (see Chapter 7, Smart Mobility Initiatives, Figure 7.17). An adapted version is illustrated in Figure 11.2.

For performance management, two additional steps have been inserted at the beginning and end of the process; at the beginning, a step has been added for defining objectives. At the end of the process, a step has been added for the evaluation of the actions taken. Note that a feedback loop has been introduced from objectives to data. This accounts for the fact that key performance indicators and analytics can only be effective if they can be measured. This requires that the appropriate data be available to allow the key performance indicators to be measured when selecting key performance indicators, it is crucial to evaluate the viability of each performance measure in terms of availability of data. It may be necessary to define a proxy measurement that can be measured in place of the actual measurement. If proxy are direct measurement cannot be achieved, then the performance measure is not viable and should not be used. If the key performance indicator is important but not viable, consideration should be given to additional data collection arrangements to make the data available and hence make the KPI viable. This process of identifying future data collection requirements is part of an overall approach to data collection and acquisition planning on the basis of a thorough understanding of the use of the data and the result of using the data. Results-driven data collection acquisition can be a crucial step in ensuring the efficiency of data collection.

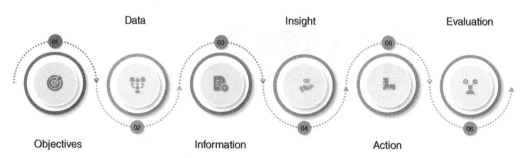

Figure 11.2 Smart mobility performance management process.

11.5.1 Objectives

Objectives consist of the following dimensions.

11.5.1.1 Expected Outcomes

Expected outcomes are broad goals to be achieved through the application of smart mobility. These include improved safety, increased efficiency, enhanced user experience, equity improvements, and decarbonization. These are the main headings which can of course be broken down into sub-goals to suit the needs of the city.

11.5.1.2 Key Performance Indicators

The terms key performance indicators, performance measures, and analytics are used interchangeably. This reflects the practice in the industry. Key performance indicators are the measures that are defined and utilized to measure performance. Ideally, there will be a distinct set of key performance indicators depending on the target for the insight and performance information.

Figure 11.3 provides an illustration of how analytics or key performance indicators can be targeted to multiple tiers of staff involved in performance management.

For example, the nontechnical decision-makers tier must feature performance measures that relate to current and future policies. The nontechnical decision-maker should be able to ascertain progress being made toward current policy goals and identify the need for new policies and regulations. The public tier must focus on outcomes that are relevant and understandable to the general public. These KPIs should illustrate how smart mobility will impact the everyday lives of citizens and visitors.

When multitiered performance measures and analytics are defined, it is also appropriate to consider the establishment of analytics-driven organizational arrangements. For nontechnical decision-makers, a specific set of performance measures can be defined and agreed the provides the maximum value for policy and regulation. Aligning performance measures with policy goals delivers the best value. In this case, performance measures should be aligned to existing policies and enable the nontechnical decision-makers to identify the need for new policies. At the

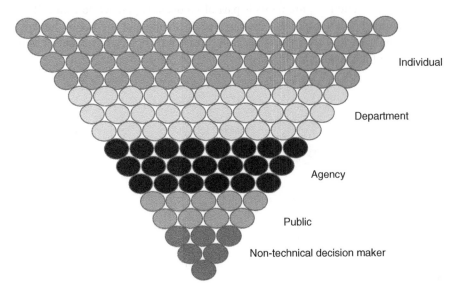

Figure 11.3 User groups.

agency level, broad performance measures can be identified, enabling the activities and resources of the agency to be focused on the achievement of the appropriate values for the performance measures. Similarly, at a departmental level, appropriate performance measures can be defined, and target values agreed across the department.

Departmental resources and activities can then be managed toward the achievement of the performance measure goals. It is important that the performance measures used for each tier are tailored to the specific needs of the target audience. Different measures are required for different purposes. Moving up through the tiers, the level of detail should increase to tailor the key performance indicators to the specific roles and responsibilities of the staff. Note that at the individual level groups of staff in teams with a specific objective or group of objectives should also be addressed in addition to the individual. Individual and team KPIs can also be related to the business processes that are being supported. More recently, performance management, and individual and team-level performance management have also been viewed as performance enablement [1]. This requires that the information regarding actual performance versus intended performance is provided to individuals and teams to enable them to take corrective action and make improvements. This of course assumes that the teams and individuals are also empowered to make the necessary changes and actions. This fits well with the concept of business process improvement using small business process improvement teams to identify challenges and weaknesses, define solutions, and then implement them.

The process classification framework is a useful tool to be applied to team and individual process performance management [2].

A high-level overview of the process classification framework is provided in Figure 11.4.

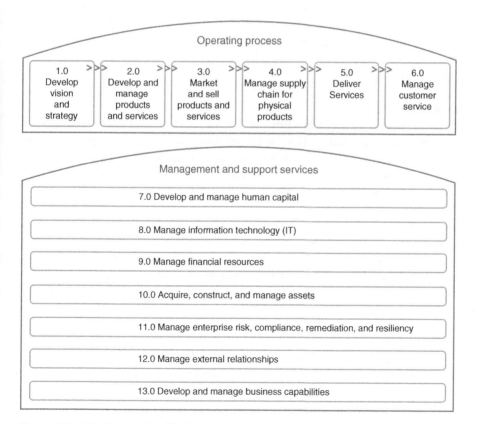

Figure 11.4 The Process classification framework.

This illustrates the typical business processes that are supported by various industries and businesses. This is a cross-industry version, which are other versions available for specific businesses and enterprises. The city service version is explained in detail in Chapter 14, Organizing for Smart Mobility Success. The process classification framework features detailed subprocesses under each of the headings shown above. This enables the process classification framework to be used as a checklist when teams or individuals and attempting to define business processes to be managed and improved.

This provides a structure and checklist to ensure that all relevant processes are considered. Since the process classification framework was derived from a cross-industry study, it also enables smart mobility performance management to be compared against performance management in other industries. It is designed as a framework and global standard that can be customized for use in any organization.

An excellent guide to key performance indicators used in smart mobility within cities was developed by one of the authors in fulfillment of the thesis requirements for a master's degree [3]. This provides a comprehensive overview of performance indicators that have been utilized in cities around the world.

11.5.1.3 Expected Key Performance Indicator Values

In addition to defining the key performance indicators to be used, it is also necessary to establish the appropriate values to be achieved if performance is to be at the required level. This requires careful consideration of the expected effects of smart mobility investments and may vary over time as performance improves. An annual review should be conducted to ensure that the values being set for performance indicators continue to be relevant to the changing performance of the city and the agencies. Emerging technologies may also provide new capabilities that could lead to revision of the expected values for the key indicators. Valuable input into the definition of key indicator expected values can also be achieved by considering lessons learned and best practices from other implementations around the globe.

11.5.1.4 Data Collection

The second step in the process for smart mobility performance management involves the collection of data from multiple sources, the verification and validation of the data, and the creation of a unified data set that can form the basis for analytics and data exchange. This step in the process will also include the application of performance management techniques to the data to ensure that the data is of sufficient quality and has been collected in the most cost-effective manner.

11.5.1.5 Information Processing

At this step in the process, the raw data is processed to provide information from which reports, trends, patterns, and insight can be determined. Information processing is where value is added to the raw data by making it suitable for the extraction of insight. The information delivered can be focused on the people aspects of smart mobility, the overall system, the individual components of the system, or the data.

11.5.1.6 Insight Extraction

Insight can be extracted from the information either automatically or by human effort. Sometimes by a combination of the two. Summaries are created as technical reports and visualizations. Often the use of business intelligence, and dashboard techniques and implemented to illustrate the geolocation of the information and highlight trends, patterns, and mechanisms. The KPIs are analytics used, the visualizations, and the dashboards would be specifically designed for the distinct levels of staff and departments discussed earlier. These include

- Nontechnical decision-makers
- The public
- Agencies
- Departments
- Individuals or teams responsible for performance management

The pyramid shape adopted in Figure 11.3 is also designed to indicate that the progression of KPIs from the point of the pyramid to the base should feature a progressive simplification and summarization. Toward the base of the pyramid, larger numbers of people are involved, and it is necessary to define simpler, easier-to-share performance measures that can be used to guide large groups of people toward the same objective. There is also a challenge to ensure that the performance measures have a level of detail to enable individual staff in teams to use the performance measures to improve job activities. It is likely that toward the bottom of the pyramid, two sets of performance measures will be required. A higher level extracted set of performance measures to be used to guide the larger group and customized performance measures for each individual job or team.

At the tip of the pyramid were nontechnical decision-makers who received performance management information, it may also be appropriate to consider the use of indices to create a simple single number to represent different performance parameters. While the detail behind these numbers will be required by other users in the pyramid, it would be most effective to provide high-level compass indices to this particular group.

Figure 11.5 shows how key performance indicators across the smart mobility service delivery spectrum are relevant to each group.

11.5.1.7 Performance Management Actions

At this point, the new insight can lead to short-term or long-term actions to improve performance. Short-term or summative performance management is the fine-tuning adjustments that can be made on a day-to-day basis in the light of the insight received in a performance management system. These would typically be changes to response plans, minor changes, organizational arrangements, or new and information control longer-term actions known as summative performance management can also be defined and implemented. These would be longer-term changes required

	Planning	Designing	Building	Operating	Maintaining
Nontechnical decision maker	⊘			⊘	
Public	⊘			⊘	
Individual	⊘	⊘	⊘	⊘	⊘
Teams	⊘	⊘	⊘	⊘	⊘
Department	⊘	⊘	⊘	⊘	⊘
Agency	⊘	⊘	⊘	⊘	⊘

Figure 11.5 KPIs by user group and process step.

to infrastructure, the network, operational management, and the deployment of new technologies. This would also include the acquisition of new data required as identified in earlier steps.

The definition of investment programs for infrastructure and significant changes to technology can be considered as results-driven investment planning, where investments are related to desired performance management improvements.

The technique introduced in Chapter 12: Mobility for People could also be implemented here. A comparison could be made between a predefined ideal trip and the trip actually experienced by the traveler. Advances in the availability of data and sophisticated data management systems enable the kind of trip-by-trip comparison that would not have been possible in the past. Insight obtained regarding the variation of the actual trip from the ideal trip can form the basis for operational changes, information advisories to travelers, and potentially long-term investments.

In this step, in addition to the definition of the actions to be taken, consideration should also be given to whose responsibility it will be to implement the actions. This is a crucial factor in the customization of the KPIs for each audience group.

11.5.1.8 Evaluation of Actions

Using the same data collection process that was used to identify, define, and measure key performance indicators, or analytics. A comparison is made between the actual effects of the actions and the intended. This can lead to revision for future strategies and the definition of the need for new data. In an equivalent manner to performance management actions. As discussed above, the responsibility for the evaluation of actions should be considered and assigned to a specific group or individual.

Figure 11.6 provides an example of how user roles for each step in the process could be assigned to distinct groups. Note that this will vary from one agency to another and from one city to another. So, this simply provides an example of how it could be done.

Group	Role
Nontechnical decision maker	Revision to policies and the introduction of new policies. Evaluating progress towards policy objectives
Public	Feedback on performance and suggestions for new services and improvements
Individual	Participation in process improvement teams and individual actions. Evaluating the effects of performance management actions
Teams	Participation in process improvement teams and individual actions. Evaluating the effects of performance management actions
Department	Collective improvement actions. Evaluating the effects of performance management actions
Agency	Collective improvement actions. Evaluating the effects of performance management actions

Figure 11.6 User roles.

11.6   Managing the Private Sector

Truly comprehensive performance management for smart mobility would go beyond the bounds of the public sector. This might seem like an impossible task, but it is necessary to coordinate public sector performance management activities with those of the private sector. It is likely the private sector enterprises involved in their own independent smart mobility initiatives will have their own internal performance management system. However, most of the systems will feature some sort of interface with the outside world. This could be at the service delivery and the process, or at some point along the way.

Many smart mobility initiatives will also require that the private sector complies with policy or regulation. For example, in many micromobility implementations around the world, the private sector operating company is required to comply with rules and regulations related to vehicle speed and vehicle parking. These are points on which the public sector can have influence over the performance measures used by the private sector. Public sector opportunities to influence private sector performance management are greater when there is some form of public–private partnership in existence. This could involve predefined goals and objectives, roles, and responsibilities for both the public and private sectors. Establishing these provides an opportunity for the public sector to influence private sector performance management. In many cases, the interface between the public and the private sector is represented as a client interface.

The public sector acquires the services of the private sector for design, build, operations, and maintenance of various smart mobility applications including public transit ticketing. This is discussed in Chapter 7, Elements of Smart Mobility. When the public sector is a client, performance measures and performance management process can be defined or heavily influenced by the public sector. They can be integrated into contractual documents that support procurement of the service. At the very least, the public sector should establish a dialogue with the relevant private enterprises to identify common ground and align activities and investments.

11.7   Organizing for Performance Management

Organizational aspects of smart mobility were covered in detail in Chapter 14, Smart Mobility Organization. As stated in that chapter, it is necessary to perform a capability assessment on the organization to determine readiness for smart mobility. With regard to performance management, the same is true. A readiness assessment should be conducted, using the principles of the capability maturity model to determine organizational readiness for comprehensive performance management.

Organizational arrangements for performance management must also consider variations in the type of user for performance management and their roles and responsibilities. At the level of individual staff, analytics-driven organizational arrangements should encompass the alignment of job objectives and job descriptions to the selected performance indicators and values. This may require a realignment of existing job objectives and job descriptions. It may be necessary to set performance measures that require the individual to work together as a team without individuals to achieve the objective. For example, a traffic signal engineer with an objective of minimizing congestion would not be able to achieve this at an individual level as multiple other staff have an influence over congestion levels. The traffic signal engineer can have an impact on congestion levels by choosing the most appropriate signal timings. However, if a congestion pricing system is in operation, then those staff responsible for operating the system will also have a considerable influence over congestion levels.

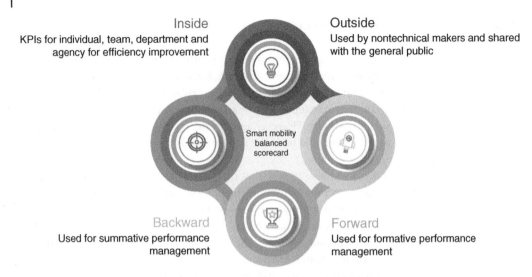

Inside
KPIs for individual, team, department and agency for efficiency improvement

Outside
Used by nontechnical makers and shared with the general public

Smart mobility balanced scorecard

Backward
Used for summative performance management

Forward
Used for formative performance management

Figure 11.7 Smart mobility performance management balanced scorecard.

Organizational arrangements for performance management also posed the question – who should be responsible for performance management. On the one hand, those responsible for operational management have access to the required data and a well-placed to conduct performance management alongside their operational management responsibilities. On the other hand, there is a lot to be said for performance management and operational management responsibilities being assigned to distinct groups. Separating performance from operations will enable a certain degree of independence and neutrality that may not be possible if the roles are combined. If a separate group is assigned the responsibility for performance management, then obviously the needs to be close cooperation and information exchange between the two, without one influencing the other.

It may be appropriate here to use the concept of a balanced scorecard approach to performance management as illustrated in Figure 11.7.

This is based on the principles of the Balanced Scorecard Institute [4], modified smart mobility performance management. It recognizes the stratification of the user group for smart mobility performance management and allocates "Inside" Key Performance Indicators for efficiency improvement by individuals, teams, departments, and agencies. It allocates "Outside" Key Performance Indicators for communication to nontechnical decision-makers and the public. "Forward" Key Performance Indicators and assigned to formative, short-term, performance management, while "Backward" Key Performance Indicators are used for summative, longer-term performance management with a retrospective perspective.

11.8 The Role of Performance Management in Operations Management

To illustrate the relationship between performance management in operations management, traffic management has been selected as a focus. In Chapter 15 Smart Mobility Operational Management, the various roles and activities within operational management for traffic were defined and described. Using this as an example, it can be illustrated that performance management is an important subset of operations management. Figure 11.8 shows the 20 business processes that are supported within a traffic management center.

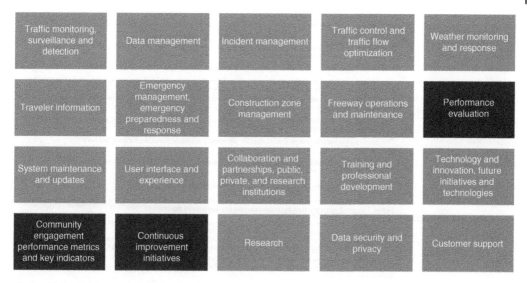

Figure 11.8 Traffic management center processes.

Three of the 20 business processes (highlighted) involve performance management: performance evaluation, community engagement, performance metrics, key indicators, and continuous improvement initiatives. Performance evaluation will include the application of performance management techniques to each individual business process and the operation of all the business processes together. The community engagement, performance metrics, and key indicators process would result in the definition of the performance measures to be used, and the target values to be attained. The continuous improvement initiatives process would be part of the performance management activity involving the identification of weaknesses and the definition of suitable solutions.

The data management process supports performance management by providing decision-quality data. The other processes will be the subject of performance management activities. The business processes are described in the following sections, along with a brief summary of the most significant key performance indicators, or analytics that can be utilized for performance management.

11.8.1 Traffic Monitoring, Surveillance, and Detection

Performance management for this process will include the use of key performance indicators like traffic monitoring, accuracy, surveillance effectiveness, and detection error rates. For a traffic signal perspective, the recent trend in the United States is to make use of the advanced communication capabilities for new traffic signal controllers to implement automated traffic signal performance measures. These include the following measures are summarized in Figure 11.9.

11.8.2 Incident Management

For this business process, performance measures would include time taken to initially detect the incident, time taken to verify the incident, time taken to respond to the incident, and the appropriateness of the response, incident clearance time, and time taken for appropriate resources to reach

ATSPM	description
Purdue phase termination	This metric summarizes the controller's phases and the reason the phase terminated
Split monitor	This measure generates separate summaries for each phase on the controller. Each plot depicts the length of the phase in seconds and the reason the phase terminated
Pedestrian delay	This time-of-day summary depicts the delay, in minutes, associated with each pedestrian actuation.
Preemption details	This measure details the pre-emption events at the signal. Either railroad crossings or emergency vehicles may pre-empt normal signal operation. This metric does not provide any options to configure the output.
Turning movement counts	Time-of-day summary showing volume for the lanes in a turning movement. Available turning movements also includes the through movement
Purdue coordination diagram	Enables what-if scenarios to be examined by varying cycle length and green times. Predicted coordination diagrams how then produced
Approach volume	This measure plots traffic approaching an intersection with advanced analysis comparing opposing directions, detector sites, and more
Approach delay	This measure plots a simplified approach delay experienced by vehicles approaching and entering the intersection The delay per vehicle and total delay are both available
Arrivals on red	This measure plots both the volume and percentage of vehicles arriving on red
Approach speed	This measure tracks the speed of vehicles approaching a signalized intersection. Phases where this data are available will be plotted. This metric requires detection set back from the stop bar that can detect speed
Yellow and red actuations	This measure plots vehicle arrivals during the yellow and red portions of an intersection's movements where the speed of the vehicle is interpreted to be too fast to stop before entering the intersection It also provides users with a visual indication of occurrences, violations, and several related statistics The purpose of this metric is to identify engineering countermeasures to deal with red light running
Purdue split failures	Purdue split failure relies on stop bar presence detection, either by lane group or lane-by-lane This metric calculates the percent of time that stop bar detectors are occupied during the green phase and then during the first five seconds of red.

Figure 11.9 Advanced traffic signal performance measures.

the incident. The US Federal Highway Administration defines the following performance measures for traffic incident management [5]:

- Time between first recordable awareness of instant by a responsible agency and first confirmation that all lanes are available for traffic flow.
- Time between first recordable awareness of incident by responsible agency and time at which the last responders left the scene.
- Number of unplanned crashes, beginning with the time of detection of the primary instant when a collision occurs either within the instant scene are within the Q including the opposite direction resulting from the original incident.

11.8.3 Traffic Control and Traffic Flow Optimization

Key performance indicators for this process include queue lengths, congestion delays, total throughput through intersections, total throughput through links between intersections, reliability of travel time, the alignment of traffic signal timings with traffic demand, and the performance of emergency vehicle and public transportation vehicle preemption systems. Measures can also include total people throughput at an intersection, and delays incurred by pedestrians and bicycles. Safety will also be an important performance measure for this business process as well as for the overall activities of the traffic management center. The automated traffic signal performance measures described above are also used for this process.

11.8.4 Weather Monitoring and Response

Performance measures will include the alignment of response. The weather conditions, the speed of response, the speed at which weather conditions are detected, and the overall reduction of traffic capacity caused by the weather condition. The US Federal Highway Administration defines 25 performance route measures related to weather monitoring and response [6].

These include performance measures summarized in Figure 11.10.

Building sustainable relationships, ensuring that road weather management investments improve Highway performance
The appropriate communities use and rely on road weather observations
Advancing the state-of-the-art, advancing the state of the practice, integrating weather-related decision-support into traffic operations
Advancing the state of the practice by raising road weather capabilities and awareness

Figure 11.10   Road weather management performance measures.

11.8.5   Traveler Information

The effectiveness of traveler information business processes related to the behavior change induced and travelers. This can include determining the percentage of travelers who move from peak to off-peak travel due to the availability of decision-quality traveler information. Utilization of each mode can also be measured to determine the effectiveness of the traveler information, especially if a mobility as a service approach is being taken. The effectiveness of travel information can also be measured by customer satisfaction surveys and regional traffic network performance. Typical performance measures used for travel information include accuracy of information, timeliness of information delivery, geographical coverage of the information, usability and accessibility of the information, response to incidents, integration of different modes, effects on travel behavior, customer satisfaction, data quality, accessibility, and cost-effectiveness.

11.8.6   Emergency Management, Emergency Preparedness and Response

This business process can be measured using the response time of emergency services, the effectiveness of emergency plans, and the speed and appropriateness of response. The typical range of performance measures applied include response time, clearance time, instant duration, traffic flow, secondary instance, resource utilization, communication effectiveness, public satisfaction, coordination and collaboration, safety and health outcomes, economic impact, training, and preparedness.

11.8.7   Construction Zone Management

Construction zone management can be measured using average travel times through the zone, safety levels before and during construction, traffic speeds for the construction zone, and safety levels achieved within the construction zone. Significant performance measures utilized include lane queues, user costs, risk exposure, safety, and public perception.

11.8.8   Freeway Operations and Maintenance

This business process includes incident management and also additional activities related to the routine operations of freeways. While incident management involves the operation of freeways during unusual operating conditions such as special events and weather events, freeway management extends to include day-to-day routine operations. Performance management can be accomplished using average travel speeds, travel time, travel time variability, total throughput of traffic, and total people throughput. Examples of significant performance measures include how many people or vehicles are using the network, where and when are they being delayed and subjected to unsafe conditions, how frequently do those delays occur, how bad other delays, what are the reasons for the delays, effective operational improvements on the delays [7].

11.8.9 System Maintenance and Updates

The major performance indicator for system maintenance and updates will be Mean Time Before Failure for the whole system and for individual components. Expenditure on system maintenance and updates can also be an important performance measure. This can be measured as a total number of dollars per annum per device or system component.

11.8.10 User Interface and Experience

This addresses the major interface between the advanced traffic management system and the user, and general public. Performance can be measured through the use of customer satisfaction surveys and tracking software used to track the progress of queries and issues.

11.8.11 Collaboration and Partnerships, Public, Private, and Research Institutions

The most effective way to measure the performance of this business process is in terms of measuring the outcomes of the partnerships. This can include smart mobility deployment, research projects, and other joint initiatives. The output from the partnerships should be measured against that which was intended when the partnership was established.

11.8.12 Training and Professional Development

Training and professional development activities can be measured using four KPIs as defined by Dr. Donald Kirkpatrick [8]. This was introduced in Chapter 6 Planning for smart mobility.

11.8.12.1 Level 1: Reaction
The degree to which participants find the training favorable, engaging, and relevant to their jobs. In other words, did the participants find the training enjoyable and useful?

11.8.12.2 Level 2: Learning
The degree to which participants acquire the intended knowledge, skills, attitude, confidence, and commitment based on their participation in the training. Did the participants learn something?

11.8.12.3 Level 3: Behavior
The degree to which participants apply what they learned during the training when their back on the job. Is the new knowledge that the participants acquired applicable to the job?

11.8.12.4 Level 4: Results
The degree to which targeted outcomes account as a result of the training including support and accountability packages. What is the bottom-line impact on the organization?

11.8.13   Technology and Innovation, Future Initiatives

The simplest way to measure this would be to count the number of technological innovations and future initiatives that were accomplished. However, effective performance management would go beyond this measure of activity and attempt to ascertain the value of each innovation and initiative. This requires that benefit-cost analysis, data, and analytics be built into each innovation and initiative.

11.8.14   Research

In a similar way to technology, innovation, and future initiatives, research performance management should be based on the value of the output. This again requires that cost-benefit data collection and analysis be built into each research initiative, along with a clearly agreed set of goals.

11.8.15   Data Security and Privacy

The range of important performance measures that can be used for data security and privacy include the following:

- Data breach frequency and impacts
- Instant response measures
- Access control measures
- Encryption and data protection measures
- User authentication measures
- Privacy compliance measures
- Data retention and disposal measures
- Third-party risk management measures
- Employee training and awareness measures
- Privacy impact assessment measures
- Customer support measures

11.9   Applying Technology to Performance Management

Considering the performance management process defined in Figure 11.8, there is a range of opportunities to apply technology for performance management.

11.9.1   Data Collection

The use of advanced sensors, both roadside and in vehicle represents a good example of the application of technology for performance management. Technology application also extends to the techniques used to create the unified data set including data validation, and verification. Technology can also be used to conduct orthogonal sensing and data coming from different sources and provide the management capabilities required to keep the data usable and accessible.

11.9.2   Information Processing

Information processing converts raw data into information from which trends, patterns, and mechanisms can be identified. Recent advances in data science and data management systems can be applied to ensure the information processing is conducted in the most efficient manner, with results readily accessible to end users.

11.9.3 Insight Extraction

A range of business intelligence tools can be used to understand trends and patterns within the information, creating reports, and visualizations. As most of the data related to smart mobility is geo-located, the use of visualizations is of particular importance. Dashboards are a particularly important element in the extraction of insight and a visual display of key performance indicators. Just like key performance indicators, they must be tuned to the needs of the audience group to be most effective. Dashboards can cover entire countries, a region, or a city, or a corridor of interest. While the elements on the dashboard are specifically tailored to the objective and the audience group, a number of typical dashboards as shown in the following figures to provide an understanding of the nature of dashboards. The first dashboard example shown in Figure 11.11 depicts European cities progress toward sustainable development goals [9].

Progress toward the sustainability goals are shown as an average value according to color-coded bands. Using the interactive dashboard is possible to select one of a number of sustainable development goals and see the progress for that city on the map. The 16 goals that can be evaluated are as follows:

1) No poverty
2) Zero hunger
3) Good health and well-being
4) Quality education
5) Gender equality
6) Clean water and sanitization
7) Affordable clean energy

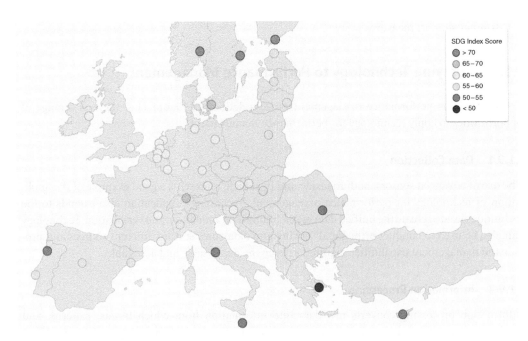

Figure 11.11 SDG goals dashboard.

8) Decent work and economic growth
9) Industry, innovation, and infrastructure
10) Reduced inequalities
11) Sustainable cities and communities
12) Responsible consumption and production
13) Climate action
14) Life below water
15) Life on land
16) Peace, justice, and strong institutions
17) Partnership for the goals

Smart mobility falls under sustainable transport which fits into goals, three good health and well-being 9, build resilient infrastructure, promote inclusive and sustainable industrialization and foster innovation, and 11, sustainable cities and communities. Smart mobility has a direct impact on the sustainability of the transportation system in terms of managing fuel consumption and emissions, which also has an impact on air quality and hence good health and well-being. Smart mobility can also have an additional impact on good health and well-being by making access to medical facilities and appointments easily available through better mobility options. This is a good example of performance management on a national level.

The next example shown in Figure 11.12 is an example of a statewide dashboard from Florida Department of Transportation in the United States. It displays the four key performance measures for traffic incident management – roadway, clearance time, incident clearance time, secondary crashes, and Road Ranger average time on the scene. The Road Ranger is the courtesy service patrol.

The next example shown in Figure 11.13 is a good example of a city or regionwide performance management dashboard. The website is known as Vital Signs and was developed by the Metropolitan Transportation Commission in San Francisco, California [10].

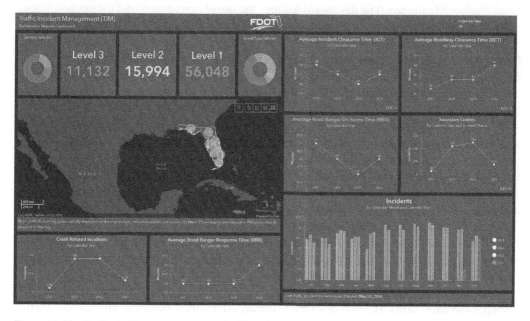

Figure 11.12 Traffic incident management dashboard. *Source:* https://gis.fdot.gov/arcgisportal/apps/dashboards/bde2398d1b2a48cfb4ff613e096d39e0

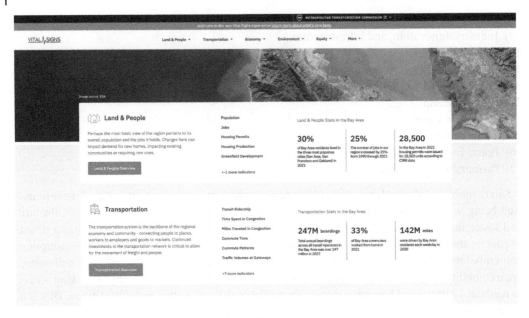

Figure 11.13 Vital Signs dashboard. *Source:* https://vitalsigns.mtc.ca.gov/

In addition to mobility, the dashboard also addresses land and people, economy, environment, and equity. The mobility performance measures used include traffic volumes at regional gateways, miles traveled in congestion, public transportation ridership, daily, miles traveled, street pavement condition, highway, pavement condition, bridge condition, commute mode choice, commute time, and time spent in congestion. The transportation are mobility performance measures are shown in rotation in the left-hand box. When the snapshot was taken Highway pavement condition was being displayed in the box. The dashboard features the use of percentages and indices as it is also a public-facing dashboard.

The next example shown in Figure 11.14 is from Utah Department of Transportation in the United States [11] https://www.udot.utah.gov/connect/business/traffic-data/traffic-operations-data/. It shows a corridor-level dashboard that can display average traffic volumes by month along a specific corridor. The dashboard allows the user to select a specific year and a specific day of the week.

The next example shown in Figure 11.14 is also from Utah Department of Transportation [12] and shows a regional view of the Cottonwood Canyon's region. In addition to displaying traffic volumes, it also displays congestion levels in terms of the current travel time divided by the free-flow travel time. This is a good example of a performance index in use. This dashboard is particularly designed for public use, hence the use of a performance index to simplify the information and make it more consumable (Figure 11.15).

The next dashboard from Cobb County in Georgia [13] shown in Figure 11.16 is good example of a nontechnical decision-makers and targeted dashboard and KPIs. The KPIs depicted can be related to policy goals and objectives, including safety, regional, travel time, airport traffic, asset management, capital improvements, and maintenance.

These dashboard examples serve to illustrate the overall design and format for performance management dashboards. They are designed to convey information and allow interaction with the

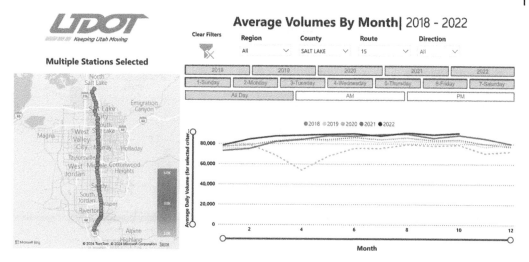

Figure 11.14 Corridor management dashboard.

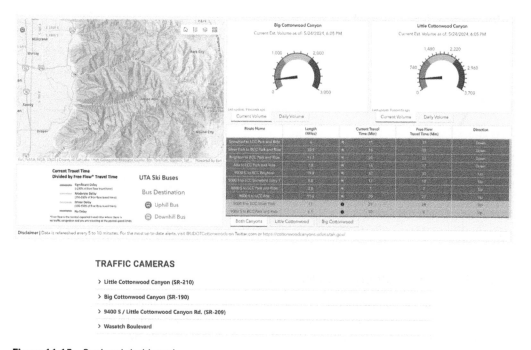

Figure 11.15 Regional dashboard.

user to look at the information from different perspectives. It is likely that more sophisticated dashboards will be used on an internal basis for agencies, departments, teams, and individuals. These are likely to have additional functionality to enable what-if scenarios to be evaluated by changing input parameters using sliders, or dials. This is particularly useful for assessing the effects of intended performance management actions and future investments.

Transportation Performance Dashboard

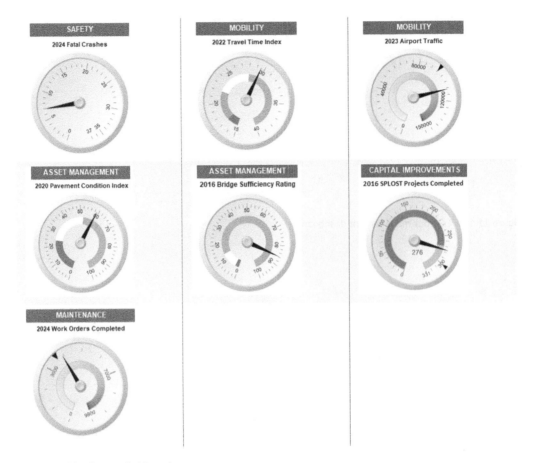

Figure 11.16 County dashboard.

11.9.4 Response and Strategy Development

The application of decision-support technologies artificial intelligence and machine learning can be used to support the development of effective responses and strategies. Simulation modeling can be used to understand the effects of response plans and strategies prior to implementation.

11.9.5 Response and Strategy Evaluation

The same technologies used to support data collection can also be used to create the raw material for evaluating the effectiveness of responses and strategies. Decision-support technologies including artificial intelligence and machine learning can also be used in the evaluation process. The results of the evaluation process can suggest improvement actions and visualizations, which are then fed into response and strategy development.

11.10   Summary

In a complete approach to smart mobility performance management, there are multiple dimensions and multiple audience groups. While current approaches tend to focus on specific applications and specific groups of people, because of new data sources and better data management, it is possible to be more comprehensive in the future. Beginning with the definition of smart mobility performance management, this chapter then takes a detailed look at smart mobility performance management and some of the challenges.

The smart mobility service delivery process as introduced in Chapter 10 was then reviewed and the application of smart mobility performance management within each step of the process was explored and discussed. Some of the issues that can occur as a result of stovepipe performance management across the service delivery process were identified, highlighting the need for detailed performance management within each step, but also an overarching comprehensive approach to smart mobility performance management across the process. The drawbacks of a fragmented approach to data collection for performance management were also discussed, highlighting the duplicate effort and greater expense caused by the lack of a centralized approach. It was also highlighted that similar fragmentation takes place between the public and the private sector, which has become an issue due to the growing importance of the role played by the private sector in modern smart mobility.

As a response to the challenges, a concept of comprehensive, complete smart mobility performance management was introduced. This provides effective and comprehensive performance management, addressing the challenges associated with stove piping while encompassing the opportunities offered by an overarching view. A simplified overview of the comprehensive smart mobility performance management approach was presented followed by a discussion of the detailed activities within each process step.

The key performance indicators, performance measures, or analytics were then discussed, making the point that the three terms are used interchangeably within the transportation industry. The users for performance management were stratified according to individual/team, department, agency, public, and nontechnical decision-makers. The importance of KPI customization for each group was stressed. The process classification framework was introduced as a tool that can be used by the various groups when implementing performance improvement actions. Each step in the comprehensive smart mobility performance management process was then discussed, explaining the activities that will be performed under each process.

The role of performance management, relative to operational management was then discussed using a high-level process diagram that explains the activities required for a traffic management center, highlighting those that are directly relevant to smart mobility performance management. Each 1 of the 15 processes for the traffic management center was discussed and sample key performance indicators that would be used for each process were described.

The importance of dashboards for visualization and for the extraction of insight from information was also stressed. Several examples of dashboards were then provided ranging from Europewide to corridor-specific applications.

There are several important points covered in this chapter. Firstly, the use of business processes as a means to structure and guide smart mobility performance management. Performance management can also be integrated with business process improvement, enabling improvements to the process to be made at the same time as other improvements to services and infrastructure. Secondly, in the course of setting key performance indicator expected values, it is extremely useful to scan for experience and lessons learned from prior implementers. This is particularly important as the availability of new data enables new key performance indicators to be applied. Thirdly, it is obvious from the process defined that complete performance management cuts across departmental boundaries, the steps in the delivery process for smart mobility, and involves more than just measurement. Fourthly, there is considerable scope to apply smart mobility technologies to the process of performance management. New sensors, new data sources, and new ways to analyze data are some of the technologies that can be brought to bear to make smart mobility performance management more efficient and effective. Artificial intelligence and machine learning can also have a significant impact in automating manual processes and more effective extraction of insight from the information.

The fifth and final point is that benchmarking for performance management could be enabled more effectively if there were guidelines or standards for data to be collected for smart mobility projects. As was seen in Chapter 5, there is a wide range of smart mobility initiatives underway around the world. However, it is difficult to use these for benchmark purposes because of the inconsistency of the before and after data.

The authors best advice for smart mobility performance management practitioners remains as stated earlier in this chapter, "if you can't measure it, you can't manage it and if you're only measuring you are still not managing it." Success in smart mobility performance management also calls for boldness and a commitment to the intended outcomes and objectives. It is essential to plan it like you mean it, communicate objectives effectively then take credit for achieving them.

References

1 BetterWorks. Six keys to better performance enablement in 2023. August 15, 2023. https://www.betterworks.com/thank-you-for-reading-6-keys-to-better-performance-enablement-ebook/, retrieved August 15, 2023, at 8:24 PM Central European Time.

2 American productivity and quality consumption, process classification framework. August 15, 2023. https://www.apqc.org/resource-library/resource-listing/apqc-process-classification-framework-pcf-cross-industry-pdf-8, retrieved August 15, 2020, at 3:26 PM Central European Time.

3 Safi, A.M. *Comprehensive key performance indicators assessment for smart city in sustainable cities transportation system.* December 2019.

4 The Balanced Scorecard Institute. What is the balanced scorecard? August 15, 2023 at 9:30 PM Central European Time. https://balancedscorecard.org/bsc-basics-overview/, retrieved August 15, 2023, at 9:31 PM Central European Time.

5 Federal Highway Administration. Office of operations, traffic incident management performance measurement: on the road to success. August 15, 2023, https://ops.fhwa.dot.gov/publications/fhwahop10009/tim_fsi.htm, retrieved August 15, 2023, at 8:30 PM Central European Time.

6 Federal Highway Administration. Office of operations, 2019 road weather management performance measures update. https://ops.fhwa.dot.gov/publications/fhwahop19089/ch1.htm, retrieved August 15, 2023, at 8:31 PM Central European Time.

7 Federal Highway Administration. Freeway management and operations handbook, Chapter 4 – performance monitoring and evaluation. August 15, 2023. https://ops.fhwa.dot.gov/freewaymgmt/publications/frwy_mgmt_handbook/chapter4_01.htm, retrieved August 15, 2023, at 8:32 PM Central European Time.

8 Kirkpatrick and Partners. The Kirkpatrick model. August 15, 2023, https://www.kirkpatrickpartners.com/the-kirkpatrick-model/, retrieved August 15, 2023, at 8:34 PM Central European Time.

9 Sustainable Development Solutions Network. European cities SDG index, prototype version 2019. https://euro-cities.sdgindex.org/#/,%20as%20defined%20by%20the%20United%20Nations, retrieved August 15, 2023, at 8:39 PM Central European Time.

10 Metropolitan Transportation Commission. Vital Signs. August 15, 2023. https://www.vitalsigns.mtc.ca.gov/, retrieved August 15, 2023, at 8:41 PM Central European Time.

11 Utah Department of Transportation. Traffic operations, data. August 15, 2023. https://www.udot.utah.gov/connect/business/traffic-data/traffic-operations-data/, retrieved August 15, 2023, at 8:42 PM Central European Time.

12 Utah Department of Transportation. Cottonwood Canyons. August 15, 2023. https://cottonwoodcanyons.udot.utah.gov/canyon-road-information/, retrieved August 15, 2023, at 8:43 PM Central European Time.

13 Cobb County Government. Transportation, transportation performance dashboard. August 15, 2023. https://www.cobbcounty.org/transportation/about/performance-dashboard, retrieved August 15, 2023, at 8:45 PM Central European Time.

12

Mobility for People

12.1 Informational Objectives

After reading this chapter, you should be able to:

- Describe how people fit within the smart mobility revolution
- Explain the background to people aspects of mobility and the need for further progress
- Describe the characteristics of the people involved in smart mobility
- Describe the characteristics of the end user for smart mobility
- Discuss the balance between satisfying and managing demand
- Describe trip characteristics including trip purposes
- Explain how to capture the needs of the customer
- Explain how to maximize multimodal choice and awareness of choice
- Describe the Mobility Modulor, mass customization and market segment of one
- Define a new human-centric approach to mobility service delivery
- Explain the role of the mobility architect and the network manager
- Define the transition from today to tomorrow
- Describe ideas based on practical experience on how to make planning and design better
- Define next steps
- Describe the role of big data and analytics in measuring mobility, accessibility, trip stage conditions, equity, and traveler characteristics

12.2 Introduction

This is a significant chapter. It builds on years of experience and practical lessons learned to evaluate our approach to people aspects of smart mobility. It is a rich subject area as witnessed by the fact that the material tended to "run off the page" as the chapter was being written. The definition of one concept led to the identification of another concept, leading to the identification of yet another concept. This chain of events brought an understanding that the subject has multiple dimensions and many connections across mobility and society. Writing the chapter took a lot longer than originally expected. While it is important to address this wider context, it is also important to stay focused on the main subject of the book – smart mobility. Consequently, this chapter introduces a wide range of concepts associated with people within the smart mobility revolution

Smart Mobility: Using Technology to Improve Transportation in Smart Cities, First Edition. Bob McQueen, Ammar Safi, and Shafia Alkheyaili.
© 2024 The Institute of Electrical and Electronics Engineers, Inc. Published 2024 by John Wiley & Sons, Inc.

with a focus on "what," and "why" rather than "how." The objective of the chapter is to inspire an innovative approach to planning, design, and delivery of smart mobility services with the more effective focus on people aspects.

The chapter introduces significant innovative approaches to incorporate the characteristics and needs of people within smart mobility planning, design, and operations. The intention of this chapter is to present the new concepts and approaches and provide sufficient knowledge transfer to enable a planning or operations practitioner to take informed decisions regarding the definition of a detailed approach plan for a specific city.

Although the main theme of the book is focused on the application of technology to transport, there is little to be gained through technology if outcomes are not delivered for people. Smart mobility could be viewed as the integrated application of technology, infrastructure, products, and services to transport, with a heavy focus on people needs. Integration is an important term here as there is a tendency to approach transport and smart mobility in highly focused areas. Transport is operated as a series of silos or stovepipes with specialist activity in each channel. Sometimes cockpits are developed for specific jobs and functions. These consist of the data and tools required to conduct a job or a group of jobs. It would be more effective to adopt a complete approach that spans the modes and combines services and infrastructure into a coherent service offering.

Therefore, this chapter focuses on the various people aspects of smart mobility. It is also recognizing that a wider range of people are involved in, and affected by, smart mobility beyond the end user, consumer, or traveler, this includes all segments of society and those responsible for mobility service delivery activities including policy making, planning, design, project delivery, operations, and maintenance for smart mobility services, systems, and infrastructure.

This chapter defines the need to do a better job of understanding the characteristics of people involved in travel and the delivery of smart mobility services. It explains how the technologies that enable smart mobility can be used to gain a better understanding of people's characteristics and the conditions experienced by travelers at various stages in the total trip. The concept of a Mobility Modulor is introduced as a way of harnessing the power of technology to provide better insight into traveler needs and trip characteristics. The central premise of the chapter is that it is possible to do a much better job of incorporating the characteristics of people within the design and delivery of smart mobility services. The basis for improvement involves a relentless drive toward mass customization and a "segment of one" approach to customizing smart mobility services. Our definition of people-centric smart mobility services focuses on a move toward customizing services for individual trips for individual travelers. Taking account of the variation and people characteristics and the conditions experienced at various stages of the trip. It is a tall order, challenging but not impossible.

The chapter begins with an exposition of the various people aspects that include both the customer and other participants involved in smart mobility. The different types of people involved are identified and their characteristics defined. In addition to the variation in human characteristics, it is also important to address the variation in the conditions experienced at different stages in the trip and the purpose of the trip. A trip from point A to point B can include multiple modes, can be preplanned or spontaneous, and can include interim stops along the way. The different conditions experienced each station on the trip have an influence on the human needs as the trip progresses. The purpose of the trip can also have a significant effect on the choice of mode and the trade-off between the cost of travel and time taken.

Having addressed human characteristics, how human needs vary by trip stage and trip purpose, and the insight required for those involved in service delivery to customize services to the traveler, the characteristics of the trip are then addressed. This includes the different stages of a multimodal trip, interchanges between modes and the nature of the trip being made including its purpose, and approach to harnessing the power of technology to create better insight and understanding regarding people and trip characteristics is then discussed.

The concept of a Mobility Modulor is introduced. Le Corbusier coined this term when he developed an anthropometric that spanned scale of proportions that spanned the metric and imperial measurement systems, with a focus on the scale of the human body [1]. The concept is adapted and extended to address variations in human characteristics, travel conditions, and trip purpose.

Building on the Mobility Modulor concept, an innovative approach to planning, design, and operations for smart mobility is presented and explained. This introduces the concept of mobility architecture in which the Mobility Modulor is used within a framework approach, by an individual identified as the mobility architect, responsible for the overall application of the framework to mobility. The steps necessary to transition from today's approach to this innovative approach to smart mobility are also discussed. The chapter concludes with a summary of the essential points contained in the chapter and some suggestions for next steps.

The need to address people aspects in smart mobility was identified a few years ago in a conversation between one of the authors and a leading US transportation practitioner. His background as a former senior Department of Transportation official and his current role, leading a major US institution, provides him with a wide perspective on the needs for transportation and smart mobility. When asked what his focus would be on smart mobility, he was expected to talk about technology or infrastructure but quite surprisingly he said that his focus and his number one concern was people. He noted that very often an extreme focus is placed on technology and the wonderful things that technology can do. Sometimes the point is missed that technology is worthless unless it has a positive impact on the people for which the solutions have been designed. Therefore, this chapter of the book has a focus on people. It explores the characteristics of the people who use the transportation system and those that do not use it but benefit because of the existence of the system. It discusses the various trip types and trip purposes that are satisfied by smart mobility and the need to maintain a balance between satisfying demand, managing demand, and managing the cost of transportation.

As the perception of the customer can be as important as the actual reality of the situation, the psychology of the smart mobility system user is also explored along with ways to improve awareness and perception. By accommodating different customer characteristics and understanding the needs of different trip purposes. The importance of understanding how choice of mode is influenced by the trip's purpose and how factors such as willingness to pay are also heavily influenced by the reason for the trip.

It has been mentioned several times in the book that is important that data is converted to information from which insight and understanding are extracted as the basis for smart decision-making, responses, and strategies. This is revisited here when addressing how big data and analytics can play a vital role in understanding the customer, the customer's travel-making behavior, and the diverse needs for different trip purposes. Uncovering trip patterns and mechanisms underlying transportation demand is key in making sure that human needs are satisfied. At the time of writing, there is a considerable focus on improving the equity associated with transportation. Therefore, the specific topic is called out for special mention when discussing how analytics can be applied to understanding customer characteristics and the variation in the demand for transportation. This also includes variations in the operating conditions being experienced related to the smart mobility system.

The chapter continues with an exposition of a people-centric, multimodal approach to smart mobility service delivery within smart cities. This explains how the characteristics of the various modes can be matched to the customer's needs and trip purpose factors to provide a much better fit between the services provided and current needs. It explains the use of the Mobility Modulor by people involved in each stage in the smart mobility service delivery spectrum. The discussion also includes some thoughts on how to transition from the current approach to this new people-centric approach.

The chapter concludes with the definition of some next steps for the application of a people-centric approach to smart mobility service delivery.

12.3   People Aspects

There are multiple dimensions associated with the people aspects of smart mobility. Obviously, primary focus is on the end user or the traveler. It should be noted that even people who do not travel benefit from improved efficiency, safety, climate change management, and equity that can be delivered by smart mobility. For example, approximately 80% of goods travel by truck, meaning that the travel conditions experienced by these vehicles have a direct impact on the cost of goods in stores. Beyond the end user, it is important to consider the wide range of people involved in planning, designing, and delivering smart mobility services. In fact, changes in thinking and approaches by this group of people are likely to have the largest impact on the end user.

To provide an effective way, this discuss the range of people involved in the delivery of smart mobility services, it is worth considering the high-level concept of operations for smart mobility services. This is illustrated in Figure 12.1.

The range of people that support the delivery of smart mobility services at various stages is split into both public and private sector enterprises. Figure 12.1 shows the private sector people in light blue and the public sector people in darker blue. The periphery of the diagram provides an overview of the job functions involved within each category. The inner core of the diagram shows the high-level roles and responsibilities for each group.

Considering the private sector first, each group can be described as follows.

12.3.1   Executives and Business Planners

These are the senior leaders responsible for guiding the organization, approving the future vision, and ensuring that the vision is implemented effectively and efficiently. They are responsible for overall direction and management of the enterprise. This group also includes people responsible for planning future business and services. Based on a detailed understanding of market needs and customer characteristics, they define a future vision for the enterprise, in alignment with current and future market needs.

12.3.2   Designers

These people are responsible for taking the plans developed by the executive leaders and business planners, turning them into designs that can be implemented, budgeted, and scheduled. This includes the design of products, services, and solutions for the smart mobility market. Designers add the detail and make specific technology choices that turn policies and plans into implementable actions. This can also be outsourced to private sector consultants for specific projects or initiatives.

12.3.3   Sales and Marketing

The people responsible for marketing and selling the products and services being delivered by the enterprise. Marketing people focus on raising awareness and generating interest in the products and services that are either currently on offer or under development. Salespeople focus on generating further interest in the products and services and offer and supporting action by the customer for the services in terms of procurement of goods and services.

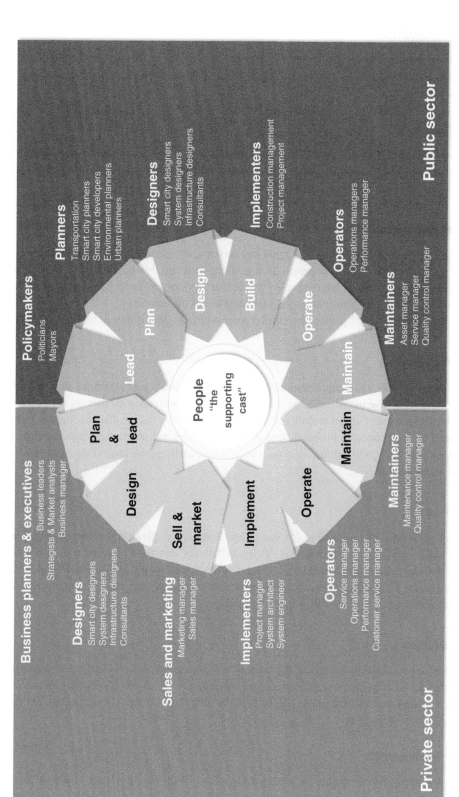

Figure 12.1 Public and private sector people involved in smart mobility.

12.3.4 Implementers

This group of people is responsible for the construction of any infrastructure required and the establishment of services. They ensure that the customer gets what they have paid for at the agreed cost and schedule.

12.3.5 Operators

For services and offerings that require ongoing support, this group of people is responsible for the operations of systems and infrastructure. Operational managers are also responsible for performance management for both long-term and short-term operations.

12.3.6 Maintainers

These people are responsible for ensuring that any systems, infrastructure, or other assets required to deliver the service are kept in appropriate condition, with suitable intervention required. This group of people is also responsible for quality control for services, products, assets, and other items that are delivered to the customer or support the delivery to the customer.

For the public sector say that there are similar roles with a slightly different focus as follows.

12.3.7 Policymakers

These are the nontechnical people responsible for setting policies and strategies that should be followed by everyone else in the public agency. Policymakers set broad goals that are to be achieved through the application of smart mobility technologies. They have the job of ensuring that policies are balanced to maximize the probability of attaining the desired outcomes while avoiding any undesirable side effects. This group also has a role in stimulating new industries and supporting economic development.

12.3.8 Planners

This group of people is responsible for short- and long-range planning at the strategic level. Planning topics can include land use, smart city, urban, the environment, and transit. Planners typically have the responsibility to adopt the policies developed by the policymakers and develop appropriate plans for policy implementation. Note that this role can often be outsourced to a private-sector consultant for specific projects or plans.

12.3.9 Designers

These people are responsible for taking the plans developed by the planners and turning them into designs that can be implemented, budgeted, and scheduled. The ad the details that turn policies and plans into implementable actions. This can also be outsourced to private sector consultants for specific projects or initiatives.

12.3.10 Implementers

This group of people is responsible for getting things done. They take the designs from the designers and manage people and other resources to ensure that the designs are implemented according to the design. They can be responsible for construction and project management for infrastructure, technologies, and services. Implementation can sometimes be transferred to the private sector under the auspices of design-build and operate projects.

12.3.11 Operators

As the name suggests, this group of people is responsible for operating infrastructure and services over the long haul. This usually includes performance management, ensuring the infrastructure and services are delivered in the optimal manner. Here again, this can be transferred to the private sector under the auspices of a design-build and operate arrangement.

12.3.11.1 Maintainers

These people are responsible for asset management and maintenance to ensure the infrastructure, devices, and services are delivered in an optimum way. They are responsible for managing the resources required to maintain all these elements including the definition of the appropriate time for intervention. This role can also be outsourced to the private sector under the auspices of a design-build and operate arrangement.

It is worth noting that for both the public and private sector roles defined above, the optimal results are achieved when there is effective communication between groups. While each group is specialized in a particular stage in the service delivery process, feedback, and feedforward across each stage are essential for optimal operation of the service, infrastructure, or product.

To do their jobs effectively within a people-centric approach, people in each group require significant insight into various aspects of the traveler and the trip. The data needed for each of these groups is summarized in Figure 12.2. The insight and information needs will be addressed in more detail in the following sections on traveler and trip characteristics.

		Traveler characteristics	Trip characteristics	Demand for travel	Variations in travel conditions	Operational performance	Regional need, issues, problems	System, device, infrastructure condition	Future needs	User feedback
Private	Business planners	•	•	•					•	•
	Executives	•	•	•	•	•	•	•	•	•
	Sales and marketing									•
	Implementers				•					
	Operators				•	•				•
	Maintainers							•		•
Public	Policymakers			•	•		•		•	•
	Planners	•	•						•	•
	Designers	•	•	•					•	•
	Implementers									•
	Operators	•	•	•	•	•		•		•
	Maintainers									•

Figure 12.2 Data needs for different people groups.

12.4   Traveler Characteristics

The most important group when it comes to people characteristics is of course the traveler. One definition the authors have often used of smart mobility is that it involves hardware, software, data, and people. Logically, if people are neglected then the investment in hardware, software, and data would be a waste of time. The outcomes for investments are also suboptimized if the characteristics and definitions of the people smart mobility services are designed for are not clearly understood.

12.5   Empathetic Design

Understanding traveler characteristics is much more than measurement of data. While comprehensive and accurate data collection is especially important that data will not be of any use and less is turned into information and the information is used to develop a detailed insight regarding the traveler. Bearing in mind that the overall objective is to improve the quality of service, product, and infrastructure delivery related to smart mobility and do a better job addressing the people aspects.

Understanding the customer is where smart mobility practitioners can learn lessons from marketing. Because of the availability of data and new abilities to turn the data into information, the marketing industry is now considering what is called a "market segment of one." This is interesting because marketing has driven the segmentation of people into distinct categories to tailor marketing approaches, solutions, and services to them. It has long been understood that there are flaws in this approach as people are miscategorized or also averaged into one bucket. Alvin Toffler foresaw this future in his book "Future shock" first published in 1971 [2]. He had the insight that advanced manufacturing, the introduction of sophisticated software, and the ability to develop systems at high quality and low price, would lead to products and services that could be individually customized to each person. Rereading "future shock" also reminded the authors that the main premise of the book was that people would be in a state of "future shock," overwhelmed by a huge array of choices and enormous waves of data, in other words too much change in too little time. There is a strong parallel to the smart mobility revolution which is the central premise of this book. The results of the smart mobility revolution could also be overwhelming if the appropriate actions, knowledge, and tools are not developed.

In addition to the real-time data described in Figure 12.2, it will be necessary to have access to anthropometric data that describes the physical characteristics of the traveler. This can be collected and stored on an archived basis as the data will not change as often as the data in Figure 12.2. The data, summarized in Figure 12.2, is a particularly important part of a people-centric approach that can address equity management in smart mobility. Anthropometry is the comparative study of the measurements and capabilities of the human body. The objectives are to provide easy access and comfortable functionality. This is complementary to ergonomics [3], ergonomic or human factors design that aims to make systems work for the human rather than vice versa. It involves the application of psychological and physiological principles to the engineering and design of products, processes, and systems. Ergonomics is focused on the interaction between people and machines, including the design and engineering of the User Experience (UX). Both sciences rely on the definition of average characteristics for average people and are also male-centric. The answer to the wide variation in people's characteristics has been the average due to the expense and difficulty of collecting data. Mass production has required the adoption of standard averages. In a people-centric approach, averages are replaced by mass customization and a "market segment of one" approach, customizing each individual trip for the individual traveler.

12.6 Customer Psychology

Customization to the individual requires more than an understanding of physical and mental capabilities, it is also important to understand psychology. This is not intended to be a treatise in psychology especially as none of the authors are qualified to produce that, however, it is necessary to point out this added dimension with respect to customization of services for people. All the authors have backgrounds in transportation and the application of advanced technology, so it is telling that none of us have the qualifications to discuss psychology in any depth. The subject will be revisited in the later discussion on education and professional development needs for people-centric approaches to smart mobility services. A good example of the psychology of the traveler lies in the difference between user perception and actual data regarding a situation. For example, a journey to a new destination can seem much longer than a journey to a previously visited destination. The element of the unknown seems to add to the perception of time. Waiting at a bus stop for the bus to arrive can also seem longer than the actual time. Transport for London addresses this by equipping bus stops cannot turn timers providing information on the time to the next bus. An unexpected side effect, related to psychology, was that if the display showed more than 10 minutes to the next bus, then passengers waiting would visit adjacent stores instead of waiting at a bus stop. This leads to the interesting thought that it is not enough to quantify the benefits to be gained from a smart mobility service application, such as the countdown timer on the bus stop, it is also necessary to understand who might benefit. In Transport for London for example, local shopkeepers benefited from the technology implementation. The identification of beneficiaries could also lead to ways to share the cost of technology implementation by partnering with those who might benefit.

Another example lies in the trade-off between privacy and service. Several years ago, a leading US transit agency introduced a smart card system for transit ticketing. They offered users two choices to acquire the card, either anonymously or as a registered user. They made it clear to people who registered that the data would be used to determine trip patterns and travel behavior, which would have a potential impact on privacy. They also informed users that if they chose to receive the card anonymously, then if the card were lost or stolen, they would lose any money they had put on the smart card. If the user chooses to become registered user, then the transit agency would guarantee balance recovery from the instant that the card was lost or stolen. More than 90% of people who chose to use the card opted for the registered user approach. This is a clear demonstration of the trade-off between privacy and service. Psychology is simple. Offer users additional benefits of services and subject to proper management of their data, they are prepared to trade privacy.

From a safety perspective, the perception of safety is as important as actual safety. People who feel they are being placed in a hostile or threatening environment are less likely to use specific modes of transportation. This perception issue could also relate to automated vehicles due to a fear of the unknown, people are reluctant to trust the automated system.

People's psychological aspects must be addressed by providing adequate information on the actual performance of the system and current statistics. This should be combined with the development of trust between the service user and the service provider. An important part in the development and maintenance of that trust lies in clear, accurate, and reliable information. Information must be delivered at the right time, in the right place, and in the appropriate format.

12.7 Trip Characteristics

In the USA Federal Highway Administration Office of transportation policy studies defines trip purposes: work, shopping, family, and social [4].

Work trips address the journey to and from work. Shopping trips address journeys to and from shopping opportunities such as malls and stores. Family trips include journeys to and from schools, church, medical, and other family, or personal business. Social includes visits to friends and relatives and other social and recreational trips. One of the difficulties in trip purpose characterization is that many trips are conducted for multiple purposes. Typically, this is resolved by simplifying the trip and assuming that the trip purpose can be characterized by the trip purpose that consumes most of the time during the trip. While this kind of simplification of trip purpose category can work very well for high-level planning analyses, it will be insufficient to support the sort of improvement in planning, design, and operations required for smart mobility. Remembering that the objective is to enable the development and evolution of empathy for the traveler, going beyond a simple assessment of the demand for travel. Another objective is the achievement of a higher degree of equity in the transportation service delivery process. A proposed trip characterization dataset to support these objectives is shown in Figure 12.3 along with a brief explanation of the need for the data.

The collection of this data on a continuous and reliable basis will require the application of advanced sensing technologies and the use of data management to create the unified dataset from multiple sources. The raw data collected can also be supplemented using trip purpose inference techniques using probe and location intelligence data along with data from electronic payment systems to infer trip purpose and trip characteristics.

Data element Trip characterization data	Notes on the need for the data
Home based	Information delivery options are different for home based
Non home based	This can also cover on trip information needs
Pre planned	May have more flexibility in timing, route and mode
Spontaneous	Places emphasis on trip chain preservation and efficient interchange management
Shopping	Purchases have significant impact on user needs and modal choice
Education and school related	Enables integration with smart school systems
Medical and caregiving	Enables integration with medical appointment systems
Leisure	May affect tolerance to delay and cost versus reliability trade off
Weather conditions	Could make some modes untenable
Time of day, week, month, year	Could influence mode choice and reliability tolerance
Single or multimodal	Trip interchange management
Transport operating conditions	Attractiveness of different modes videos according to operating conditions
Interchanges used	Interchange management, trip time reliability management
Trip time reliability	Critical to mode attractiveness
Visitor or resident	Information need vary
Short or long term	Luggage size and weight
Local trip	Multimodal trip design
National trip	Multimodal trip design
International trip	Multimodal trip design

Figure 12.3 Trip characteristics data.

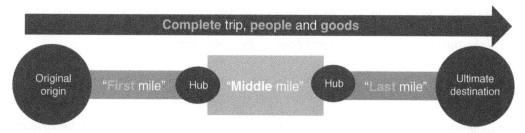

Figure 12.4 A complete trip.

It is also important to collect data and gather insight on the complete trip made by the traveler in addition to data on the individual modes and the interchanges. Figure 12.4 depicts a simplified model of a complete trip.

Note that the most advanced models in use today are tour-based not trip-based. A tour is a chain of trips made by an individual that begins and ends at home without any intermediate stops at home, whereas a trip is a single movement from an origin to a destination. A good example of this is the SF-CHAMP model developed by the San Francisco County Transportation Authority [5]. The simplified single trip is adopted here to enable clear communication of the principles.

An advanced approach to understanding trip characteristics would also involve the establishment of an ideal trip and comparison between the ideal and the actual trip experience. This is depicted in Figure 12.5. This shows in terms of journey time and journey distance how an ideal trip was defined and the variation between the ideal and that experience monitored by the system. In addition to feedback from the traveler, this kind of feedback can also be extremely valuable in better planning and service delivery for smart mobility.

The lower journey trajectory depicts the ideal preplanned trip. The notes represent interchanges between modes or services. The darker-colored upper trajectory represents the actual expedience traveling conditions and the shaded area represents the variation from ideal. This type of ideal versus actual journey comparison would be conducted on a large sample of trips within the region to be managed. This type of trip variation approach can also be customized to the individual, supporting a higher level of equity in transportation service delivery.

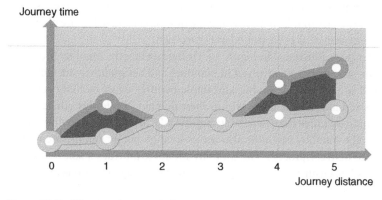

Figure 12.5 Trip experience variation.

As it has been defined, the data needed for a people-centric approach to smart mobility seems daunting. However, with the advent of new sensing and data collection technologies and the ubiquity of smartphones, there are new possibilities for data collection.

12.8 The Mobility Modulor

The complicated and comprehensive need for data as described in the early part of this chapter could present a significant challenge. The data is required to take an innovative approach that will mold mobility services much more carefully to people characteristics and the characteristics of the trip. The answer that has been defined for the data challenge is what is known as the Mobility Modulor. The second part of the name, "Modulor" has an interesting origin. The famous (some might say infamous) French architect le Corbusier, wanted to define a scale of measurement that related to the human body and spanned imperial and metric measurement systems. He found that the metric system, and which units are related to the circumference of the earth, did not do a good job of addressing the people characteristics of design. He believed that if there are machines for flying and machines for driving, there should also be machines for living. His ideas were quite radical, and misuse of the concepts led to some of the most disastrous housing projects in the world. While the authors do not endorse le Corbusier's style of architecture, the concept of a Mobility Modulor is interesting when trying to accommodate people's characteristics and evade conditions experienced in travel from original origin to ultimate destination. The original Modulor was a scale of measurement related to the height of a man with an outstretched arm. It uses averages and takes no account of gender or other variations in the population. Consequently, it is quite ironic that the concept is now used as the basis for an innovative approach for people-centric design and equity.

The mobility as a service Mobility Modulor that is proposed here is an extensive knowledge base of anthropometric and other design factors that can be used as the basis for better smart mobility design. Instead of describing an average man, the data describes the range of human characteristics across genders and capabilities. It provides the basis of human dimensions and capabilities that must be considered when developing effective and successful smart mobility services.

Such a concept is now possible because of advances in data science and data analytics and the emergence of a wider range of data available regarding smart mobility services and travel conditions. It is also analogous to the digital twin techniques now been used to evaluate potential designs and policies, prior to implementation. The idea is to replicate the real world in which designs will operate in an artificial, digital representation in which concepts can be tried and evaluated before implementation. The intent of the modular is to inform a range of factors involved in the planning and design of smart mobility services, this is illustrated in Figure 12.6.

One of the most important aspects illustrated in Figure 12.6 is the fact that the Mobility Modulor acts as a bridge that spans the silos of people involved in planning and design for smart mobility. The range of people involved in planning designing and implementing smart mobility services tend to specialize and others to improve efficiency within their domain. While this efficiency is vital, the nature of smart mobility requires that there be strong cross-fertilization and commonality of design tools and datasets. The Mobility Modulor also serves to align public and private sector entities involved in the service delivery process. The Mobility Modulor provides a sole source of accurate data for each of the silos, ensuring that policy, planning, and design decisions are made within a consistent context. An interesting byproduct of the development and use of the Mobility Modulor will be an increase in the efficiency of data acquisition and collection. The authors

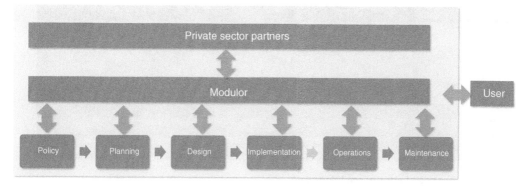

Figure 12.6 The Mobility Modulor in context.

suspect that a great deal of investment is involved in collecting data for specific purposes, rather than developing a central repository. A central repository also enables effective configuration control, with changes and updates reflected across the dataset.

Figure 12.2 provides an initial view of the data required for the various groups involved in the mobility delivery process. However, the Mobility Modulor also needs to contain the data from which information can be derived to assist the end user or traveler. Therefore, the Mobility Modulor dataset would be expanded to include a wider set of data. The data is described in Figures 12.7–12.10.

Data element Trip characterization data	Notes on the need for the data
Home based	Information delivery options are different for home based
Non home based	This can also cover on trip information needs
Pre-planned	May have more flexibility in timing, route, and mode
Spontaneous	Places emphasis on trip chain preservation and efficient interchange management
Shopping	Purchases have significant impact on user needs and modal choice
Education and school related	Enables integration with smart school systems
Medical and caregiving	Enables integration with medical appointment systems
Leisure	May affect tolerance to delay and cost versus reliability trade off
Weather conditions	Could make some modes untenable
Time of day, week, month, year	Could influence mode choice and reliability tolerance
Single or multimodal	Trip interchange management
Transport operating conditions	Attractiveness of different modes videos according to operating conditions
Interchanges used	Interchange management, trip time reliability management
Trip time reliability	Critical to mode attractiveness
Visitor or resident	Information need vary
Short or long term	Luggage size and weight
Local trip	Multimodal trip design
National trip	Multimodal trip design
International trip	Multimodal trip design

Figure 12.7 The need for trip characterization data.

Travel conditions data	
Stress	A factor in mode choice
Vibration	A factor in mode choice
Temperature	A factor in mode choice
Noise	A factor in mode choice
Crowding	A factor in mode choice
Smell	A factor in mode choice
Safety	A factor in mode choice
Perception of safety	A factor in mode choice
Uncertainty and unreliability	A factor in mode choice
Interchnage difficulty	A factor in mode choice

Figure 12.8   The need for travel conditions data.

People characteristics data	
Height	Input required for great design
Reach	Input required for great design
Weight	Input required for great design
Circumference	Input required for great design
Mental capabilities	Input required for great design
Physical capabilities	Input required for great design including hearing, site
assistance tools	wheelchair, walker, crutches, service animal, traveling companion
Gender	Input required for great design
Race	Input required for great design
Social demographic	Input required for great design

Figure 12.9   The need for people characteristics data.

Mobility service delivery data	
Demand for travel	Assessing market for services and service design
Operational perofrmance	The performance of the service for the entire trip chain, devices, infrastructure, and systems
Regional need, issues, problems future needs	Providing the context for service delivery evaluating future markets and designing future services
User feedback	Assessing the user experience

Figure 12.10   The need for mobility service delivery data.

The inclusion of this wider set of data is an extremely valuable aspect of the Mobility Modulor. Since all users are drawing on a common dataset and from a single context, then there is close alignment between the information extracted from those in the mobility delivery process and the end user or traveler.

This view of the data content of the Mobility Modulor is certainly not complete at this stage and further development work will be required to detail the exact contents of the Mobility Modulor to meet the needs of each of the people groups.

It will also be necessary to define a data acquisition plan that explains how each of the datasets will be measured either directly or by inference. While it is beyond the scope of this chapter to provide a data acquisition plan, since the plan must be customized to the individual implementation, Figures 12.11–12.14 provide some ideas on how the various data elements might be collected or acquired.

The possible data collection acquisition methods summarized in the figures serve as examples of how the required data may be collected. The methods will either involve direct measurement of the data or inference of the data from other data sources.

Data element	Data collection and acquisition methods
Trip characteristics	
Home based	Mobility as a Service, mobility as a service, or social media
Non home based	Mobility as a Service, crowdsourcing, or social media
Pre-planned	Mobility as a Service, mobility as a service, or social media
Spontaneous	Mobility as a Service, mobility as a service, or social media
Shopping	Mobility as a Service, mobility as a service, crowdsourcing, social media, or retail systems
Education and school related	Mobility as a Service, mobility as a service, crowdsourcing, social media, or education systems
Medical and caregiving	Mobility as a Service, crowdsourcing, social media, or medical appointment systems
Leisure	Mobility as a Service, mobility as a service, social media, ticketing systems
Weather conditions	National weather services, road weather information systems, crowdsourcing
Time of day, week, month, year	Mobility as a Service, mobility as a service, social media, crowdsourcing
Single or multimodal	Mobility as a service, social media, crowdsourcing
Transport operating conditions	Social media, crowdsourcing
Interchanges used	Social media, crowdsourcing, ticketing systems
Trip time reliability	Social media, crowdsourcing
Visitor or resident	Social media, crowdsourcing
Short or long term	Mobility as a Service, social media, crowdsourcing
Local trip	Mobility as a Service, social media, crowdsourcing
National trip	Mobility as a Service, social media, crowdsourcing
International trip	Mobility as a Service, social media, crowdsourcing

Figure 12.11 Modulor trip characteristics data: data collection and acquisition methods.

Data element	Data collection and acquisition methods
Travel conditions data	
Stress	Crowdsourcing and travel monitoring apps
Vibration	Crowdsourcing and travel monitoring apps
Temperature	Crowdsourcing and travel monitoring apps
Noise	Crowdsourcing and travel monitoring apps
Crowding	Crowdsourcing and travel monitoring apps
Smell	Crowdsourcing and travel monitoring apps
Safety	Crowdsourcing and travel monitoring apps
Perception of safety	Crowdsourcing and travel monitoring apps
Uncertainty and unreliability	Crowdsourcing and travel monitoring apps
Interchange difficulty	Crowdsourcing and travel monitoring apps

Figure 12.12 Modulor travel conditions data: data collection and acquisition methods.

Data element	Data collection and acquisition methods
People characteristics data	
Height	Anthropometric databases, adapted security,or ticketing portals
Reach	Anthropometric databases
Weight	Anthropometric databases
Circumference	Anthropometric databases
Mental capabilities	Medical systems, crowdsourcing
Physical capabilities	Medical systems, crowdsourcing
Assistance tools	Mobility as a Service, mobility as a service, social media, crowdsourcing
Gender	Mobility as a Service, mobility as a service, social media, crowdsourcing
Race	Mobility as a Service, mobility as a service, social media, crowdsourcing
Social demographic	Mobility as a Service, mobility as a service, social media, crowdsourcing

Figure 12.13 Modulor people characteristics data: data collection and acquisition methods.

Mobility service delivery data	
Demand for travel	Assessing market for services and service design
Operational performance	The performance of the service for the entire trip chain, devices, infrastructure, and systems
Regional need, issues, problems future needs	Providing the context for service delivery evaluating future markets and designing future services
User feedback	Assessing the user experience

Figure 12.14 Modulor mobility service delivery data: data collection and acquisition methods.

12.9 Why Is It Valuable?

The Mobility Modulor has several values:

- Single data repository
- Providing the raw material for decision support for travelers
- Support for advanced analytics
- Support for data exchange
- Support for analytics exchange
- Enablement of an empathetic approach to mobility service delivery
- Support for a higher level of equity in transportation

12.9.1 Single Data Repository

The Mobility Modulor acts like glue that links existing data sources and innovative data sources together in a unified dataset. Data also includes that obtain from feedback from the traveler. This has the advantage that all data users are using a single version of the truth when it comes to data. It also makes it more efficient to update the data in a single location. This also provides a higher level of cybersecurity with fewer data access points. The creation of analytics and decision-making is happening within a single data context supporting uniformity and coherence.

12.9.2 Providing the Raw Material for Decision Support for Travelers

The Mobility Modulor will support a higher level of situational awareness for all those involved in the mobility service delivery process. The data acts as the raw material on which better decision support can be provided. Decision support for each stage of the trip from original origin to ultimate destination, delivered in a timely manner, is of vital importance to achieving safety, efficiency, user experience, equity, and decarbonization objectives. This also addresses one of the major barriers to effective modal choice – the easy option is to take a private car. The Mobility Modulor can play a key role in making it just as easy to make other, better, transportation mode choices.

12.9.3 Support for Advanced Analytics

The Mobility Modulor will support advanced analytics across a wider data horizon. These analytics will reveal trends, patterns, and causal mechanisms that can be fundamental to a greater understanding of transportation demand and operating conditions.

12.9.4 Support for Data Exchange

The availability of a central data repository not only supports processing for advanced analytics but also allows data exchange between users. This leads to a higher level of situational awareness across all user groups and affords the possibility for future features such as support for a data market. This would enable data suppliers and data users to establish an agreed value for both data and analytics.

12.9.5 Support for Analytics Exchange

In the same way that data exchange is supported as described above, data analytics can also be exchanged. The Mobility Modulor can feature analytics exchange and library where previously created analytics can be made available for others to use, or to form the basis for their own customized analytics. This enables the Mobility Modulor to serve as an additional function to capture best practices.

12.9.6 Enablement of an Empathetic Approach to Mobility Service Delivery

A critical value of the Mobility Modulor is the ability to support empathetic delivery of mobility services in policy, planning, design implementation, operations, and maintenance activities. Empathy can be defined as:

Cognitive empathy: the ability to understand another's perspective. ...

Emotional empathy: the ability to physically feel what another person feels. ...

Empathic concern: the ability to sense what other people need from you.

[https://www.linkedin.com/pulse/understanding-three-types-empathy-emotional-christy-kennedy]

To develop, empathy requires that all involved in the mobility service delivery process can walk in the shoes of the traveler. This in turn requires a detailed understanding of traveler

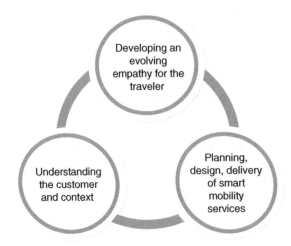

Figure 12.15 An empathetic approach to design.

characteristics, trip conditions at each stage of the trip, and traveler reaction to the various features of smart mobility services.

Therefore, the goal must be to support empathetic approaches through a detailed understanding based on scientific data collection on a continuous basis. This must be accompanied by the development of insight based on that data. The insight should be such that a level of empathy for the traveler can be developed and evolved, as depicted in Figure 12.15.

The process starts with a better understanding of the customer and the context within which the customer abides. In the case of smart mobility, it requires a detailed understanding of traveler characteristics including capabilities and trip purpose. This leads to the development of an evolving empathy for the traveler based on this increased understanding. The empathy created is then applied to the planning design and delivery of smart mobility services.

12.9.7 Support for a Higher Level of Equity in Transportation

Another important value of the Mobility Modulor is an innovative approach to the establishment and preservation of equity within the transportation delivery process. Note that practitioners talk about establishment **and** preservation. Having made things better, it is important that things stay better, and equity is maintained on a continuous and ongoing basis. Equity can be defined as fair and equal access to the needed transportation services by all segments of the community regardless of the people characteristics described in Section 12.3 on People aspects. Equity is an essential attribute of smart mobility, providing equal opportunities for access to a range of services including medical and educational. Since most of our current transportation infrastructures are also funded by public sector funding, it is only fair that these opportunities are made available across as a wide range of the community as possible. The Mobility Modulor supports the management of better equity in transportation by raising situational awareness regarding traveler characteristics and the conditions experienced. The Mobility Modulor can support advanced analytics such as an accessibility index that measures the ease or difficulty of travel from one point of the city to another in terms of the time taken for the trip, the reliability of the trip, and the cost of the trip as a proportion of household income.

The Mobility Modulor serves a key role in providing a channel for feedback from the end user regarding the effectiveness of policies, plans, and designs. This provides a kind of continuous feedback loop that is vital to improving the effectiveness of smart mobility and maintaining it. This can be taken one step further by providing the ability to treat end users as prosumers. This concept was first introduced by Alvin Toffler in his book *The Third Wave*, published in 1984 [6], and is more relevant today than ever. The end user and consumers are much more informed than they have ever been because of the Internet. Instead of treating the end user as a spectator in the design process, it would be much more effective to include the end user within the design process as a professional consumer – "prosumer." Within the auspices of this approach, the end user will play a vital and active input and participation in the planning and design process. The end user will also provide continuous feedback on the pinion and service levels. This can be a two-way channel as the Mobility Modulor can also be used to support traveler education. It could be argued that a vital component in the effectiveness of the traveler's input to the planning and design process is the traveler's knowledge regarding smart mobility services and technology capabilities. In the book *Transport for Humans* [7], it is noted that to have the best experience a range of travel skills are required and that it is important to develop and maintain the skills. This is especially true when smart mobility services involve sophisticated technologies and provide a wider range of choices to the traveler. This can also address the question of how today's travelers are informed about making the best use of services and the travel options available. Transport psychologist at TRL once told one of the authors that "if you drive by habit, you are immune to new information" this could even be extended to general life.

The range of technologies available today including virtual-reality, augmented reality, sophisticated visualizations, and dashboard tools can all be used as enablers for an informed traveler. However, these are just tools that are only useful if a decision is made to use the tools for a different and more effective approach. The authors particularly like the approach taken by the John Deere company in the United States. About 10 years ago, they decided to improve the way in which they involve end customers in product planning and design [9]. John Deere is one of the leaders in manufacturing agricultural and construction equipment. They announced their most important machine ever, which turned out to be a mobile recording studio. At exhibitions, instead of fair and extreme focus on current products and services, John Deere set up the recording studio and invited customers to give their feedback. The customers were invited to sit in a modified backhoe seat and select a range of questions to answer. Two miniature HD cameras were used to record their responses. This moves the focus away from John Deere's current product line to the needs of the customer – a customer-centric approach. At the same time, they also announced a new articulated dump truck and two new diesel-electric hybrid wheel loaders that were engineered from start to finish with active input from real-world customers. The innovative and successful marketing approach taken by John Deere suggests that the industry should not be afraid to open the doors to real-world customers in the planning and design process. The main caveat to this is that the customers need to understand the context and the technology being applied to the solution, requiring that effective customer education be also supported.

12.10 Mobility Modulor Operation and Management

How many of us have had an email from the Chief Executive Officer of a large corporation promising undying love and promising better customer service than before? Only to discover that email came from a "do not reply" address. So how much did that customer service initiative really mean? One of the authors had this very experience recently. Having subscribed to an online service which let the author down during two or three crucial times, the author received an email from the Chief Executive Officer apologizing profusely and promising to make things better. The email came from a "do not reply" email address so one of the authors took the trouble to track down the actual email address for the chief executive and sent him an email complaining about the service and pointing out that sending messages from a "do not reply" email address is not good customer service. Guess what? The author is still waiting for a response from the CEO. The signal is clear, the enterprise wants to talk but is not interested in hearing feedback.

The Mobility Modulor must be a combination of hardware, software, data, and people. It needs to have staff who will monitor the automated signals and automated responses, providing the human touch where required. The Mobility Modulor could function as a smart switch in this respect with specific issues and challenges related to mobility services being directed to the operator of that service. This should be one way to distribute the load and enable a higher degree of customer service than make be provided by a centralized approach.

The roles of planning, designing, implementing, operating, and maintaining the Mobility Modulor can be assigned to either the public or private sector or a combination of both under a public–private partnership arrangement. The consumer-facing services delivered by the Mobility Modulor would seem to be more suited to a private-sector approach, while universal access to data could be better supported by the public sector. Since the private sector will require a revenue-generating mechanism, it is likely that traveler services would be accessible on a paid basis. This raises the question of equity in transportation, balancing this with efficient implementation.

12.11   The Functions of the Mobility Modulor

As described, the Mobility Modulor is a unified dataset to support the end user and all the people involved in the activities related to delivering mobility services. However, as stated several times throughout this book, data does not provide solutions. Data must be converted to information from which insight must be extracted to form the basis for responses, and advisories. To support this, the Mobility Modulor must go beyond a data repository and support information processing, analytics development, data sharing, and analytics sharing. This is illustrated in Figure 12.16.

Stage 1: Data is obtained from multiple sources including roadside and infrastructure-based sensors, other transportation management systems that already exist, connected vehicles, location intelligence, and crowdsourcing. It is assumed that data sourcing and sharing agreements have been put in place to enable the transfer of the data.

Stage 2: Data is ingested, verified, made compatible, and entered the centralized data repository.

Stage 3: One or more data engines are used to convert the raw data into information including summaries and visualizations.

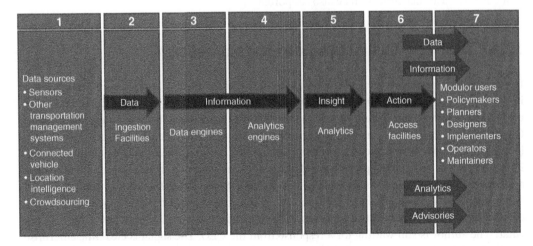

Figure 12.16   Mobility Modulor functionality.

Stage 4: One or more analytics engines are used to produce a range of analytics that define current transportation operating conditions, trip characteristics, user characteristics, and other analytics that will be of use to both sets of people – those involved in service delivery and the end user.

Stage 5: The analytics are stored in such a way that they are accessible to meet the needs of both user groups.

Stage 6: Access arrangements are utilized to enable both user groups to access information, exchange data, share analytics, and in the case of the end user gain access to travel advisories.

Stage 7: Data and analytics are made available to all user groups, delivering a higher level of situational awareness and better possibilities for traveler decision-support.

This is a very high-level simplified view of the functionality of the Mobility Modulor and does not take account of design choices on the use of technology. There are many possible technological approaches to enabling the defined functionality.

12.12   Current Service Delivery Approach

There are several aspects of the current mobility service delivery process that should be considered. The first is "stove piping" or silos of activity.

This usually occurs by mode. For example, those responsible for freeway management and traffic signals do not routinely talk to the transit people who in turn do not talk to the emergency management people, who in turn do not talk much to the payment systems people. Of course, this is exaggerated for effect because there is some degree of cooperation or regional level in many cities around the world. A spectrum can be defined that characterizes the level of cooperation as depicted in Figure 12.17.

Level 1: Conflict: mutual distrust and outright animosity between the different agencies within a region.

Level 2: Peaceful coexistence: understanding the other agencies exist but no data sharing and information sharing between them. There are also no attempts to align objectives between

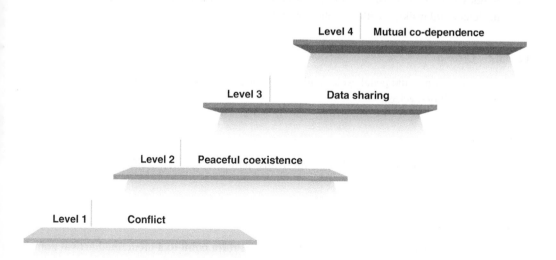

Figure 12.17   Levels of integration.

agencies, in fact, there are cases where different departments in the same agency have not aligned their objectives and operate in a suboptimal way.

Level 3: Data sharing: there are agreements in place and data is being shared across the agencies to minimize duplication of data collection and maximize the value obtained from data. At this level, there may also be some preliminary discussions on alignment of objectives.

Level 4: Mutual codependence: processes within one agency are being conducted for the benefit of another agency and there is an interlocking arrangement between agencies in terms of data collection, information processing, and objectives. It is good alignment between objectives and cost-sharing as required to implement new strategies and innovation.

To characterize the current situation, it is likely that most agencies around the world are somewhere between level II and Level 3. In fact, there is some concern that Level 3 may not be achievable unless there is a unitary organization responsible for all transportation within a city. This is because agencies do not want to get anybody coming between them and a customer and do not want any constraints on applying their own strategies and own policies. Examples of unitary transportation agencies were identified and discussed in Seoul and London as part of the study of global progress in Chapter 5.

12.13 Proposal for a New People-centric Approach

A new people-centric approach to the various activities involved in the delivery of smart mobility services would include the use of the Mobility Modulor to inform activities and decision-making for all groups of people involved and for the entire trip from original origin to ultimate destination. Existing tools and applications can be used to enable the use of the Mobility Modulor. Six example modes of transportation can be used to illustrate the use of the Mobility Modulor for the complete trip.

12.13.1 Walking

This includes walking unassisted, walking with the aid of a walker, crutches, or walking stick, use of a wheelchair and walking with a stroller or shopping cart.

12.13.2 Cycling

Involves the use of manual pedal cycles and the use of electrically assisted cycles, e-scooters, and other micromobility services.

12.13.3 Private Car

Include the use of private cars including minivans, light-duty pickup trucks, SUVs, and other private vehicles.

12.13.4 Transit

The use of mass transit services from a predefined departure point to a predefined arrival point. For example, taking a bus from the departure bus stop to the arrival bus stop. This can also include light rail and trams.

12.13.5 Transportation Network Company

Services such as UBER and LYFT offer on-demand road transport between original origin and ultimate destination using a specially developed smartphone app to summon the service and make the payment.

12.13.6 Paratransit

Specialized transit services specifically developed for those people who cannot use normal mass transit due to physical capability or mental capability constraints. Typically requires preregistration to use the service.

There are more modal choices depending on the people characteristics described above, these examples have been selected merely to illustrate the usefulness of smart mobility services at different stages in the trip. Figure 12.18 illustrates how these six mode choices could fit into the complete trip.

The first three mode choices walking, cycling, and private car can also support the first mile, the middle mile, and the last mile within the complete trip. This assumes that the car or bicycle is parked close to the original origin and can be parked close to the ultimate destination at the other end of the trip. It is assumed that transit will cover only the first mile as there is a need to travel from the original origin to the point of departure for the transit service. Similarly, at the other end, there is a "last mile" need from the point of arrival of the transit service to the ultimate destination. Transportation service providers could cover all three segments of the complete trip; however, for the purposes of this analysis, it is assumed that there will be a need to walk to the pickup point for the transportation service and from the drop-off point to the ultimate destination. Similarly, with paratransit. Most paratransit services will also help with the first mile and last mile. Having defined six example transportation modes within the complete trip framework is possible to consider how the Mobility Modulor and smart mobility applications can deliver value to travelers at each stage of the journey.

A sample of six smart mobility solutions or services has been identified to explain how smart mobility can affect the traveler during each stage of the complete trip. Of course, there are many

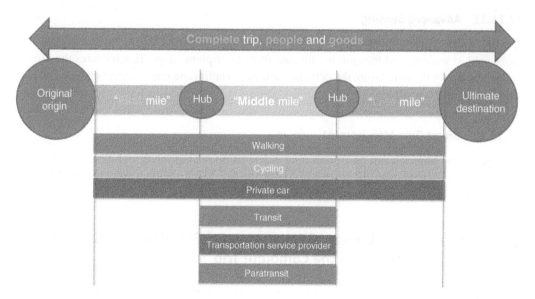

Figure 12.18   Complete trip.

more possibilities for solutions and services but these six have been chosen to demonstrate the effect that smart mobility services can have on each stage of the complete trip and in fact the complete trip from original origin to ultimate destination. The six smart mobility solutions and service examples are described as follows.

12.13.7 Advanced Transportation Management

This category of solutions and services covers the management of freeways and limited access roads, expressways, urban arterials, and urban surface streets. Using a variety of telecommunications, sensing, information delivery, and data management technologies.

12.13.8 Smart Vehicles

This category encompasses automated vehicles, connected vehicles, and driver support systems. Automated vehicles are assumed to be capable of driving without a driver, while connected vehicles are assumed to have a two-way communication link between the vehicles and the back office. Driver support systems encompass applications like automated lane keeping and distance keeping as well as self-parking vehicles.

12.13.9 Electronic Payment

Services enable electronic payment for tolls, parking, and transit fares.

12.13.10 Freight and Urban Logistics Management

In a comparable manner to advanced transportation management, but with a focus on goods movement and urban logistics.

12.13.11 Advanced Sensing

A range of sensing techniques including Internet of Things roadside infrastructure-based sensors, probe vehicle sensors, and location intelligence from smartphone apps. This includes the use of opt-in crowdsourcing to measure the demand for transportation and trip-making behavior and the use of social media analysis.

12.13.12 Smart Data Management

The application of a range of data processing, information management, and data science hardware and software to manage the process of converting data to information, extracting insight, and developing appropriate action plans.

12.14 The Possible Effects of Smart Mobility Solutions and Services on Different Modes Within the Complete Trip

The six example solutions and services are mapped to the different trip stages in modes within the complete trip as shown in the following figures.

12.14.1 Walking

Figure 12.19 provides some examples of how the walking experience can be improved by the application of smart mobility solutions and services, within a people-centric design approach.

12.14.2 Cycling

Figure 12.20 provides some examples of how the cycling experience can be improved by the application of smart mobility solutions and services, within a people-centric design approach.

12.14.3 Private Car

Figure 12.21 provides some examples of how the private car use experience can be improved by the application of smart mobility solutions and services, within a people-centric design approach.

12.14.4 Transit

Figure 12.22 provides some examples of how the transit use experience can be improved by the application of smart mobility solutions and services, within a people-centric design approach.

Within a people centric design approach Advanced transportation management	Adaptive and coordinated traffic signal control will improve time allocations to pedestrians. More effective and efficient parking should improve the pedestrian experience by enabling drivers to park at their destination. This will be especially valuable in inclement weather
Smart vehicles	Automated vehicles can minimize collisions with pedestrians. Increased safety will also be provided through onboard systems that can detect pedestrians and the probability of collision. The solutions and services could remove a volume of vehicles on the road making it safer for pedestrians
Electronic payment	Removing the need to use cash can improve overall safety and security for pedestrians
Freight and urban logistics management	Rerouting heavy vehicles avoiding urban areas could improve safety for pedestrians. Better parking management for urban logistics can also improve pedestrian safety by maintaining sightlines and appropriate safety distances
Sensing	Better traffic signal timings that take account of the needs of pedestrians. This can include the use of walking speed sensors to customize traffic signal timings to the characteristics of different pedestrians. A two-way dialogue between the pedestrian andt he back office will provide better information to the pedestrian and better data on operating conditions. Social media analysis and crowdsourcing data on an opt in basis will also provide feedback on trip making patterns, travel behavior and user sentiment
Smart data management	Information derived from data will assist pedestrians with choice of the best assessment of the risks involved. They can also provide additional services that can be accessed via a smart phone app to enhance the pedestrian experience

Figure 12.19 Improvements in the walking experience.

Advanced transportation management	Adaptive and coordinated traffic signal control will improve time allocations for cyclists. Smart parking also applies to cycles for modal interchange
Smart vehicles	Automated vehicles can minimize collisions with cycles, using collision avoidance technology. Could also reduce the volume of vehicles on the road making it safer for cyclists
Freight and urban logistics management	Rerouting heavy vehicles avoiding urban areas could improve safety for cyclists
Electronic payment	Can assist a cyclist to pay for parking or pay for modal transfers when taking a multimodal approach including cycling. The use of electronic payment for road users charging could also improve conditions for cyclists
Sensing	Better traffic signal timings that take account of the needs of cyclists. A two-way dialogue between the cyclist in the back office will provide better information to the cyclist and better data on operating conditions. Social media analysis and crowdsourcing data on an opt in basis can provide feedback on cyclist user sentiment and operating conditions is feedback to transportation service operators
Smart data management	Information derived from data will assist cyclists with the choice of the best route and assessment of the risks involved. They can also support additional services that can be accessed via a smart phone app to enhance the cycling experience

Figure 12.20 Improvements in the cycling experience.

1	Advanced transportation management	Transportation management will make commuting more dependable, while also making the driver aware of alternative transportation choices for that time of day. It will be easier to identify open parking spaces and more efficient to pay for the spaces
2	Smart vehicles	Will make travel safer, more cost-effective, and reduced omissions and fuel consumption. Better driver information and better traffic management from probe vehicle data. Improved safety through driver decision support and automation. Faster and more reliable commute times
3	Electronic payment	Within a mobility as a service approach this will enable easier payment for tolls and other transportation services including parking
4	Freight and urban logistics management	Better freight in urban and logistics management will help to reduce congestion on the roads
5	Sensing	Better traffic management because of richer, larger sample size data. A two-way dialogue between the vehicle and the back office will provide better information and better data on operating conditions. Social media analysis will provide data on trip making patterns, travel behavior and traveler sentiment. Crowdsourcing data from smart phones and in opting basis will also add to the insight and improve transportation management as well as providing continuous and real-time data on travel conditions
6	Smart data management	Information derived from data will be used to support decisions and provide a customized travel experience. Data on performance management and prevailing transportation conditions will be used to fine-tune transportation services and traffic management strategies

Figure 12.21 Improvements in the private car use experience.

12.14.5 Transportation Service Provider

Figure 12.23 provides some examples of how the transportation service provider experience can be improved by the application of smart mobility solutions and services, within a people-centric design approach.

Advanced transportation management	Traffic signal priority for transit vehicles will improve transit commutes. More effective fleet management and better passenger information will lead to efficiency and user experience gains. Smart parking will also support modal interchange what a driver can park a vehicle and then board the transit system
Smart vehicles	Automated vehicles can replace the driver, reducing the cost of operations. Also enables better travel information and better traffic management from probe vehicle data. This will also lead to reduced costs through eventual removal of the transit operator. Faster and more reliable transit trip times
Electronic payment	Will make it easier to pay for transit and provide more flexible fare structures. Also improve security by removing cash from the point of service
Freight and urban logistics management	Will have an impact on the cost of goods. Buses will benefit from reduce congestion because of better freight and urban logistics management
Sensing	Better transit management because of richer data as the basis for better travel information. Improved fleet management and schedule adherence. Smart phone apps will support a two-way dialogue between the traveler and the back office which will provide better data regarding operating conditions and better information for the traveler. The use of social media analysis will provide additional insight on trip making and travel behavior as well as traveler sentiment. This also includes the use of crowdsourcing data from smart phones on an opt in basis
Smart data management	Information derived from data will be used to support decisions and provide a customized travel experience. Data on performance management and prevailing transportation conditions will be used to fine-tune transportation services and traffic management strategies

Figure 12.22 Improvements in the transit experience.

Advanced transportation management	Better transportation management will result in shorter and more reliable journey times. smart parking Can also be extended to resolving congestion conflicts caused by drop-off and pickup for transportation network companies
Smart vehicles	Automated vehicles can reduce the cost of operations by enabling better fleet management can be delivered through more effective communications with the driver and data from the vehicles. Automated vehicles could remove a substantial cost component from transportation network company operations and reduce costs. in future transportation network operators may use flying cars which would reduce travel times
Electronic payment	Travelers can make use of the integrated payment system to pay for transportation network company services
Freight and urban logistics management	Better freight and urban logistics management will reduce congestion prevailing advantage and cost savings to transportation service providers
Sensing	Smart phone apps are already in use to support reservations and electronic payment; the technology can be expanded to include advanced data collection on all aspects of operations. This will include social media analysis and crowdsourcing data on an opt in basis to provide additional data on trip making, travel behavior and traveler sentiment
Smart data management	Better overall fleet and network management through the availability of real-time data regarding current vehicle locations on prevailing transportation conditions, traveler sentiment feedback as input to improving services and for quality control

Figure 12.23 Improvements in the transportation service provider experience.

12.14.6 Paratransit

Figure 12.24 provides some examples of how the paratransit experience can be improved by the application of smart mobility solutions and services, within a people-centric design approach.

12.15 Transitioning to a New People-centric Planning Design and Delivery Approach

Having defined the future vision, it is now necessary to address how the industry can get there in manageable steps. A transition plan from today's situation to tomorrow's desired vision, at its simplest level consists of three components.

Advanced transportation management	Better transportation management will result in shorter and more reliable journey times
Smart vehicles	More efficient paratransit can be achieved with autonomous vehicles replacing the driver. Can deliver better fleet management through two-way communications with the driver and providing a data stream from the vehicles. The use of automated vehicle technology could also provide a more effective and efficient service to get from home to healthcare opportunities, as an example
Electronic payment	Electronic payment services and solutions can be used to make the administration of paratransit more efficient, manage registration and any associated costs
Freight and urban logistics management	Better freight in urban logistics leads to reduced congestion urban areas, increasing the efficiency of paratransit services
Sensing	Better data leads to better information leads to better decisions leads to better paratransit management. Smart for apps can provide support for reservations and provide feedback on trip making patterns and travel behavior as well as traveler sentiment
Smart data management	More efficient and effective paratransit journeys through better management of the transportation process and the overall paratransit actuation and management system. Integration of the paratransit management system with medical appointments systems can also lead to more efficiency and better coordination of services

Figure 12.24 Improvements in the paratransit experience.

12.15.1 A Definition of Where the Industry Is Today

This will define the current state of technology, the services currently available, the business models, and the organizational arrangement currently in place.

12.15.2 A Definition of the Desired Future Vision

A typical future vision will define the desired outcomes such as:

- Safety: reduction in accidents, reduction infidelities, rejecting and injuries, and reduction in property damage. Improved safety reduces the impact of mobility services and improves travelers' perception of safety.
- Efficiency: reduce congestion, reduce delays, reduce stops, improvement in travel time reliability, more efficient choice of mode for travel
  - enhanced user experience: did the user have a pleasurable experience, where any difficulties managed appropriately, did the user have the appropriate choice of travel options.
- Decarbonization: are we minimizing fuel consumption and emissions through managing unnecessary travel and variations in travel speed. Is the most appropriate mode being used for each trip?
- Equity: ensuring that every citizen and visitor has the appropriate level of access to travel opportunities within the city.

These elements are discussed in detail in Chapter 6 on Planning for Smart Mobility. In addition to the vision, a range of enabling activities will also be required, described as follows.

12.15.3 Defining and Implementing Mobility Architect and Network Manager Roles

The effectiveness of the Mobility Modulor lies in the adoption of a multimodal, complete trip approach to all aspects of smart mobility service delivery. This will require the definition and implementation of two new roles for transportation practitioners. The first of these is what the authors call the mobility architect. This role supports the coherent application of the Mobility Modulor across all modes and all trip stages. The role would also involve coordination on data sharing agreements and data sourcing. The second is the role of the network manager. In other segments such as computer networks, energy, and water supply, the network manager serves to coordinate all activities related to service delivery and monitor the demand for services and variations in supply conditions. Typically, this role is fragmented across various modes within an urban area. Together these two roles will ensure that the Mobility Modulor is used effectively and that service delivery is coordinated across the city.

12.15.4 Changes in Education

It is obvious that insufficient attention is currently being paid to the characteristics of the traveler and the traveler conditions being experienced. Part of the solution to this would be the introduction of people-centric subjects within existing educational and professional development frameworks. This would enable new planners and engineers to be equipped to adopt the people-centric approach and enable mid-career professional development opportunities for those already working in the industry. Educational opportunities should also be designed in a way that encourages a higher number of women to become involved in planning and design processes for mobility services.

This education should also be extended to improving the traveler's knowledge by educating and informing the travelers knowledge on how best to use mobility services and take advantage of new decision-support possibilities.

Education and professional development activities can also include activities specifically designed to develop empathy with the traveler. A notable example of this is the bus company that requires drivers to experience cycling near fast-moving buses [10]. Empathy development activities – the bus driver on bicycle example. Ideas and techniques can also be adapted and adopted from other industries such as the design industry where 3D modeling, prototypes, and digital twins are all used to explore the end-user experience and provide feedback into the design process. This can lead to a multisensory experience that develops a greater understanding for the designer and the end user. A dialogue with a leading design company (not involved in transportation), led to the new insight that useful design should go beyond providing the basic functions and services and deliver a pleasurable experience. The ultimate objective should be to design future mobility services to meet the needs of the user, rather than have the user adapt to the service.

12.15.5 Collaboration on Data Collection

The Mobility Modulor at its core is a unified dataset that can form the basis for efficiency and coherence in planning and operations for mobility services. It will be important to increase the level of collaboration between public agencies and private sector companies regarding data collection. Public agency data-sharing arrangements will be required, and suitable public–private arrangements should be defined to support the most effective collection, acquisition, and use of data. This also extends to the use of the latest in data science approaches to turn the data into information and insight using advanced analytics and inference models.

Current data should be supplemented by new and emerging data sources including probe vehicle data, connected vehicle data, and location intelligence data from smartphones. These are all required to be subject to appropriate data governance arrangements to safeguard privacy and preserve the security of the data.

Data collection acquisition can also go beyond traditional sources to provide a larger and continuous data supply. For example, anthropometric data could be sourced from retail operations that use special scanners to support an augmented reality approach that allows people to see what clothes look like before trying them on, or even at home with her entering the store. There are considerable opportunities for creativity and innovation in collecting and acquiring the data required for the Mobility Modulor. It may also be possible to form partnerships with travelers for them to opt in to provide anthropometric data using a smartphone.

12.16 Summary

The task of describing how people fit within the smart mobility revolution proved to be a bit more challenging than expected. Research into the subject provided several revelations regarding current weaknesses in the planning design and implementation of mobility services. This indicates a lack of viable data and the inability to develop a suitable level of situational awareness regarding mobility services, operating conditions, and the conditions experienced by the end user. This also extends to the characteristics of the end user.

An oversimplified approach is being taken to planning and modeling leading to the needs of significant parts of the traveling public to be poorly addressed or neglected. It is obvious that the needs and characteristics of the traveling public exhibit such are range of variation that simple averages are insufficient to provide the basis for the design parameters. Variations include height, weight, reach, physical and mental capabilities, and the use of additional equipment such as strollers, walkers, crutches, support from traveling companions and service animals. While it will be important to balance the cost-of-service delivery and the need to satisfy demand compared to managing it, the industry can do a lot better than it does today.

The variation in traveler characteristics is accompanied by variation in trip characteristics including trip purpose. The choice of available transportation modes can be severely limited by the trip purpose. For example, a traveler would not attempt to travel to a retail outlet to buy a widescreen television, using a micromobility electric scooter. Weather conditions have a similar effect.

There are new data collection acquisition possibilities that can inform and improve our planning and design efforts for mobility services. The facilities that enable the collection and transmission of this data can also be used to support a two-way dialogue with the traveler, enabling a much larger sample size of data regarding traveler sentiment and conditions being experienced. There are also new opportunities using such facilities to treat the traveler as a prosumer with a much more interactive role in the planning and design process.

One of the fundamental goals and mobility service delivery is to ensure that travelers have an appropriate level of choice regarding mode of transportation. This should also be extended to awareness of the choices and sufficient knowledge to make effective use of them. The emerging mobility as a service solutions take a great first step toward informing travelers.

The concept of the Mobility Modulor is introduced as a single repository of data capable of supporting a new people-centric approach to the delivery of mobility services. This also supports progress toward "mass customization" and a "market segment of one" approach to maximizing equity by tailoring services to the individual, using the new insight as a foundation. The role of big data and analytics has also been explored to convert the data into information from which insight can be extracted and actioned.

The Mobility Modulor has also been set in context of a proposed new people-centric approach to mobility services, using the Mobility Modulor. Two new roles for transportation practitioners have also been defined in support of the use of the Mobility Modulor, the mobility architect and the network manager. These roles will play an important part in the successful operation of the innovative approach.

To support a smooth transition from the current approach to the proposed people-centric approach, a set of proposed activities has been defined and described.

A new people-centric approach to mobility services will not be achieved overnight as significant momentum associated with the current approach must be overcome. This includes the delivery of new knowledge and information to practitioners and travelers. It will not be an easy task, but the potential outcomes make it a worthy cause.

References

1 Arellano, M. On the Dislocation of the Body in Architecture: Le Corbusier's Modulor [Sobre la dislocación del cuerpo en la arquitectura: El Modulor de Le Corbusier] 27 Sep 2018. ArchDaily. (Trans. Johnson, Maggie). https://www.archdaily.com/902597/on-the-dislocation-of-the-body-in-architecture-le-corbusiers-modulor ISSN 0719-8884, accessed December 30, 2022, at 5:45 PM Central European Time.

2 Toffler, A. 1971, Future shock, Bantam Books. https://www.amazon.com/Future-Shock-Alvin-Toffler/dp/0553132644, accessed May 19, 2023, at 8:55 PM CET.

3 Wikipedia. 2022, Human factors and ergonomics. Wikipedia en.m.wikipedia.org/wiki/Human_factors_and_ergonomics, accessed May 19, 2023, at 10 PM CET.

4 Federal Highway Administration, Office of Transportation Policy Studies. 2023, https://www.fhwa.dot.gov/policy/otps/nextgen_stats/chap7.cfm, accessed May 18, 2023, at 1 PM CET.

5 San Francisco County Transportation Authority. 2023, SF-CHAMP modeling, San Francisco County Transportation Authority. https://www.sfcta.org/sf-champ-modeling, accessed May 17, 2023, at 10 PM CET.

6 Toffler, A. 1984, The Third Wave, Bantam. https://www.amazon.com/Third-Wave-Alvin-Toffler/dp/0553246984, accessed May 16, 2023, at 5 PM CET.

7 Dyson, P. *Transport for humans (perspectives)*. London Publishing, 2021. https://www.amazon.com/Transport-Humans-Perspectives-Pete-Dyson/dp/1913019357/ref=sr_1_1?crid=32NQMCSSFVU64&keywords=transport+for+humans&qid=1684423042&s=books&sprefix=transport+for+humans%2Cstripbooks-intl-ship%2C153&sr=1-1, retrieved May 19, 2021, at 9:30 PM CET.

8 Engineering News Record. March 22, 2011, John Deere's Newest Machine Is A Recording Studio, ENR. https://www.enr.com/blogs/13-critical-path/post/15691-john-deere-s-newest-machine-is-a-recording-studio, retrieved May 19, 2021, at 9:30 PM CET.

9 The Logical Indian. *Mexico 19 November 2022, Mexico: company puts bus drivers and bicycles to demonstrate risks for cyclist riding past them*. The Logical Indian. https://www.youtube.com/watch?v=pa4qpLos3xs, retrieved May 19, 2021, at 9:45 PM CET.

13

Smart Mobility Policy and Strategy

13.1 Informational Objectives

After reading this chapter, you should be able to:

- Provide a definition of policy and strategy
- Explain the role of policy in smart mobility success
- Describe the need to balance industry encouragement and growth with regulation and enforcement policy
- Discuss the implications of policy leading technology
- Discuss the implications of technology leading policy
- Explain the most effective means of engaging the private sector
- Describe an approach to manage disruption
- Describe how to incorporate policy and strategy into an effective smart mobility planning approach
- List challenges and opportunities associated with policy and strategy
- Discuss how to optimize policymaking for smart mobility

13.2 Introduction

Policy and strategy are often considered as the softer side of smart mobility. However, the presence of effective policy and good strategy are the absence can have a significant impact on the success of any smart mobility initiative. In this chapter, both policy and strategy are considered, with a detailed discussion on what exactly they are, the influence that they have in smart mobility, challenges, and opportunities associated with policy and strategy and some ideas on how to manage them for good effect. Managing policy and strategy as part of a wider issue related to the alignment of different levels of management within a smart mobility planning and implementation organization.

Policy and strategy are treated together in this chapter as both relate to the definition of broad directions and outcomes, leaving the detail to operational and tactical levels. Policy and strategy also define desirable outcomes providing guidance to the overall smart mobility initiative. It could also be understood that this guidance also provides a definition of outcomes to be avoided or undesirable side effects of the smart mobility implementation.

Smart Mobility: Using Technology to Improve Transportation in Smart Cities, First Edition. Bob McQueen, Ammar Safi, and Shafia Alkheyaili.
© 2024 The Institute of Electrical and Electronics Engineers, Inc. Published 2024 by John Wiley & Sons, Inc.

13.3   Smart Mobility Policy

According to the Merriam-Webster dictionary [1], policy can be defined as follows:

Prudence or wisdom in the management of affairs

Management or procedure based primarily on material interest

A definite course or method of action selected from among
alternatives and in light of given conditions to guide and determine present
and future decisions

A high-level overall plan embracing the general goals and
acceptable procedures especially of a governmental body

All of these definitions are applicable to smart mobility. Prudence or wisdom is required to ensure that both public and private expenditure are invested appropriately to deliver outcomes. Management and procedures based on material interest relate to the need to deliver outcomes in smart mobility related to safety, efficiency, enhanced user experience, climate change management, and equity management. It can also be assumed the material interest also includes the actions and motivations of the private sector engaged in the delivery of smart mobility solutions and services. The third bullet relating to a definitive course or method of action selected from among alternatives seems to be the most apt for the smart mobility situation. As stated earlier, this relates to policy as the definition of a chosen path for both current and future decisions. Taking account of current conditions includes the current status of technological capabilities and constraints.

Current conditions can also relate to the organizational ability of the smart mobility implementation organization including appropriate skills and organization configuration to support the desired approach. The fourth bullet reinforces the thought that policy and strategy are typically high-level, broad-brush plans that guide more detailed action, rather than providing a definitive plan of attack. It is likely that there are multiple tactical plans available that would fit within the policy and strategy framework. For example, if a strategy is focused on congestion reduction, then detailed approaches that would fit under the strategy would include those that attempt to manage demand and those that attempt to manage supply through better infrastructure and service management.

13.4   Smart Mobility Strategy

Starting again with a definition as provided in the Merriam-Webster dictionary [2]

The science and art of employing the political, economic,
psychological, and military forces of a nation or group of nations to afford
the maximum support to adopted policies in peace or war

The science and art of military command exercised to meet
the enemy in combat under advantageous conditions

An adaptation or complex of adaptations (as of behavior,
metabolism, or structure) that serves or appears to serve an important
function in achieving evolutionary success

These definitions provide an indication of how strategy plays an important role in the success of a smart mobility initiative. Although couched in military terms, the first definition describes strategy as a means of aligning and coordinating all available resources and directing them toward the desired outcome. The second bullet, again described in military terms, explains how strategy is used to create advantageous conditions for success. In this case, the enemy can be considered to be unacceptable levels of safety in mobility, inefficiency, poor user experience, undesirable impact on climate change, and inequity.

The third part relating to the achievement of evolutionary success suggests that strategy can be an important tool in making incremental progress toward the desired goals and outcomes. This may seem counterintuitive when the existence and emergence of a smart mobility revolution have been described earlier in the book. However, the two concepts can be reconciled in this case as radical and accelerated change caused by the smart mobility revolution can form part of a structured and coherent plan that enables progress toward outcomes to be made in measurable increments.

The revolution does not need to be chaotic, just radical, and accelerated. While the definitions discuss military forces as the resources being aligned and coordinated toward a successful outcome, the principles are equally applicable to the resources available for smart mobility namely:

13.4.1 Hardware

Hardware technologies are applied within the framework of smart mobility services. These include processing, data storage, and telecommunications infrastructure.

13.4.2 Software

Software and solutions are also applied within the framework of smart mobility services.

13.4.3 Services

The services are driven by hardware and software that deliver value to the users of our smart mobility system.

13.4.4 Data

Data is being recognized as an important resource within the smart mobility realm along with the availability of a wide range of data, it is also necessary to have the capability to turn the data into information, extract insight, and develop action plans and responses based on the new insight.

13.4.5 People

People involved in smart mobility include a wider range than just the end user. They include operators, planners, managers, and staff who populate the planning and implementation agencies and private sector enterprises that support smart mobility initiatives.

It could be quite useful to consider the things to avoid through the implementation of smart mobility as the enemy. Lower than acceptable levels of safety costs lives and inefficiency costs time and money. A poor user experience can put people off using certain modes of transportation and may even affect the longevity of politicians. Effective climate management affects our future and our children's future and in equity in transportation as a disproportionate effect on certain segments of society. These could all be considered as "enemies."

Figure 13.1 The relationship between policy and technology.

13.5   What Should Come First, Policy or Technology Application?

Figure 13.1 illustrates the relationship between policy and technology. It is depicted as a circular relationship because in an ideal world, policy would be developed and technology application would follow, adhering to policy guidelines. In reality, technology applications typically lead to the development of policy. The example described above advanced air mobility, automated vehicles, and Mobility as a Service are great examples of technology application that have driven the need for policy to be developed.

It is also a circular relationship because policy can be influenced by technology and vice versa. Policies will be informed, and good policy will be informed by a strong understanding of what can and cannot be done in smart mobility [3]. This requires a knowledge of technological capabilities and constraints, preferably obtained from practical experience from prior deployments. This makes it particularly important now that an effective mechanism for sharing practical experience and lessons learned from field trials and from full-scale implementations around the globe is available. Technology application can also be influenced by policy which defines what can and cannot be allowed. For example, a technology application that results in disproportionate equity across the population may well be in violation of policy.

13.6   The Influence of Policy and Strategy on Smart Mobility Success

One of the ways in which policy and strategy influence smart mobility success relates to the high-level nature of these things. By operating at a wide level with a low level of detail, many more alternatives can be explored. This wider span of investigation raises the probability that the correct choices will be made. Of course, it is still necessary to develop detailed designs and approach plans that fit within the framework of policy and strategy. However, these come after the broad choices have been evaluated and decided allowing the higher level of resources required for detailed design to be focused on a fewer number of options.

Figure 13.2 illustrates this concept, which is an upside-down triangle. At the top, strategy and policy explore a wide range of options and consider the best alternatives, then define broad directions and frameworks. A high-level what will be done, but what be done, what is desired, and what is undesirable is defined at this level. A wide range of options can then be investigated and narrowed down to a smaller number that can be the subject of detailed design and evaluation.

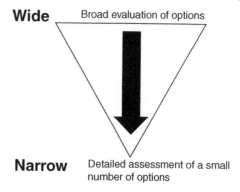

Wide — Broad evaluation of options

Narrow — Detailed assessment of a small number of options

Figure 13.2 From wide to narrow, broad evaluation to detailed assessment.

This broad evaluation of options at the strategy and policy level also allows a wider range of input from stakeholders to be considered. The likely effects of the strategies on a wide range of people and organizations can be assessed as part of the process to narrow down the options, which can then be subject to detailed assessment. Policy and strategy are important aspects of resource management. Even with the most sophisticated technology available and high-speed computing, there is still a need for careful stewardship of resources, especially the people involved in the process.

People can be thought of as having a certain amount of bandwidth, just like a communication network. To get an effective level of attention from people, it is necessary to focus on the amount of information being provided. This is especially true as more data is available and that data can have a swampy effect on the human attention span, and less is turned into effective information. Policy and strategy provide tools to assess options at a higher level, deferring the application of additional resources for detailed evaluation once an initial start has been conducted on the options policy and strategy also service communication vehicles that enable the assessment of the high-level options to include the number of people, especially those important to the subsequent success of the implementation.

13.7 Examples of Smart Mobility Technology Applications

In this section, a few examples of smart mobility technology applications are defined with policy and strategy aspects relating to implementation of smart mobility applications discussed. The intention here is to show how policy and strategy can have both positive and negative impact on the success of smart mobility applications.

13.7.1 Micromobility

Micromobility is an umbrella term that covers the use of electric scooters and electric bicycles on a pay-as-you-go basis. These days such vehicles are provided on a "dockless" basis with no need for special vehicle stands or parking. The vehicles are simply left at strategic locations around the city. Users who have preregistered for the use of the service can use a smartphone app to rent a vehicle. The smartphone app manages the reservation and appropriate electronic payment for the service. The smartphone app can also apply geolocation techniques to the vehicle, enabling the vehicle's approximate location to be determined any given time. This is especially useful for micromobility service operators to track vehicle locations and decide if vehicle repositioning is required to maximize the use of the service and revenue.

Several cities around the globe have been at the forefront of the introduction of micromobility services. Most of these micromobility services have been driven by private sector investment which has made services available at an early stage but has often left policymakers flat-footed. It is estimated that more than $5.7 billion has been invested by the private sector and micromobility startup since 2015 [4]. In some cases, the availability of the service and hence the technology has led to the definition and development of suitable policies for micromanagement.

This has led to the discovery of opportunities and challenges associated with micromobility services. The obvious opportunities of micromobility lie in the ability to offer flexible transportation services designed for short-distance use within an urban environment. Powered vehicles offer a way to address urban topology barriers but could make nonmotorized transportation options less attractive. For example, cities with a considerable number of steep hills such as San Francisco, San Diego, and La Paz may offer less attractive topologies to nonmotorized transport options such as pedal-powered bicycles, and scooters. Micromobility services may also offer at least a partial answer to the issue of ambient weather conditions.

Thunderstorms in Orlando, extreme high or low temperatures, and sandstorms in Dubai can make nonmotorized transportation options less attractive. Micromobility services may also be unsuitable under such environmental conditions but may represent a better option than traveling using nonmotorized transportation or on foot. The question of ambient weather conditions as a factor in the choice of mode of transportation, along with other factors such as young, mobility-impaired travelers, and old travelers addressed in more detail in Chapter 12 where modes of transportation and customer characteristics were discussed.

Early experience in the introduction of micromobility services in cities has also uncovered number of challenges. One of the more significant challenges is speed enforcement. Micromobility vehicles can use roadways but can also use pavement and sidewalks shared by pedestrians. To safely share the space, it is necessary to develop and apply policy and regulation describing the appropriate speed for micromobility vehicles by location. The average speed of an electric scooter (eScooter) is 25 km/h, and some high-powered versions can reach speeds of 160 km/h [5]. Compare this to average pedestrian walking speed of 5 km/h [6], average pedal cycle speed of 22 km/h [7] and an average manual scooter speed of about 12 km/h [8].

The speed differential is illustrated in Figure 13.3 which shows the speed increase for each micromobility mode as a percentage of average walking speed. The high differential between the different modes indicates a potential safety issue unless higher-speed modes are carefully regulated.

Fortunately, the smart mobility application has an integrated ability to locate micromobility vehicles. This can be used for speed enforcement purposes and also to set variable speed limits depending on the type of location where the vehicle is being operated. This will however require the definition of policy.

Another challenge with micromobility services is the indiscriminate placing of vehicles after use. Electric scooters and electric bicycles are simply discarded in the middle of a sidewalk or close to other pedestrian facilities and can create a hazard for other side users. Here again, the geolocation capability of the micromobility service application can be used to define virtual corrals when it is appropriate to leave the vehicle or even fine users who do not park the vehicle appropriately at the end of the trip.

Other challenges in which policy definition can make a significant improvement to safety include use of helmets, writing under the influence of alcohol or drugs, protection for vulnerable writers, and the need for data regarding use and safety of micromobility.

Figure 13.3 Speed increases for micromobility modes.

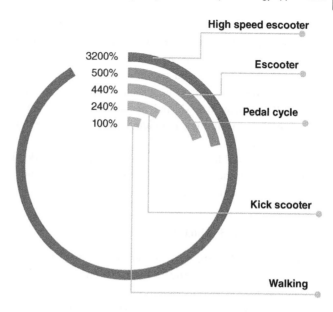

3200%
500%
440%
240%
100%

High speed escooter

Escooter

Pedal cycle

Kick scooter

Walking

13.7.2 Advanced Air Mobility (AAM)

Advanced air mobility, also referred to as Urban Air Mobility (UAM), or Future Air Mobility (FAM) involves the application of electric Vertical Takeoff and Landing (eVTOL) and other types of vertical takeoff and landing vehicles to offer short-range services from mobility hubs or Vertiports. Here again, the introduction of these smart mobility applications has been driven by the private sector. It is estimated that more than $7 billion in new investment was made in 2021 in the future air mobility sector [9].

The opportunities associated with advanced air mobility are clear and significant. Effectively this technology application can add 1/3 dimension to urban transportation systems, alleviating congestion on surface streets by offering an aviation alternative for both people and goods. AAM systems are very efficient at short ranges between 3 and 500 miles. The efficiency of the mode can only really be fully realized if Vertiports has successfully integrated into the surface transportation system. There is a great opportunity to integrate Vertiports operation into a mobility hub that supports interchange between multiple such mobility hubs can significantly improve the effectiveness of multimodal transportation within a city by offering travelers the option to interchange smoothly and efficiently between modes. Urban mobility hub can also serve as a focus point for business operations and perhaps even serve as hotels and conference centers.

Challenges that have been discovered this early stage in the emergence of advanced air mobility applications relate to a disconnect between private sector investment in the aerial vehicles and public sector policy and regulation related to the land infrastructure that is the Vertiports While substantial effort has been made by the various national aviation administrations to develop policy and regulations for this type of service and define how it interacts with commercial aviation, there is significant work still required especially related to the location, design, and operation of the Vertiports. Policy and regulation governing land use should be adapted to accommodate land-based infrastructure for AAM.

Another challenge for advanced air mobility is the need for the policy definition and regulatory authorities to extend current knowledge and capabilities to include AAM systems. Electric vertical takeoff and landing vehicles and the associated Vertiports infrastructure is something that is relatively new to local agencies responsible for policy, planning, land use, and regulation. It will be necessary for these agencies to acquire the knowledge and experience necessary and other to develop effective policies and regulations. To achieve this, it will be necessary for national and central governments to provide sufficient support for agencies to acquire such knowledge and successfully implement AAM policies and regulations Based on such knowledge these agencies must develop institutional, regulatory, and architectural frameworks for a competitive AAM services market.

It is also essential to develop suitable policy frameworks that achieve the objectives of government, and protect the interests of travelers, while also avoiding the possibility of stifling growth of new industries and innovation.

13.7.3 Automated Vehicles

Automated vehicles are also known as autonomous vehicles and sometimes bundled with connected vehicles as connected and autonomous vehicles automated vehicles (CAV). The term automated vehicles has been used in this chapter and indeed throughout this book to avoid any confusion and automated vehicles can be completely autonomous. In many cases, automated vehicles require a relationship between the vehicle and road infrastructure. The terms connected and autonomous vehicles have been avoided as connected vehicles have quite different characteristics from automated vehicles. Connected vehicles have a two-way communication interface between the vehicle and the roadside and with other vehicles but are not necessarily automated.

Automated vehicles involve the replacement of the driver with a combination of sensors, processes, communications, and software on board the vehicle. These are capable of sensing traffic conditions and understanding the environment around the vehicle to fulfill the functions normally supported by the human driver. There are more than 1400 field trials involving the use of automated vehicles on public roads in the United States alone [10]. Most of these trials are funded and implemented by the private sector. The field trials are highly focused on private sector objectives with little data sharing between the trials. This fragmentation poses a challenge to the development of consistent policy and regulatory guidelines.

There are many policy challenges related to automated vehicles and perhaps the most significant relates to the effects on the economy, industry development, safety, and data.

The emergence of automated driving technology offers the possibility of new business opportunities, the establishment of new industries, and the application of innovation on a large scale. However, the bottom-up approach that is currently adopted with regard to the development of localized policy and regulation represents a potential barrier to effective implementation. The automobile manufacturing business operates on a global basis. It would be exceedingly difficult to comply with a patchwork quilt of localized policies and regulations. This serves as a resistance factor to the attainment of the full benefits that can be achieved from vehicle automation. Another potential policy challenge is related to automobile insurance. What will the policy be for driverless vehicles involved in a collision? Will the vehicle honor be responsible, or the automobile manufacturer, or perhaps even the organization that has leased or rented the vehicle?

Also, if an automobile mechanic works on an automated vehicle, then does he or she take responsibility and liability for anything that goes wrong with the vehicle? The state of Michigan has been pioneering in this respect by establishing policy specifically to cover the automobile mechanic issue. Senate Bill 998 provides an exemption for automobile mechanics and repair facilities from product liability as a result of repairing automated vehicles. This is based on the condition that the repairs are performed according to manufacturer's specifications [11].

In fact, the Michigan legislation, passed on December 9, 2016, is a great example of the type of framework of policy and regulation required to preserve safety, while stimulating the development of the automated vehicle industry. The legislation allows the operation of driverless vehicles on public roads, the operation of on-demand driverless taxi networks, and platooning of electronically connected trucks and other vehicles. Prior to this, legislation only permitted the testing of automated vehicles with a human driver present at all times. The new legislation removes these limits. The legislation also protects automated vehicle manufacturers from any liability arising out of modifications to a vehicle on the vehicle systems without the knowledge or consent of the manufacturer.

In common with the other example cited above, there is obviously a need for a framework approach to be taken to policy and other to make sure that all significant aspects are addressed by the policy development. As noted earlier, this policy should be at least at a national level if not at a global level because of the global nature of automobile manufacturing.

13.7.4 Mobility as a Service

What is Mobility as a Service? This turns out to be a great question as there does not seem to be a single agreed definition of what mobility is service really is. These are a few popular definitions. The first centers around the provision of a smartphone app to travelers to support better decision-making. The act provides information to travelers about the complete range of options to get from point A to point B. These options can include a single mode such as a private car, or a multimodal trip involving interchanges between different modes or different services.

Decision support information on the various choices is made available to the traveler and provides expected travel time including any time required for interchanges, travel time reliability, and the total cost of travel as a percentage of household income. Ideally, it will also provide the traveler with the opportunity to make a single electronic payment and reservation for the entire trip. This can be one way or return depending on the travelers' requirements. An important dimension for Mobility as a Service adopting this definition is the ability to provide for spontaneous travel decisions as well as those that are planned in advance. Support for spontaneous travel decisions places a particular burden on the system to ensure that interchange possibilities are preserved. This in turn poses a challenge to service operations. Mobility as a Service will provide decision support both before the trip and during the trip.

Another popular definition of Mobility as a Service is a payment of a monthly fee in return for on-demand access to a private car. When this is combined with automated vehicle technology, then there is a possibility of automated on-demand transportation. Variations on this theme include Robo taxis and a flexible rental car fleet that provides pay-as-you-go, and on-demand access to rental vehicles. This is often referred to as Mobility on Demand (MOD).

An important aspect in improving accessibility and mobility in an urban area is not just the availability of services, but the travelers understanding and perception that these options are available and are viable. Mobility as a Service provides a can of decision-quality information to enable travelers to be aware of the various travel options and also the impact of those options. The impact of the travel option choice could be temporal, financial, or environmental. For this reason, it is likely that Mobility as a Service in addition to providing information and the cost of the various options will also provide carbon footprint information.

One of the biggest challenges with Mobility as a Service is bridging the stovepipes that currently exist around the various modes of transportation within an urban area. Many of these modes have sophisticated information and management systems that are focused only on the mode and are not suitable for sharing information across modes. Another challenge with Mobility as a Service relates to who will operate the service. These are viable business models for either the public or the private sector to operate or even a hybrid with the public and private sector working together in partnership. A good example of a privately operated Mobility as a Service application would be UBER or LYFT. Examples of Mobility as a Service type applications currently operated by the public sector include the various paratransit or transit on-demand services operated for those who cannot use conventional transit systems.

A significant opportunity related to Mobility as a Service is the ability to even out the use of various modes of transportation in a city. Moving toward an ideal situation with the best route and the best mode and the best timing for the journey is always selected by the traveler. As discussed earlier, this also includes making sure the travelers have a full awareness of the transportation options available at the time of the journey and the consequences of choosing that particular mode. There is also a good opportunity for cooperation between the private and the public sector under the auspices of appropriate business models. An interesting current example of this is cooperation between cities and Uber with respect to paratransit service. When the public sector paratransit service closes for the evening, a subsidized UBER trip can be made instead.

13.8 Significant Smart Mobility Policy Areas

There is a wide range of policy areas that can be considered related to smart mobility. The most significant of these areas are depicted in Figure 13.4.

Each of these areas can be described as follows:

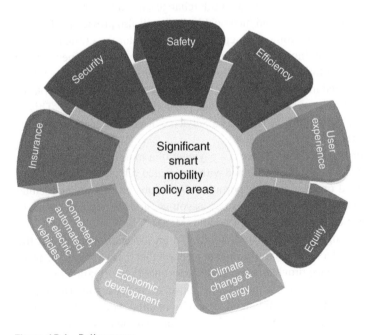

Figure 13.4 Policy areas.

13.8.1   Safety

Safety is always of paramount importance with respect to smart mobility and transportation activity. Policies are typically focused on the reduction of fatalities, injuries, and property damage as a result of the transportation service delivery process. It should also be noted that traveler's perception of safety is an important aspect of the overall safety picture.

13.8.2   Efficiency

Efficiency policy can address and process several aspects including travel time, time lost due to congestion sub optimization of the overall city and regional transportation system. The latter can occur as a result of a mismatch between supply and demand, or travelers choosing the wrong mode for the trip, the time of day, the origin, and the destination.

13.8.3   User Experience

User experience is an important aspect of policy as most users of the transportation system will also be part of the electorate to vote for political leadership. It could be argued that all significant policy areas have an element of user experience and user satisfaction. Effective policy must also address the user's perception of the value and quality of smart mobility services as well as the measured quality.

13.8.4   Equity

Policy must ensure that investments in smart mobility services are fair to all segments of society. Technology has the potential to customize mobility services to a much higher degree than has been previously possible. There is a need to ensure that technology is applied in an unbiased way to deliver the values and benefits that await is possible range of people in the population.

13.8.5   Climate Change and Energy

Climate change is a policy area with significant current activit. Policymakers are making bold promises regarding greenhouse gas reduction and decarbonization over the next 20–30 years. This includes a move away from fossil fuels and the internal combustion engine toward electric vehicles. This also includes the use of a wider range of modes of transportation including micromobility, automated first mile last mile shuttles, and nonmotorized alternatives.

13.8.6   Economic Development

Many policies related to smart mobility and smart cities have a focus on making the city more attractive to inward investment, the relocation of industry, and businesses in the generation of new job opportunities. The desire to create locally focused policies also provides an opportunity for cities and regions to shine through the proper application of policies for smart mobility and smart cities.

13.8.7   Connected, Automated, and Electric Vehicles

This was discussed previously as an example of a technology application area, with a discussion on policy and strategy aspects. However, this policy area is of such significance in smart mobility and smart cities that it is worth adding further information here. Policy development has been challenged through the emergence of automated vehicles and their use within the field trials and open roads and public roads in many parts of the world. Policymakers are being forced to address the safety of allowing fully automated vehicles with no drivers to operate on public roads. Connected vehicles also provide policy challenges, especially related to the use of the data that can be extracted from the vehicles using the connected communication technologies. Electric vehicles also pose challenges with respect to a new demand for electricity in new locations. While there is a desire to stimulate and encourage electric vehicle manufacturing and deployment, it is important to have the appropriate policy framework to manage possible distortion in the demand pattern for energy.

13.8.8 Insurance

This particular policy area has mainly been driven by private-sector insurance companies. These companies have understood the possibility of smart mobility technologies enabling a better approach to risk management. They have also understood that automated vehicles may also pose additional risks that have to be quantified and assessed. In particular, the question of liability in the event of a crash is to be analyzed and understood. If an automated vehicle has a collision, and whose fault is it – the owner, the car manufacturer, or the solution provider that developed and supplied the automated vehicle system?

13.8.9 Security

Security is becoming an increasingly important aspect in smart mobility. As automation and telecommunication technologies are applied to vehicles, then the possibility increases of cyber-security attacks. The risks associated with automated control of vehicles must also be assessed and appropriate policies defined. An interesting example of a policy that has been developed for cybersecurity and transport is United Nations Regulation Number 155 – Cybersecurity and Cybersecurity Management System [12]. This requires that vehicles are developed and manu-factured within the framework of a cybersecurity management system (CSMS). The framework includes the identification of risks to vehicles, the assessment categorization and treatment of risks, the definition of processes used for testing cybersecurity, and arrangements for detection and response to cyber-attacks.

In addition to cybersecurity, it is also important to consider the physical security of smart mobility systems and applications. There will be little point in having a high-integrity cybersecu-rity approach if physical security is weak. For example, a computer can have sophisticated soft-ware to protect it against attacks and malicious actions. However, this would be of little effect if physical security arrangements were poor enabling a bad actor to access the device directly. The authors are aware of at least one instance where a public sector transportation agency lost track of a hardware asset due to the lack of an effective asset management system. The device in ques-tion was protected from a cybersecurity point of view but was simply relocated without prior permission or approval.

An interesting way to view security and safety is in terms of two sides of the same coin. It can be argued that security revolves around protecting the system from the user. This includes users who may inadvertently cause damage to the system or bad actors with malicious intent. It can also be argued that safety is simply protecting the user from the system. This could be intended actions of the system or unexpected consequences of system operation. Figure 13.5 illustrates this.

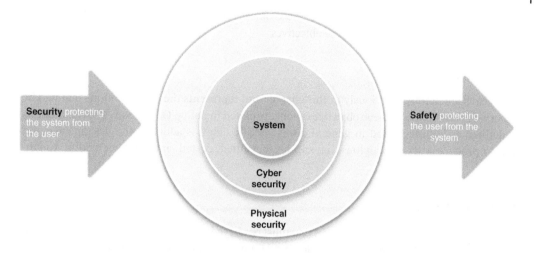

Figure 13.5   The relationship between safety and security for smart mobility systems.

13.9   The Path from Policy to Operational Management

Ideally, there would be a clearly defined path between defined policy and the application of smart mobility technology during operational management. Operational management is where smart mobility services, powered by technology, deliver the values, and benefits expected.

Figure 13.6 depicts the ideal path between policy and operational management.

There are multiple elements along the path from policy to operational management, therefore Figure 13.6 also helps to explain the relationship between the following elements: policy, strategy, regulation, best practice, standards, planning, design, and operational management. These can be described as follows:

13.9.1   Policy

This element has been defined early on in this chapter and represents a clear statement of what is desired and what must be avoided in the implementation of smart mobility within a city or region. Policy encapsulates the objectives of political leadership. Note that it is typical that a range of

Figure 13.6   The ideal path between policy and operational management.

objectives are encapsulated in policy and these objectives are not always mutually compatible. Policy must, of necessity, balance these objectives.

13.9.2 Strategy

It has also been defined previously in this chapter and represents the broad motions that will be supported and others to achieve objectives as encapsulated in policy. Of necessity strategic actions are defined as high-level and in general terms specific detail is added at the design stage and other to produce a set of activities that can be scheduled, costed, and implemented in a managed way.

13.9.3 Regulation

Regulation defines the roles and processes that must be obeyed to conform to policy. Regulations also form part of the law of the land. For example, speed limits are regulations that define the appropriate speed for different road types and different geographic locations. The consequences of not complying with speed limits are also specified as part of regulations. It could be argued that the likelihood of people complying with regulations is directly proportional to the consequences of being caught and the probability of being caught. This leads the whole discussion on the effectiveness of enforcement and how tire performance management can be applied to enforcement. This will be addressed in Chapter 15 when various aspects of smart mobility performance management are discussed.

13.9.4 Best Practice

Commercial or professional procedures that are widely accepted as being correct or most effective. Typically, best practices are formed and agreed on the basis of practical experience and lessons learned. Best practice forms an important part of knowledge sharing across field trials and full-scale implementations of smart mobility.

13.9.5 Standards

Smart mobility systems are inherently distributed and have complex data and information requirements. Multiple systems in smart mobility applications are expected to work together requiring the framework of standards to be in place to support a high level of interoperability and connectivity. It is typical in the early days of smart mobility technology application standards that have not yet been agreed and defined. In the early days, private sector solution providers invested substantial amounts of their money and investors' money motivating them toward a defensive posture that avoids standardization.

As the market for the solution of service grows, it becomes obvious that further market growth will depend on the existence of appropriate standards, These help to achieve economies of scale, and manage the risk for the smart mobility solution deploments. Iit is at this point in the market where standards become extremely applicable. Standards can be local or regional but are most effective when global in scope and developed by international standardization bodies such as CEN and ISO. Regional standards are typically developed across geographically or politically connected regions. In many cases, a preexisting national standard can be adopted as a starting point for an international standard. The main objective of standards is to harmonize products, services, and solutions.

13.9.6 Planning

The process of evaluating and assessing multiple strategic-level approaches as a way of identifying those that will not work and those that show promise for further examination and analysis. Planning should take full account of policy objectives and strategic directions that have been defined earlier. The planning process can also be informed by practical experience and lessons learned from prior deployments. It is also in the planning process where smart mobility is integrated and harmonized with other means of transportation, across modes, taking full account of economic and workforce development needs.

13.9.7 Design

The design process has sufficient detail to enable service, product, or a solution to be specified and procurement. A highly detailed analysis can be conducted at this stage in terms of the likely capital and operating cost of the smart mobility solution and the expected values, benefits, and outcomes. In some cases, preliminary designs can be developed with less detail, to enable initial cost-benefit analysis to be performed.

13.9.8 Operational Management

The stage of the process where smart mobility solutions and services are applied to deliver values and benefits. Operational management involves day-to-day operation of the transportation systems in a region, detection of threats and opportunities, and the formulation of appropriate actions in response to changes in transportation demand and operating conditions. Operational management takes place in typical, predictable recurring transportation conditions but it plays a particularly important role in unpredictable nonrecurring transportation conditions associated with special events, either synthetic or natural. These include sporting events, road construction, hurricanes, sandstorms, and other adverse weather conditions.

Operational management is undoubtedly at the sharp end of the service delivery process for smart mobility requiring that the appropriate decisions are made quickly and accurately. To be successful, operational managers require sophisticated decision support that defines the current situation, provides reliable predictions on what will happen next, and data-driven advice on response plans. This will be covered in more detail in Chapter 15 when smart mobility operational management is discussed.

13.10 General Challenges Associated with Smart Mobility Policy and Strategy

The three examples described earlier, advanced air mobility, automated vehicles, and Mobility as a Service are examples of specific opportunities and challenges associated with each of the solutions. It is also useful to consider a set of general challenges associated with smart mobility policy and strategy. The first of these is the need to balance the preservation of government objectives and the safety of the traveler and population at large. This must be balanced against the desire to stimulate economic development within the city or region by promoting the growth of new industries and innovation associated with the transportation process: the smart mobility industry.

The second general challenge, which has already been identified in the policy aspects examples, is a need to manage the sequence in which policy and technology applications are developed. When policy leads to technology application, it may not take full account of technology capabilities and limitations that are only fully proven when applied in the field. Policies defined ahead of technology implementation run the risk of being too theoretical and not practical.

The third general challenge associated with policy and strategy relates to effective engagement of the private sector. The world is witnessing a sea change in investment and leadership of smart mobility deployments. Where previously the public sector had a dominant role, the private sector has now come to the fore in terms of investing, deploying, and growing the market for smart mobility services. The role of the public sector has changed to that of an advocate and influencer that can benefit from private sector investment and activity. This requires a strong partnership approach to be taken by the public sector, a role which is relatively new and for which the public sector may not yet have acquired the appropriate skills.

13.11   Communications Between Policymakers and Technologists

As noted earlier, strong effective policies based on a good understanding of the capabilities and limitations of technology in their current shape and form. Policymakers can be considered as the experts in "what" is required. Solution providers and technology companies can be considered to be the experts in "how" it can be done, based on an intimate understanding of current technology capabilities. It is necessary to support a strong interaction between "what" and "how," between policymakers and technology experts, to arrive at the appropriate policy that will take full account of current technology capabilities. Policy should also account for emerging technologies and predicted technology capabilities in the near future.

13.12   An "Ideal" Policy

The concept of applying performance management to policy definition and development, leads to the question "what exactly is an ideal policy?" After all, performance management is all about planning and providing resources to ensure the goals and objectives are achieved and this requires that such goals and objectives are predefined and agreed. It is likely there is no such thing as a universal ideal policy as policy needs vary between different groups in different types of people. There may also be aspects of the policy that are simply not achievable in practical situations. However, it is still valuable to consider a checklist that defines the ideal policy. The objective is to provide a starting point for those involved in smart mobility policy development and not to provide a catalog, or complete list of aspects of an ideal policy. Based on discussion with policymakers and practitioners in the field of smart mobility and smart cities, the following factors have been identified in relation to an ideal policy:

- Encompasses all predictable future scenarios
- Addresses all stakeholder needs
- Fosters innovation
- Gets the best from technology capabilities
- Ensures that what is wanted is achieved
- Ensures that what is not wanted is avoided
- Stickiness – avoidance of reversal

- Informed by the capabilities and constraints of current technology
- Incorporates an effective knowledge of lessons learned and practical experience from prior implementations

The factors are discussed in the following section:

13.12.1   Encompasses All Predictable Future Scenarios

This is probably the most difficult factor to address. History has shown that progress in technology is not linear and cannot be extrapolated from previous experience. Destructive technologies emerge that had not been anticipated or forecasted. It could be argued that the Internet falls into this category. However, difficulty should not be used as an excuse for not addressing this factor.

Diligence in understanding current technologies and researching technologies that are about to emerge can result in a series of predictions regarding future scenarios. This could be considered as risk management for policy as the longevity required from policy can only really be provided if an effective approach is taken to encompassing predictable future scenarios. An analysis can also be conducted to consider the sensitivity of the policy to changes in scenarios and errors of prediction.

13.12.2   Addresses All Stakeholder Needs

This is another difficult factor. There is a wide range of stakeholders involved who will be impacted by smart mobility policy. These include the following:

End users – the effectiveness of the policy in enabling the delivery of the required outcome, while avoiding undesirable side effects has a significant impact on the end-user experience.

Solution providers and technology developers – enterprises involved in developing solutions for the smart mobility market make use of technologies either developed in-house or by others. Both of these groups are affected by policy. In some cases, policy can stifle or disrupt the market for products and services.

Practioners, operators, and services suppliers – those involved in the delivery of smart mobility services and the efficient operation of mobility in a city and across regions can be affected by policy considerations.

Nontechnical decision-makers – this group includes the policymakers themselves and can be affected by industry and public reaction to the policy. If the policy is perceived as not effective or too draconian, it can affect the longevity of nontechnical decision-makers and policymakers.

13.12.3   Fosters Innovation

In addition to providing protection from what is not desired, policy is also expected to foster innovation. Economic development and the creation of job opportunities are usually fundamental components of policy and strategy. This requires an effective balance between guidance and flexibility. The guidance provided by policy should clear and concise, balanced by some degree of flexibility to accommodate innovation and the emergence of better approaches.

13.12.4   Ensures That What Is Wanted Is Achieved

Policy should be based on a clear definition and understanding of the intended outcome and the objective to be achieved.

13.12.5 Ensures That What Is Not Wanted Is Avoided

Closely related to the previous factor, policy should ensure the avoidance of undesirable side effects. For example, emissions from internal combustion engines.

13.12.6 "Stickiness" – Avoidance of Reversal

This is related to policy longevity; it is especially important to policymakers that short-term reversals in policy are avoided. The policy should be effective enough and suitably informed so that it "sticks."

13.12.7 Locally Driven

Discussions with policymakers suggest that most have a desire for policy to be driven at a local level. This ensures that policy is appropriate to local needs and takes full account of the characteristics of local travelers. This does not however take account of the global nature of mobility technology applications such as automobiles and smartphones.

While local adaptations are possible, it is not possible to design a car or a smartphone specifically for the needs of the city, while remaining economically viable. This requires a delicate balance between locally driven policies and those policies required at national, and international level.

13.13 Applying Performance Management to Policy

Having defined the ideal policy, it is now possible to apply performance management to policy through the use of a policy scorecard. Figure 13.7 shows a suggested template for such a scorecard.

The scorecard contains header information that describes the policy number, the overall theme of the policy, a policy description, and the policy statement. This is then followed by a policy evaluation which, as would be expected, uses the factors described above within an evaluation framework that determines how well the policy performs.

The resulting policy score can be a simple total of the scores from the individual factors or a weighted total. In addition to evaluating the performance of a particular policy, this enables comparison with the performance of other policies.

13.14 Thoughts on Maximizing the Effectiveness of Smart Mobility Policy

This is perhaps one of the most complex areas of smart mobility, mainly because of the many and varied forces that shape policy. There is also a subjective aspect of policy that is difficult to quantify and measure. Another important aspect of smart mobility policy is that it often lags the introduction of technology applications. While this is not ideal, it is probably necessary as policy must be informed by technological capabilities and constraints. These are often not defined or discovered until the technology applications have actually been implemented in practice. Policymaking involves navigating a balanced path between varying objectives that are sometimes not compatible.

Policy scorecard	
Policy number	
Policy theme	
Policy	
Policy statement	
Policy evaluation	
Evaluation factor	Score
Encompasses all predictable future scenarios	
Addresses all stakeholder needs	
Fosters innovation	
Gets the best from technology capabilities	
Ensures that what is wanted is achieved	
Ensures that what is not wanted is avoided	
Stickiness–avoidance of reversal	
Informed by the capabilities and constraints of current technology	
Incorporates an effective knowledge of lessons learned and practical experience from prior implementations	
Locally driven	
Total score	

Figure 13.7   Policy scorecard template.

13.14.1   Bridging Policy and Operations

Policy involves the definition and communication of bold strategic objectives. Ideally, these objectives will not be unfunded mandates – directives with no associated resources but will be supported by the appropriate budget allocations. It is also a tendency for a gap to emerge between these bold high-level strategic objectives and the practical planning and operations required for smart mobility. This is an opportunity to apply technology toward the optimization of policymaking and implementation. These are significant examples of the application of data and information technologies to the management of traffic and transportation. Such systems can be extended to include a bridging role between nontechnical decision-making, planning, and operations. Figure 13.8 illustrates how such a technology application could serve this role.

There are five elements in the diagram as follows:

1) Operational management
2) Policymakers

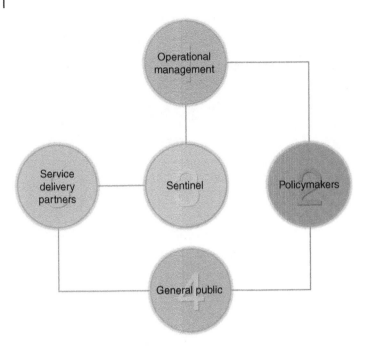

Figure 13.8 Bridging policymaking and operational management.

3) Sentinel
4) General public
5) Service delivery partners

Considering each of these in turn:

13.14.1.1 Operational Management
In this element, data coming from the Sentinel is assessed and evaluated. The data is converted to information from which insight and understanding are extracted. On the basis of the insight and understanding strategies and responses are formulated, which in turn form the basis for technical decision-making. Operational management also informs engage the action of the sentinel, while also providing information and guidance to the service partners. Operational management also provides guidance to the policymakers and in turn, receives guidance back from the policymakers while also receiving feedback from service partners. It is expected operational management will possess a range of hardware and software tools to enable effective decision support.

13.14.1.2 Policymakers
In close coordination with the general public, policymakers define the bold statements of direction that are necessary to set strategic and coherent path to the future. Policymakers receive insane feedback from the general public about needs, issues, problems, and objectives, while also receiving information from operational management regarding technical decisions and situational awareness.

Ideally, policymakers would take a framework approach to the definition of policy needs. Perhaps making use of the policy scorecard discussed earlier. It is also important for policymakers to be informed regarding the capabilities and constraints of current technology. It is not necessary for policymakers to understand exactly how technologies work; however, they should have a clear and concise understanding of what the technologies are and what they can do. A clear understanding of outcomes is critical to this.

13.14.1.3 Sentinel

This is not a typical term that would be used for network monitoring and surveillance. It is used here to differentiate between the monitoring and operational management roles. The role of the Sentinel is to continuously detect and assess threats and opportunities, suggesting appropriate action to operational management. The Sentinel also supports a continuous assessment of public sentiment and feedback from the general public. Finally, the Sentinel uses surveillance and detention capabilities to measure the effectiveness of plans and responses that have been implemented by operational management Stands and keeps watch. It is expected that the Sentinel will make use of advanced detection, sensing, and telecommunication technologies to maintain a sufficient level of situational awareness to detect threats, opportunities, changes in the demand for transportation, and changes in transportation operating conditions.

13.14.1.4 The General Public

Consists of travelers, non-travelers, and of course voters. The general public and the primary consumer of services delivered by both the public and private sector. The public received the benefit of the smart mobility services and provided feedback to the other elements of the framework regarding user experience and perception on the value of the benefits received.

13.14.1.5 Service Delivery Partners

These can be both public and private sector organizations responsible for the delivery of smart mobility services. Smart mobility services are supported by a combination of technological applications. Service delivery partners can also be regional stakeholders and other regional and city transportation agencies that deliver a variety of services to the general public, taking appropriate action, applying appropriate strategies, and receiving feedback from the public.

The approach takes advantage of transportation management and decision-support software that is already being applied in practice. The innovative element in this approach involves the creation of bridges between policymaking, private sector service partners, and the general public.

The overall approach involves a complete and structured framework for the definition of policy based on the identification of needs. Policy goals and operational key performance indicators are then related to each other to ensure that policy objectives are being addressed at the operational level.

13.15 Structured Management of Policy and Strategy

A formal mechanism should be defined and established for informing policymakers regarding technological capabilities and constraints. They should be based on a scanning approach to understand current technology status based on implementation experience and practical lessons learned around the world. This should be incorporated within the wider context of an effective planning approach described in Chapter 6. It should be noted that policy and objective statements could be both inputs and outputs from the planning process. It is likely that preexisting objectives have been identified before the planning process. It is also likely that additional objectives and policy goals will be identified in the course of the planning process.

It is also particularly important for policymakers to have a clear understanding of the opportunities that can be unlocked through effective policy for smart mobility. Effective policy will ensure the achievement of desirable outcomes while navigating around undesirable side effects. Policy making, however, does not stand in isolation and it is not sufficient to define bold policy goals without also making provision for the achievement of the goals. This requires that supporting financial arrangements are made that are complementary to the policies and enable the objectives to be achieved.

Throughout the policymaking process, it is clear that effective communication is an essential ingredient. This includes communicating technology capability and constrained information to policymakers. Policymakers must also support effective communications to partners, private sector, and the general public. An interesting aspect of smart mobility and transportation is that many projects and initiatives will span across more than one generation.

This also needs to be considered when developing an appropriate communication strategy, which should be effective in the short term, the medium term, and the long term. A balanced approach to technology must also be taken that allows flexibility for emerging and disruptive technologies. This can be enabled by placing a focus on what is to be achieved rather than how it can be done. What has to be achieved is typically more stable and less variable than how to do it.

13.16   Summary

This chapter addresses an extremely complex subject within the world of smart mobility. That are many factors that must be considered in the development of effective policy and strategy for smart mobility, and this often involves balancing mutually exclusive objectives. The chapter includes the definition of policy and strategy and an explanation of the role that policy can play in smart mobility success. It discusses the need to balance industry encouragement and stimulate growth while also providing an appropriate level regulation and enforcement policy.

Early experiences in the application of smart mobility have shown that in many cases technology capability and implementation happen in advance of effective policy formulation. The advantages of technology-leading policy and policy-leading technology are also discussed. An important need to ensure that policymaking is informed by a thorough understanding of technology capabilities and constraints is also discussed, highlighting the need to engage private-sector solution providers and share knowledge and information with practitioners responsible for prior implementations. Accommodating emerging and disruptive technologies is also discussed, suggesting that a focus on "what" is required rather than "how" it can be done is a good approach for future-proofing against emerging and disruptive technologies. What is required remains relatively stable compared to how it can be done, as the pace of technology change increases.

A structured framework approach is defined and described showing how policymaking, planning, and operations can be linked together effectively. The relationship between policymaking and effective planning processes for smart mobility is also discussed. A number of challenges and opportunities associated with smart mobility policy have also been identified, described, and discussed in the chapter. The chapter concludes with some thoughts on the best way to optimize policymaking for smart mobility.

Like many of the chapters in the latter half of the book, this chapter is closely related to the subject matter covered in other chapters of the book. Chapter 3 Smart Mobility a Problem Statement provides an explanation of many of the needs, issues, problems, and objectives that must be addressed by policymaking. Chapter 4 and the value and benefits of smart mobility provides an explanation of the potential outcomes from the successful implementation of smart mobility applications, defining a set of objectives for policymaking.

Chapter 5 on smart mobility progress around the world provides a foundation of knowledge that can be used to inform policymaking based on practical experiences and lessons learned around the world. Chapter 6 addresses an accelerated and highly effective approach to planning for smart mobility. This can provide the channels for the input of smart mobility policy into the planning process and the output of new policy and objectives as a result of the process. Chapter 8 provides an overview of the technologies that are used in smart mobility, enabling policymaking to be informed regarding the characteristics and constraints of technologies.

Having read this chapter, the authors recommend that it is treated as an information source that provides a number of models that can be applied to improving policymaking for smart mobility. It can be taken as the basis for an approach that can be customized to a specific city and specific contexts. It is not intended to be a one-size-fits-all solution to policymaking for smart mobility. It is intended to provoke thought and careful tailoring of the knowledge and information to specific city context.

References

1 Merriam-Webster Dictionary. https://www.merriam-webster.com/dictionary/policy, accessed November 17, 2022, at 10:00 PM Central European Time.

2 Merriam-Webster Dictionary. https://www.merriam-webster.com/dictionary/strategy, accessed November 17, 2022, at 10:09 PM Central European Time.

3 Compass master class with Darren Capes, mobility policy lead, Department for transport, UK, linked to be added, accessed November 17, 2022, 10:11 PM Central European Time.

4 Micro mobility's 15,000 mile checkup. McKinsey and Company, January 29, 2019. https://www.mckinsey.com/industries/automotive-and-assembly/our-insights/micromobilitys-15000-mile-checkup, accessed November 17, 2022, at 10:15 PM Central European Time.

5 How fast do electric scooters go?(What is the max speed on an electric scooter?)Holy He, LinkedIn, September 6, 2022. https://www.linkedin.com/pulse/how-fast-do-electric-scooters-gowhat-max-speed-scooter-holy-he/, accessed November 17, 2022, at 10:20 PM Central European Time.

6 Fit and proper: what is the ideal walking speed for you? Business Standard, Thursday, November 17, 2022. https://www.business-standard.com/article/current-affairs/fit-proper-what-is-the-ideal-walking-speed-for-you-115100900029_1.html#:~:text=In%20the%20absence%20of%20significantsmall%20range%20within%20these%20speeds, accessed November 17, 2022, at 10:25 PM Central European Time.

7 Average cycling speed – what is it? BikeLockWiki, https://www.bikelockwiki.com/average-cycling-speed/, accessed November 17, 2022, at 10:30 PM Central European Time.

8 Bassett, E. How fast can kick scooter's go? (Plus suggested models for speed), two wheel better, March 17, 2021. https://twowheelsbetter.net/how-fast-kick-scooter/#:~:text=This%20may%20be%20over%2015,6%2D7%20mph%20is%20sustainable, accessed November 17, 2022, at 10:35 PM Central European Time.

9 Esque, A.; Reidel, R. A milestone year for future air mobility. McKinsey and Company, February 8, 2022. https://www.mckinsey.com/industries/aerospace-and-defense/our-insights/future-air-mobility-blog/a-milestone-year-for-future-air-mobility, accessed November 17, 2022, at 10:45 PM Central European Time.

10 McQueen, B. Unsettled issues regarding policy aspects of automated driving systems, EPR 202-0016, Society of Automotive Engineers International, August 31, 2020. https://www.sae.org/publications/technical-papers/content/epr2020016/, accessed November 17, 2022, at 10:50 PM Central European Time.

11 O'Melveny, Michigan enacts new autonomous vehicle legislation. Alerts and Publications, December 12, 2016. https://www.omm.com/resources/alerts-and-publications/alerts/michigan-enacts-new-autonomous-vehicle-legislation/#:~:text=Senate%20Bill%20998%20exempts%20auto,performed%20according%20to%20manufacturer%20specifications, accessed November 17, 2022, at 10:59 PM Central European Time.

12 UN regulation number 155 – cybersecurity and cybersecurity management system. UNECE Sustainable Development Goals, published 4th of March 2021. https://unece.org/transport/documents/2021/03/standards/un-regulation-no-155-cyber-security-and-cyber-security, accessed November 17, 2022, at 11:05 PM Central European Time.

14

Organizing for Smart Mobility Success

14.1 Informational Objectives

After you read this chapter, you should be able to:

- Describe the purpose of the organization
- Understand how the organization can support the smart mobility revolution
- Describe the difference between performance and productivity
- Describe organizational performance management
- Describe how to conduct an organizational capability assessment
- Explain the role of change management in organizational alignment
- Describe a real-life example of organizational analysis

14.2 Introduction

The authors' experience with the application of advanced technology to transportation is that often, the technical solution is adapted to fit within existing organizational structures. There seems to be a momentum behind legacy organizational structures and reluctance to apply significant change. However, when addressing the needs of the smart mobility revolution, it would not make sense to adopt a "business as usual" approach to organization. Smart mobility needs different skills and a more integrated approach to organization as many of the applications cut across existing organizational silos. It is understood that from a practical perspective, a complete renovation of existing organizational structures may not be feasible. However, at the very least, it is necessary to modify and adapt the organization to the tasks involved in supporting smart mobility.

The authors believe that substantial change will be required if the world is to get the best from the smart mobility revolution. There must be a willingness to consider some changes to the existing organizational fabric. It is a bit like changing clothes to suit changing weather conditions. Sudden and dramatic changes are not useful, but business models and the way things are done will require changes from current practice. It is important to organize for success. It is also important to plan for success. The authors have worked with some organizations that do not seem to believe that they can be successful and do not plan for success, but for something less and suboptimal instead. Organizational change must support the development of a corporate belief in organizational goals.

Smart Mobility: Using Technology to Improve Transportation in Smart Cities, First Edition. Bob McQueen, Ammar Safi, and Shafia Alkheyaili.

14.3   Organizational Goals

A good organizational realignment or fine-tuning will be based on a clear and agreed understanding of the purpose and goals of the organization. From a smart mobility perspective, these goals are likely to relate to the problems to be addressed as discussed in Chapter 3: a smart mobility problem statement and the opportunities identified in Chapter 9, where smart mobility opportunities and challenges were discussed. A guiding principle is that any objective that is adopted must be measurable. If it cannot be measured directly, then a proxy must be defined. If a proxy cannot be identified, then the objective should be reevaluated as perhaps it is not a valid object of our outcome if it cannot be measured. For example, public perception and sentiment are typically areas where it is difficult to define measurable parameters.

However, did this possible to use social media and other survey techniques to measure outcomes. It may be necessary to adjust to KPI his heart parameters to be measured to ensure that they can indeed be measured. Measurability is an important aspect of measurement. It is also important to develop a clear relationship between agreed objectives and the things to be measured. This also includes an understanding of the target customer groups including planning, operations, and end users. Smart mobility customer characterization will be addressed in Chapter 12 when smart mobility modes and customers are discussed. This is important to ensure that there is a full understanding of how objectives and measures vary according to the target customer, which in turn requires a thorough understanding of customer characteristics.

In addition to an alignment to the anticipated outcomes and objectives, the organization must also be designed to support the business processes required to achieve the objectives and outcomes. After understanding of the business processes that are currently supported and those that will be required for effective smart mobility implementation must be achieved through a rigorous analysis of business processes, to ensure that the organization can be aligned to the objectives and outcomes required.

14.4   Organization Alignment to Business Processes

So, what is the best way to approach organizing for a successful smart mobility revolution? The authors understand that one solution will not be good for everybody as an organization must reflect the objectives, nature, and characteristics of the enterprise and be designed to attain specific smart mobility goals. It is also impossible to specify a detailed organizational structure without considerable information on the nature of the organization and the context within which it operates. However, what can be done is the creation of a model instead of a recipe. The important characteristics of an effective and successful organization can be defined, and a framework for achieving the best organizational structure can be created and explained.

That is what is provided in this chapter, the definition of a model from a number of different perspectives that illustrates what the authors think would be a successful organizational structure for smart mobility. The intention is that this model be reviewed, the major principles understood, and then applied to the specific situation, with the required adaptation and customization. To assist with this, in addition to describing the model, some advice is also provided on putting it into practice.

An important part of our approach is to develop a performance-based organization that is focused on outcomes. To begin, let us discuss the definition of performance. Performance measures that show if the original predefined stated objectives were attained. Sometimes performance and productivity are confused. The authors define productivity as the amount of work that can be done within a given time. So, the focus for this exercise should be on performance. The aim is to design and define a new organization that maximizes the probability of attaining the objectives of the organization. While there will be smart mobility objectives, that can be considered as outcomes, there will also be organizational objectives that are used to guide progress toward outcomes. The organizational objectives vary from one organization to another, but some standards can be defined. From a business process point of view, an organization should be optimized to support the business processes required. In this case, the business processes are those required to deliver smart mobility. Business processes are the activities that are supported and resourced by an organization to bring raw materials in and create value in the form of solutions, products, and services. The process classification framework, or PCF, was first introduced in Chapter 11, Smart Mobility Performance Management. The cross-industry version was introduced, but there is also a city services version illustrated in Figure 14.1, tailored to city government business processes.

The city services process classification framework defines a total of 13 essential business processes, 6 of which are defined as operating processes and 7 defined as support or enabling processes. The six operating processes can be described as follows.

14.4.1 Develop Vision and Strategy

A city requires a vision and a strategy for achieving that vision. The vision will explain and communicate the long-term goals of the organization and as such represents the major objectives of the organization.

	Develop vision & strategy	Develop & manage city services	Promote the city	Deliver city products	Deliver city services	Engage constituents
	1.0	2.0	3.0	4.0	5.0	6.0
7.0	Develop and manage human capital					
8.0	Manage information technology					
9.0	Manage financial resources					
10.0	Acquire, construct, and manage assets					
11.0	Manage enterprise risk, compliance, remediation, and resiliency					
12.0	Manage external relationships					
13.0	Develop and manage city capabilities					

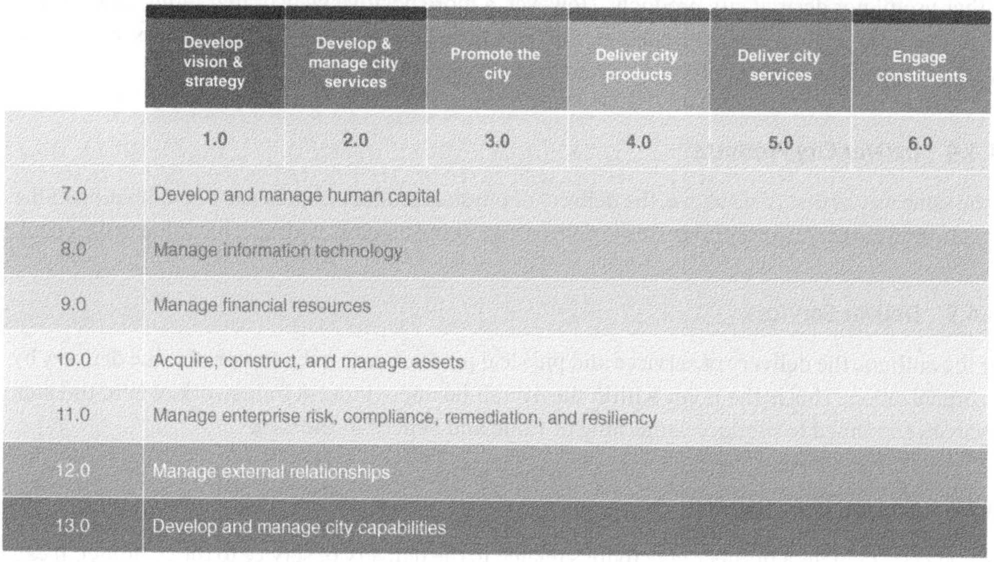

Figure 14.1 The process classification framework for city governments from APQC. *Source:* Adapted from [1].

14.4.2 Develop and Manage City Services

The process classification framework was written for organizations that develop products and those that develop services. These days the lines between products and services are blurring because the same organization can produce both products and services. While in the past, it was easier to differentiate between products and services, products such as software that may be offered as a service making use of cloud services. So, what is the difference between a product and a service.

A product is something tangible that is usually purchased on a one-off basis using capital expenditure funding. On the other hand, the service is something that can be delivered in return for a regular subscription and become part of operating expenses. A product can also be considered as a result of a series of actions and the application of a range of resources within the framework of business processes. In either case, value delivery will be by way of the service. In the case of products, products are applied to deliver services that in turn deliver value and benefits. For example, a traffic signal controller is obviously a product. It consists of a steel box containing a variety of electronics that enable traffic signal timings to be adjusted in line with traffic flows and operating conditions. The combination of steel and electronics that comprise the traffic signal controller does not deliver value in its own right. However, when applied to deliver a traffic management service, such as adaptive traffic signal control, then value is delivered in terms of reduced stops, lower emissions, lower fuel consumption, and smoother traffic.

14.4.3 Promote the City

This can also be considered to be marketing the city. It is also worthwhile to consider the difference between marketing and sales. Our definition is that marketing covers the activities required to make the user aware of the products and services and develop interest in learning more or acquiring them. The sales process is defined as the activities and resources required to build on awareness and interest to create a desire and cause a buying action for the end user. Sales and marketing can often have negative connotations of unscrupulous people trying to sell things to the user that are neither useful nor desired city products. However, a more positive view of marketing and sales as efforts can be taken to inform the customer of the capabilities of products and services and ensure that these capabilities are carefully matched to client needs.

14.4.4 Deliver City Products

In the same way as described above, the delivery of physical products represents a combination of the outputs from earlier business processes to deliver something tangible with specific value to the client.

14.4.5 Deliver Services

For the authors, the delivery of services and physical products are at the center of value delivery by the organization. This is the point within the overall business process framework, where the area processes combined to produce something of value and benefit to the user.

14.4.6 Engage Constituents

This process includes a number of activities related to the delivery of service to the customer. It can range from the provision of information to enable the customer to use the product to the fullest or take the best advantage of the service. It can also involve upgrades and replacements. It covers a vital interface between the organization and the customer. In many cases in transportation, this interface is weak. Consider, for example, the construction of a new road. If it is a major project, there is significant customer service employed to consult those affected by the growth construction. However, once the road is open but is little interaction between the service operator and the end user. Of course, the exception to this is a toll road, or a transit agency, where the service operator has a continuing dialogue with the customer regarding ongoing use of the facility, billing, and customer service.

There are also seven enabling or supporting business processes that can be described as follows.

14.4.7 Develop and Manage Human Capital

This business process category is concerned with the acquisition and retention of human resources and people. This can also include development and growth of people through the use of appropriate professional development and training programs. This is very intricately linked to business process 14, develop and manage business capabilities as in many cases an important part of organizational capability lies in the capability of the people. This business process and the following three are focused on managing various types of organizational resources.

Develop and manage human capital ensuring that the organization contains the right people and that those people have appropriate skills. This also includes ensuring that skills are aligned to changing needs.

14.4.8 Manage Information Technology

Information technology for data and information processing has become a crucial part of the organization. Managing information technology includes identifying and harnessing the appropriate data sources, turning the data into information, extracting the maximum insight, and understanding from the information, including the identification and definition of trends and patterns. This process should culminate in the development of response plans and action strategies that are appropriate to the prevailing conditions.

14.4.9 Manage Financial Resources

Both public and private sector organizations are required to ensure good stewardship of financial resources. This includes revenue and investments and should address short-term and long-term reporting requirements. Reports will be required for nontechnical decision-makers, shareholders, and the government for tax purposes.

14.4.10 Acquire, Construct, and Manage Assets

In addition to people, organizations must also support the business processes required to acquire construct and manage assets. Assets can include physical infrastructure and buildings and other non-intangible assets such as intellectual property.

14.4.11 Manage Enterprise Risk, Compliance, Remediation, and Resiliency

The organization must also support the business process required to implement effective risk management. This includes the identification and prediction of risks, the definition of strategies to avoid or mitigate risks, and protocols for managing risks that should be manifest this business process area also includes the management of compliance with regulations and any remedial works required to keep on track. Resiliency management including disaster recovery will also be part of this.

14.4.12 Manage External Relationships

In addition to the internal relationships between people and departments, it is necessary for an organization to manage external relationships including those with investors, customers, and end users.

14.4.13 Develop and Manage City Capabilities

A successful organization must also be able to support the business processes required to fine-tune business capabilities as markets and business operating conditions change. Capabilities can include new staff, new software tools, new equipment, and new assets.

14.5 Case Study – The Ministry of Infrastructure Development, United Arab Emirates

In the same way that smart mobility needs different organizational arrangements, then the way in which the framework or model is developed can be more effective if a new approach is taken. The authors: realized that recent experience provides a wonderful opportunity to explain an organizational optimization approach from a real-world, practical perspective.

An organizational optimization analysis was conducted for the then Ministry of Infrastructure Development (now Ministry of Energy and Infrastructure in the UAE). It was designed to understand the organizational change needs for the Ministry with respect to three technology application areas: Artificial Intelligence, The 4th Industrial Revolution, and Smart cities. It was part of a wider project to prepare staff and organization for new technologies and approaches. The authors were involved on both sides of the project, as specialist consultants and as part of the client group. This was known as the Advanced Scientific Assessment project and comprised separate assessments for each of the three application areas: Artificial Intelligence, The 4th Industrial Revolution, and Smart cities. The case study illustrates an innovative approach to organizing for smart mobility success and provides further detail on organizational elements and strategies.

14.5.1 Approach Adopted

The approach methodology that was adopted for the Advanced Scientific Assessment was based on adaptation of the Capability Maturity model [2]. The overall process adopted for the assessment is illustrated in Figure 14.2.

A total of five steps were supported in this process as follows:

14.5.1.1 Step 01
The original Capability Maturity model as developed by Carnegie Mellon University [2], was configured, and adapted to the needs of the three topics namely Artificial Intelligence, 4th Industrial

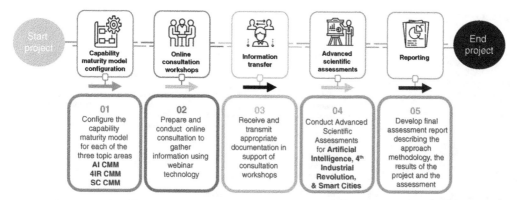

Figure 14.2 Analysis approach methodology.

Revolution, and Smart cities. This resulted in the development of three separate Capability Maturity models:

- **AI CMM** for Artificial Intelligence
- **4 IR CMM** for the 4th Industrial Revolution
- **SC CMM** for Smart cities

14.5.1.2 Step 02
A series of three online consultation and information gathering sessions were prepared and delivered, during which the three configured versions of the Capability Maturity model were used as a framework for information exchange and discussion.

14.5.1.3 Step 03
Subsequent to the online consultation and information-gathering sessions, additional documentation was transferred from the Ministry group to SARI and vice versa.

14.5.1.4 Step 04
The information received from the online consultation and information gathering, combined with additional documentation and research were used to conduct the Advanced Scientific Assessment on each of the three topics, making use of the adapted Capability Maturity models.

14.5.1.5 Step 05
A final assessment report was prepared (this document) describing the work conducted and the results. This document also contains some recommendations for improvement actions to raise the Capability Maturity level for each topic.

14.6   The Capability Maturity Model

The Capability Maturity model was originally designed for software development management by the Software Engineering Institute at Carnegie Mellon University in 1987. The model consists of two dimensions, Capability Maturity levels and capability criteria. Figure 14.3 shows the Capability Maturity levels originally adopted along with a brief description of each.

Level	0	1	2	3	4
Label	Initial	Repeatable	Defined	Managed	Optimizing
Description	• Unmanaged • No process is defined	• Configuration manage • Risk management • Measurement & analysis • Monitoring & control • Planning • Process management • Requirements management	• Decision analysis • Integrated project management • Organizational process definition • Organizational training • Process integration • Requirements management	• Organizational process performance • Quantitative project management	• Causal analysis and resolution • Organizational performance management

Figure 14.3 The original capability maturity model.

To meet the needs of the three topic areas, this original model was adapted by selecting new Capability Maturity levels and five capability criteria. The Capability Maturity levels were customized for each of the three topics, while the five capability criteria were common to all three topics. Figure 14.4 provides an overview of the Capability Maturity model adaptation for Smart cities.

14.7 Smart Cities Assessment

Smart Cities is a concept that is emerging in many parts of the globe. The number of people living in urban areas, combined with the need to improve the quality of living requires the effective application of technology to deliver services.

The range of smart city services being delivered worldwide is extensive and includes the following:

• Economy	• Emergency response	• Housing
• Education	• Governance	• Waste
• Energy	• Health	• Telecommunications
• Environment	• Recreation	• Urban planning
• Finance	• Safety	• Water and sanitation
		• Mobility

Mobility is last on this list of services; however, it is inordinately important in terms of the effect that smart mobility has on the quality of life and the efficient and effective use of energy.

Although this has been an emerging concept, the Ministry has been visionary and initiative-taking in defining the opportunity and taking some early steps toward harnessing the value and benefits of sparsity technologies that have been deployed elsewhere in the globe. The Ministry has also become a thought leader in the identification of suitable technologies and their practical application. Building on this existing platform of progress, it is necessary to develop a coordinated approach to the application of smart city technologies through a framework of projects, guided by key performance indicators linked to the intended effects and objectives.

Smart city Capability Maturity model	0	1	2	3	4
Capability criteria	Ad hoc	Opportunistic	Purposeful & repeatable	Operationalized	Optimized
Strategy	No overall roadmap for digital transformation, investment in discrete areas only	Strategy and investment at the department level, managing sharing of strategy and business case, some initial shared service transformation	Shared vision, strategy, and roadmap, multiple partners, business cases established, shared investments	Vision, strategy, and roadmap established citywide level, improved service outcomes	Strategy is optimized and evolving based on results, smart investments of clear impact on city's strategic priorities
Organization	Traditional organizational models, traditional client, provider/supplier, user relationships	Organization adapted to new ways of engaging partners, share the current ability for some initiatives	Organizational models evolved to shared accountability, partnership, and inputs	Transparent, multi-partner governance and organization, codevelopment and performance-based delivery	Organizational model stimulates an innovation system, promoting your combinations of service delivery, greater effectiveness
Data	Limited data reuse and integration, range of separate systems, issues with data integrity, quality, privacy & security, server specific data use	Optimization barriers being discussed, some advanced data analytics in place, some data sets are open to the public	Data management and optimization strategy agreed, investments in smart data management, extensive range of open data	Data used to provide actionable information, extended data capture, and analytics for decision-making	Data analytics used for dynamic and automated prediction and preventative improvements
Technology	Information communication technology architectures are predominantly designed to support individual business applications, limited investment in sensor networks and applications	Some sharing of integrated architectures, integration barriers are being discussed, some shared use of sensor networks	Investment in integrating architectures, joint investment plans	Cross organizational information, communication technology architectures in place, accelerated service delivery, citywide deployment of connected assets	Continuous review, adaptation and investment in information and communication technology, architectures defined, networkable environment across the city
Operations	Each operating unit has separate goals, objectives, and operating protocols	Some sharing of goals, objectives, and operating protocols	Some integrated information sharing, goals, & objectives, coordination and operating protocols	Extensive integrated information sharing, goals and objectives sharing, advanced decision support	Complete integrated information sharing, automated control, full integration of goals, objectives, and investments

Figure 14.4 Capability maturity model adapted for Smart cities.

The Advanced Scientific Assessment clearly shows that initial progress has been made across five capability criteria: strategy, organization, data, technology, and operations. The overall results of the assessment for Smart cities are shown in Figure 14.5.

The capability assessment dashboard, Figure 14.6, shows an average score of 10% for capability maturity regarding Smart cities.

This is an excellent position considering the short time and limited resources that the Ministry has applied to the subject. It also provides considerable scope for progress toward higher Capability Maturity levels. Each capability criterion is discussed as follows:

14.7.1 Strategy

There is a smart city strategy in place for the Ministry and evidence of investment and strategy sharing at the department level. This also includes business cases and some initial shared service transformations. However, shared vision, strategy, and roadmap do not yet exist across multiple

Capability Maturity Model criteria	Capability Maturity Model level		Score
Strategy	0	Ad hoc	10%
Organization	0	Ad hoc	10%
Data	0	Ad hoc	10%
Operations	0	Ad hoc	10%
Technology	0	Ad hoc	10%
Average	0	Ad hoc	10%

Figure 14.5 Capability maturity score summary for Smart cities.

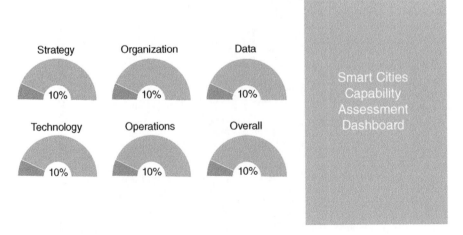

Figure 14.6 Smart cities capability assessment dashboard.

partners, and work is still required on the establishment of a business case based on key performance indicators. Vision, strategy, and roadmap have not yet been established at the national level, although larger cities such as Dubai and Abu Dhabi have a defined strategy and have partially implemented it. The relationship between investments and improved service outcomes has yet to be defined and there is not a clear understanding of how investments will have a direct impact on strategic priorities for the Ministries and for the cities. There is a need for closer coordination across all involved parties at the strategic level.

14.7.2 Organization

The existing organizational structure is not optimized for Smart cities, although a purpose-specific team has been formed. The Ministry has been quick to recognize the need to enable progress through adapting organizational structures. However, the current organization has not yet adapted to new ways of engaging partners including shared accountability for selected initiatives. Organizational and procurement models are not yet established for effective public–public and public–private partnership. An organizational structure for transparent multi-partner governance with extensive sharing and codevelopment, incorporating performance-based delivery has not yet been established. The organization is not yet able to fully support the stimulation of innovation and the delivery of new combinations of services with greater effectiveness.

14.7.3 Data

The Ministry has a sophisticated approach to the management of data and the development of solutions for specific applications. The IT department has equipped itself over the past few years to go beyond the management of internal computers and networks, to address external requirements associated with Smart cities. Barriers to system optimization and more extensive use of data are being discussed including concerns related to data sharing, data quality, data governance, and the creation of a Ministrywide data catalog. A Ministrywide data management and optimization strategy is in preparation but is not yet complete. Investment in data management and solutions tends to be focused on individual applications rather than enterprisewide solutions. There is a clear opportunity to make better use of existing data by making it accessible within appropriate cost and time resources. More extensive use can be made of analytics and use cases on a Ministrywide basis.

14.7.4 Technology

The Ministry has been adept in the application of smart city technologies across a range of subjects. The Ministry is particularly advanced in the application of smart mobility technologies such as automated vehicles and drone taxis. This initial implementation experience has given the Ministry a strong background and strong progress toward the skills required to support a full multimodal, multiproject smart city investment program. Further work is required to develop integrated architectures across smart city applications and develop a coordinated approach to smart city project implementation. A coordinated investment program that supports migration toward a total smart city is also required.

14.7.5 Operations

There is evidence of some sharing of goals, objectives, and operating protocols with respect to Smart cities. The Ministry has a strong role in ensuring coordination across departments and across cities.

Further integrated information sharing including goals and objectives, coordination, and operating protocols is required. There is currently a lack of extensive integrated information sharing and no work to investigate the possibility of automated control of smart city functions. The various smart city pillars have yet to be integrated from a planning and deployment perspective. Significant work is required to understand the anticipated effects of investments in smart city technologies.

14.7.6 Capability Criteria Summary

Significant progress has already been made in the focused application of smart city technologies both at the Ministry level and at the city level. To take advantage of this progress and to fully realize the benefits of Smart cities, it will be necessary to develop a framework approach to future projects and ensure that such projects are effectively related to intended effects, values, and benefits. A Smart cities governance framework will also be required to coordinate activities at various levels of government from federal, to Ministry to city. Progress toward a completely smart city can also be accelerated through the establishment and management of a series of public–public and public–private partnerships. In addition to the baselining delivered by this project, it would also be useful to conduct a benchmarking exercise to compare the Ministry's progress against other thought leaders and leading practitioners, on a global basis.

14.7.7 Smart Cities – Detailed Capability Criteria Assessment

14.7.7.1 Smart City Assessment Results

An Advanced Scientific Assessment for Smart cities was conducted on five capability criteria: strategy, organization, data, technology, and operations. As a result of input received from the online consultation sessions and further documentation, the following assessment results were determined. Overall, this is a subjective assessment of current capability. However, it is possible to justify the assessment scores based on evidence of progress made. Specifically, an inventory of initiatives, programs, and documents can be related to the assessment under each capability criterion. Figures 14.7–14.11 show the capability assessment score, and the relationship between such initiatives, programs, and documents.

14.7.7.1.1 Strategy: Current Maturity Level

There is a smart city strategy in place for the Ministry and evidence of investment and strategy sharing at the department level. This also includes business cases and some initial shared service transformations. However, shared vision strategy and roadmap does not yet exist across multiple partners, and work is still required on the establishment of a business case based on key performance indicators. Significant work has already been applied to develop performance

Smart Cities: Strategy	Assessment	Evidence	Status	Notes
Score	10%	Interactive smart cities are emerging	At the design stage only	
Level number	0	Intelligent mobility strategies	At the design stage only	Good progress has been made with some strategies, but the overall strategy and roadmap is incomplete. There is no overall roadmap for digital transformation, investment industry it is only
Level description	Ad hoc	Progress towards the optimal use of technical resources	Has been implemented unmeasured	

Figure 14.7 Current maturity level for Smart City strategy.

Smart Cities: Organization	Assessment	Evidence	Status	Notes
Score	10%	Policies and procedures	Implemented unmeasured	Very little organizational realignment or fine tuning, apart from some development of policies and procedures. Traditional organizational models, traditional client provider, supplier, user relationships
Level number	0			
Level description	Ad hoc			

Figure 14.8 Current maturity level for Smart cities organization.

Smart Cities: Data	Assessment	Evidence	Status	Notes
Score	10%	Infrastructure-based sensor data	Implemented	Progress has been made with data management applications which are specific to certain services and functions. There is limited data reuse our overall data management. There is a range of separate systems, issues with data integrity, quality, privacy and security. Data uses service specific
Level number	0	Geospatial infrastructure platform	Implemented	
Level description	Ad hoc	Progress towards the optimal use of technical resources	Has been implemented unmeasured	

Figure 14.9 Current maturity level for Smart cities data.

Smart Cities: Technology	Assessment	Evidence	Status	Notes
Score	10%	Intelligent connection of LIDAR data with the Road Asset System to activate self driving vehicles on the roads (automated)	In the implementation phase	Good progress is being made with the implementation of specific technologies. Overall architectures are still to be developed and the focus is still on individual business applications
Level number	0			
Level description	Ad hoc			

Figure 14.10 Current maturity level for Smart cities technology.

Smart Cities: Operations	Assessment	Evidence	Status	Notes
Score	10%			No overall operational plan, goals, objectives of smart cities have been defined. Some progress has been made with individual projects. Progress is also been made at the city level, not under the auspices of the ministry. Each operating unit has separate goals, objectives and operating protocols
Level number	0	The Digital transformation platform for project management system infrastructure projects	It has been implemented	
Level description	Ad hoc			

Figure 14.11 Current maturity level for Smart cities operations.

indicators across the Ministry [3, 4]. Vision, strategy, and roadmap have not yet been established at the national level, although larger cities such as Dubai and Abu Dhabi have a defined strategy and have partially implemented. The relationship between investments and improved service outcomes has yet to be defined and there is not a clear understanding of how investments will have a direct impact on strategic priorities for the Ministries and for the cities. There is a need for closer coordination across all involved parties at the strategic level.

14.7.7.1.2 Organization: Current Maturity Level

The existing organizational structure is not optimized for Smart cities, although a purpose-specific team has been formed. The Ministry has been quick to recognize the need to enable progress through adapting organizational structures. However, the current organization has not yet adapted to new ways of engaging partners including shared accountability for selected initiatives. Organizational and procurement models are not yet established for effective public–public and public–private partnership. An organizational structure for transparent multi-partner governance with extensive sharing and codevelopment, based on performance of the base delivery has not yet been established. The organization is not yet able to fully support the stimulation of innovation and the delivery of new combinations of services with greater effectiveness.

14.7.7.1.3 Data: Current Maturity Level

The Ministry has a sophisticated approach to the management of data and the development of solutions for specific applications. The IT department has equipped itself over the past few years to go beyond the management of internal computers and networks, to address external requirements associated with Smart cities. Barriers to system optimization and more extensive use of data are being discussed including concerns related to data sharing, data quality, data governance, and the creation of a Ministrywide data catalog. A Ministrywide data management and optimization strategy is in preparation but is not yet complete. Investment in data management and solutions tends to be focused on individual applications rather than enterprisewide solutions. There is a clear opportunity to make better use of existing data by making it accessible and appropriate cost and time resources. More extensive use can be made of analytics and use cases on a Ministrywide basis.

14.7.7.1.4 Technology: Current Maturity Level

The Ministry has been adept in the application of smart city technologies across a range of subjects. The Ministry is particularly advanced in the application of smart mobility technologies such as automated vehicles and drone taxis. This initial implementation experience has given the Ministry a strong background and strong progress toward the skills required to support a full multimodal, multiproject smart city investment program. Further work is required to develop integrated architectures across smart city applications and develop a coordinated approach to smart city project implementation. A coordinated investment program that migrates toward a totally smart city is also required.

14.7.7.1.5 Operations: Current Maturity Level

There is evidence of some sharing of goals, objectives, and operating protocols with respect to Smart cities. The Ministry has a strong role in ensuring coordination across departments and across cities. Further integrated information sharing including goals and objectives, coordination, and operating protocols is required. There is currently a lack of extensive integrated information sharing and no work to investigate the possibility of automated control of smart city functions. The various smart city pillars have yet to be integrated from a planning and deployment perspective. Significant work is required to understand the anticipated effects of investments in smart city technologies.

Smart Cities Scoring Summary Scoring was applied to the assessment by assuming the values shown in Figure 14.12 for the different Capability Maturity levels:

Based on the relationship between the level and the percentage score, Figure 14.13 summarizes the scoring for each capability criteria and overall, for Smart cities:

Figure 14.12 Capability maturity level scoring for Smart cities.

Capability Maturity Model level	Score
Level 0	10%
Level 1	25%
Level 2	50%
Level 3	75%
Level 4	100%

Capability Maturity Model criteria	Capability Maturity Model level		Score
Strategy	0	Ad hoc	10%
Organization	0	Ad hoc	10%
Data	0	Ad hoc	10%
Operations	0	Ad hoc	10%
Technology	0	Ad hoc	10%
Average	0	Ad hoc	10%

Figure 14.13 Capability maturity scoring summary for Smart cities.

Smart cities capability assessment dashboard, Figure 14.14 indicates that while some good progress has been made with respect to the development of strategy, data, technology, and operations, organizational aspects of lagging behind. There is a significant amount of work to be done to increase maturity for organizational aspects.

Smart City Recommendations for Improvement Actions   In summary, a total of 29 recommendations for improvement actions regarding Smart cities applications were defined. These have been grouped under the five capability criteria. To illustrate the five capability criteria with reference to Smart cities, the following examples have been defined:

Strategy: Smart city strategy involves the definition of goals and objectives, the creation of a future vision, and the development of a roadmap to move from today's situation to the desired future vision. Strategies can also include public–private partnership arrangements and the definition of a broad range of activities designed to achieve the objectives.

Organization: This measures the degree to which citywide governance and multimodal system operation can be achieved. This also measures the degree of alignment between objectives and the degree to which the organization is integrated.

Data: This measures the effectiveness of bringing data together from multiple sources, making it available in a cost-effective manner. This also measures the ability to take a smart data management approach that supports the use of large datasets and advanced analytics, while also enabling the distribution of appropriate information across the Ministry.

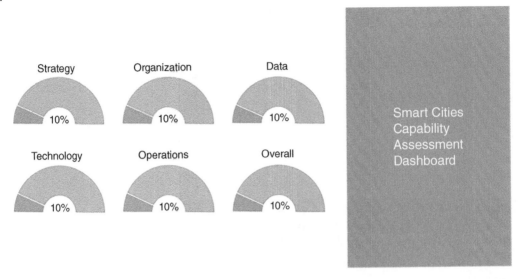

Figure 14.14 Smart cities capability assessment dashboard.

Technology: Measures the success of identifying and applying appropriate technologies to achieve the goals of the smart city. These can include sensors, telecommunications, back-office, and command-and-control systems. The degree to which technology application is coordinated for the private sector is also measured. The integrated application of the technology within an overall framework architecture is also an important measure.

Operations: The use of advanced technologies to support transport operations across multiple modes. This measures operations in routine and nonroutine transport conditions, taking account of variations in the demand for transport and operating conditions. The degree to which transport operations are connected, adaptable, and monitorable is also measured.

Strategy

1) The development of a nationwide strategy for Smart cities that includes the various Ministries and the cities
2) The development of a shared vision, strategy, and roadmap across multiple partners, driven by key performance indicators
3) Develop a clear relationship between proposed investments and anticipated service outcomes
4) Optimize the strategy based on implementation experience and ensure that smart investments have clear impact on strategic priorities

Organization

5) Fine-tuning of the current organization to adapt to smart city needs
6) Adaptation of governance models to be more effective with public–public and public–private partnerships
7) Establishment of a range of key performance indicators that define the intended outcome for the smart city program
8) The development of a shared vision and roadmap based on the key performance indicators
9) The identification of early winner projects that can deliver clear and immediate value, while also preparing the organization for larger-scale investment

10) Optimization of the strategy and the investment program based on experience gained during early initiatives

Data

11) Continued efforts to identify optimization barriers
12) The development of a plan for smart data management system that spans data silos and delivers an enterprisewide view of data
13) Incorporation of smart city data sources both public and private into the overall data management plan
14) Steady migration from application-specific solutions to an enterprisewide approach to data management
15) The development of a data valuation and monetization strategy
16) Study the potential for innovative business models involving private sector roles

Technology

17) The development of citywide smart city action and investment plans
18) A coordinated approach to the application of technology across multiple areas
19) The development of key performance indicators that can guide implementation across multiple sectors
20) The development of an integrated architecture at technology, organizational, and business model levels
21) The development of cross-organizational information sharing to support accelerated service delivery
22) Develop mechanisms for reviewing smart city strategies in the light of practical experience
23) Develop the capability to scan internationally for best practices

Operations

24) The development of a plan for extensive integrated information sharing including goals and objectives. This should include the Ministry and major cities
25) The definition of decision support requirements for smart city projects
26) Developing understanding of the capital and operating cost of smart city services
27) Develop the capability to engage the private sector in operations for Smart cities
28) Develop service-level agreements to support smart city operations
29) To find ways to integrate planning, design, project management, operations, and maintenance or Smart cities

14.8 Smart Cities Roadmap

Figures 14.15 and 14.16 show a roadmap that has been defined to bridge the gap between the current situation and the desired future situation in incremental stages, taking account of all capability criteria. Figure 14.15 depicts strategy, organization, and data, while Figure 14.16 depicts technology and operations. All criteria have been considered as a single roadmap to enable synergy to be released between efforts to improve the five criteria. This allows each criterion to be improved and improvement activities on one criterion, to improve others. The roadmap also shows a suggested priority for each action in terms of short, medium, and longer-term actions.

Fig 15	Short term (6–12 months)	Medium term (12–18 months)	Longer term (18–24 months)
Strategy	The development of a nationwide strategy for Smart Cities that includes the various Ministries and the cities		
		The development of a shared vision, strategy, and roadmap across multiple partners, driven by key performance indicators	
		Develop a clear relationship between proposed investments and anticipated service outcomes	
			Optimize the strategy based on implementation experience and ensure that smart investments have clear impact on strategic priorities
		Adaptation of governance models to be more effective with public–public and public–private partnerships	
Organization	Fine-tuning of the current organization to adapt to smart city needs		
	Establishment of a range of key performance indicators that define the intended outcome for the smart city program		
		The development of a shared vision and roadmap based on the key performance indicators	
		The identification of early winner projects that can deliver clear and immediate value, while also preparing the organization for larger scale investment	
		Optimization of the strategy and the investment program based on experience gained during early initiatives	
Data	Continued efforts to identify optimization barriers		
	The development of a plan for smart data management system that spans data silos and delivers an enterprisewide view of data		
		Incorporation of smart city data sources both public and private into the overall data management plan	
		Steady migration from application-specific solutions to an enterprisewide approach to data management	
			The development of a data valuation and monetization strategy
			Study the potential for innovative business models involving private sector roles

Figure 14.15 Roadmap for capability maturity for Smart cities: strategy, organization, data.

Fig 16	Short term (6–12 months)	Medium term (12–18 months)	Longer term (18–24 months)
Technology	The development of citywide smart city action and investment plans		
		A coordinated approach to the application of technology across multiple areas	
	The development of key performance indicators that can guide implementation across multiple sectors		
	The development of an integrated architecture at technology, organizational and business model levels		
		The development of cross organizational information sharing to support accelerated service delivery	
		Develop mechanisms for reviewing smart city strategies in the light of practical experience	
	Develop the capability to scan internationally for best practices		
Operations	The development of a plan for extensive integrated information sharing including goals and objectives. This should include the Ministry and major cities		
	The definition of decision support requirements for smart city projects		
	Developing understanding of the capital and operating cost of smart city services		
		Develop the capability to engage the private sector in operations for Smart Cities	
		Develop service level agreements to support smart city operations	
			Integrate planning, design, project management, operations, and maintenance or Smart Cities

Figure 14.16 Roadmap for capability maturity for Smart cities: technology and operations.

14.9 Capability Assessment Within the Construction Context

As construction is a particularly important activity for the Ministry, a specific focus on how the assessment relates to construction is provided in this section of the report. The assessment was conducted using the following capability criteria:

- Strategy
- Organization

- Data
- Technology
- Operations

These criteria serve the purpose of supporting a broad sweep assessment of current capabilities. To build on this, it is also necessary to address the specific needs of the Ministry. One way to look at the broader range of activities supported by the Ministry is in terms of the transport service delivery process. The process is depicted in Figure 14.17.

At a high level, each of the five steps represents a number of business processes supported by the Ministry to develop a balanced infrastructure that is sustainable and integrated, fully supporting global competitiveness. In addition to assessing current capability against each of the five capability criteria, it is also necessary to define the relationship between Artificial Intelligence, 4th Industrial Revolution, and smart city capabilities, with the business processes represented above. This illuminates the technologies from a different perspective that is related to the activities of the Ministry. While it is necessary for the Ministry to be efficient and effective in planning, design, building, operations, and maintenance, building, or construction is of particular importance because of the resources employed by the Ministry in this step. This is an area where there is significant opportunity for the application of digital technologies. The Infrastructure and Urban Development industry is lagging behind in the application of advanced technology, with a focus on primarily manual methods. The 4th Industrial Revolution, with its focus on a coherent and coordinated approach to the application of digital technologies, presents opportunities and challenges to this industry. Building on the early experiences in Ministry aside with some technologies, consideration should be given to the development of a coordinated plan for applying the following digital technologies to the construction process [5].

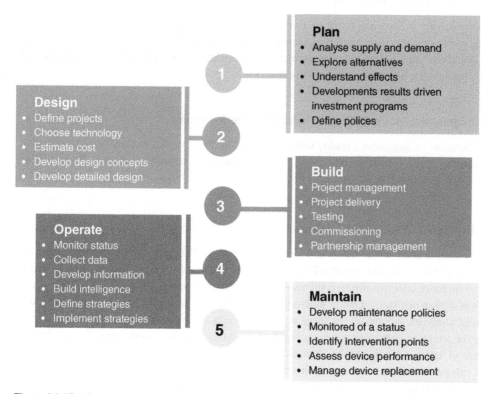

Figure 14.17 Construction processes.

14.9.1   Prefabrication and Modular Construction

This involves a shift from custom and ad hoc construction on site to a focus on prefabrication and modular construction in factories. It is likely that in a transition. Both existing construction techniques and new prefabrication and modular techniques will be implemented. The overall objective is to increase the size of the building blocks to enable resources to be applied in a quality-controlled manufacturing environment, while minimizing customized on-site construction.

14.9.2   Advanced Building Materials

In addition to the development of solutions and techniques that apply digital technologies to building, there is also great potential for advances in building materials. Building materials can take advantage of advanced technology but can also take a locally sourced approach to incorporation of materials at a raw level into construction components.

14.9.3   3D Printing and Additive Manufacturing

In addition to prefabrication and modular construction, the advent of 3D printing technologies effectively allows the factory to be brought to the construction site. This will enable extremely high quality and rapid construction within the environment of the construction site.

14.9.4   Autonomous Construction

In a similar manner to the emergence of automated driving systems on roads, such technologies can be applied to construction. Automated construction vehicles and robotic tools, driven by digital models and the appropriate automation technology are used to supplement manual labor and eventually replace the human role in construction.

14.9.5   Augmented Reality (AR) and Visualization Technologies

The incorporation of augmented reality techniques into all aspects of construction. Planners and designers can visualize concepts and designs, construction crews can see detailed instructions and visualization of the intended facility.

14.9.6   Big Data and Predictive Analytics

These can be applied across the whole service delivery process. Planning can be improved through the use of analytics to provide feedback regarding the success of previous projects and designs. Design can be more effective through the use of advanced analytics to understand the effects and outcomes of design decisions. Project delivery and construction can be improved through the use of advanced analytics to harness the large volumes of data available regarding the design of the project and current construction progress. Big data analytics can also enable operations to have a new degree of agility with respect to changes in demand and operating conditions. Maintenance activities can be focused on the use of observed data to monitor the performance of assets and infrastructure and manage maintenance activities in a scientific manner.

14.9.7 Wireless Monitoring and Connected Equipment

The Internet of Things comes into play for this type of technology. Low-cost, high capability sensors can be connected wirelessly to a back-office to establish a framework of connectivity that provides the raw data for both short-term and long-term management of activities.

14.9.8 Cloud and Real-time Collaboration

Building on the connectivity potential for wireless monitoring and connected equipment, people can network together, and processing power made available through the use of wireless networks and data management in a cloud environment. This will reduce latency on available information and support a much more collaborative approach across the construction process.

14.9.9 3D Scanning and Photogrammetry

In addition to advanced sensors, emerging 3D scanning and photogrammetric techniques can be applied to provide a large volume of data to guide design and construction. The use of LIDAR and other sensing technologies can create a high-resolution picture of the building site and the planned facility.

14.9.10 Building Information Modeling (BIM)

The development and use of digital models to support design, engineering, and construction activities. This can be expanded to include operations and maintenance to provide a lifecycle approach to building management.

14.9.10.1 Digitalizing the Process from Conception to Planning, to Design and Implementation

In addition to these individual technologies, it is also appropriate to use the 4th Industrial Revolution technologies across the entire process from planning, through implementation, to operations and maintenance. There is great potential for reducing the time and cost involved in getting from initial concept to completion of the construction or building. The coordinated application of these technologies along the business process and value chain could have significant impact on speed, cost, and quality of construction.

These technologies can be combined into a plan for the coordinated application of digital technologies to construction. Building on this assessment and the subsequent development of a strategy framework, a detailed plan for the application of the 4th Industrial Revolution technologies can be developed specifically for the construction process. One of the outcomes of the plan will be improved capability under the strategy, organization, data, technology, and operations capability categories used in this assessment.

14.10 Change Management

Over several decades of experience, the authors have learned an important lesson. The development of an appropriate organizational structure to support the technology and business elements of smart mobility is actually a major change management consideration for the organization. Significant changes to be required and this must be done to ensure that staff are motivated to work, not threatened and the resources within the organization are properly harnessed. Recognizing this is the first step toward effective change management. To be good at change is more than a technical or business consideration, it involves winning the hearts and minds of the staff and appropriately motivating them through a clear communication of the intended outcomes and intended actions, the closer this can get to letting each staff member know what their potential role and responsibility will be in the new organization, the better. The development of a good change management plan is essential, and it should contain the following features at a minimum:

- A clear definition of the value proposition that is being sought by smart mobility implementation and consequential organizational change.
- A description of the ways in which the value proposition will be communicated. They should include definition of the messages and the delivery channels as well as the intended audience.
- An explanation of the impacts on both the people in the organization. This should include the entire organization, the departments that make up the organization and the individual staff members that make up the departments.
- A clear explanation of the proposed change management arrangements – what will be done and how long will it take. This should also include an explanation of first staff members can provide feedback and input to the plan.
- A definition of the keys to success – does the department or person driving the change perceive as the keys to success? This is more than confirmation to the person driving the change, it is also an opportunity for departments and staff to attain ownership in the success.

14.11 The Need for Change Management

Mainly because there is a lot at stake. The use of smart mobility to optimize transportation requires application of innovative technology and the smart use of data. This cannot be achieved by an organization that is fine-tuned to deal with asphalt, concrete, and steel. While the organization must continue to be excellent in these areas, it is essential to add new capabilities to get the best of hardware, software, data, and people in an effective and efficient configuration. It is also important that the process implemented to optimize the organization must itself be efficient and effective. That's why a plan is important. The authors believe that there are two different ways to motivate a group of people and it is quite simple. They can be told it if they do not change and increase the level of cooperation required by new skills, and approaches, then something bad will happen. They can also be told that if they do change in the level of cooperation in alignment with the needs of smart mobility then something good will happen. Often using the two approaches at the same time can be highly effective. The "something bad" relates to past experience in project management for projects that have a significant information technology component, just like smart mobility. Ambiguous business objectives, uncoordinated stakeholders, and excessive rework mean that 75% of project participants lack confidence that their projects will succeed. On average, large IT projects run 45% over budget and 7% over time, while delivering 56% less value than predicted. The biggest barriers to success are listed as people factors: changing sets and attitudes, 58%, culture corporate culture barriers 49%, and lack of senior management support, 32% [6]. The "something good" is easier to define, it is all of the opportunities discussed in Chapter 9 on opportunities and challenges and the value that can be delivered by smart mobility as addressed in Chapter 4 – the value and benefits of smart mobility. At the highest level, these include safety, efficiency, enhanced user experience, climate change management, and equity management.

There are several aspects to consider when developing a successful change management approach. These are primarily the development of a future vision, a detailed understanding of the current organizational situation, and a roadmap to get to the future vision stage by stage. An effective approach for this is covered in Chapter 6 on planning for smart mobility. Effective change management is really about something simple and that is how to maximize the probability that what is wanted is achieved while avoiding or managing the undesirable side effects, in other words, what is not wanted. A robust approach to this has some specific planning needs which include the following [7].

14.11.1 Inspiring Members of the Current Organization

Making use of the vision that has been developed, inspire members of the current organization by communicating the vision, setting the expectations by describing the future, and explaining how to get there as a team or organization. This should also include a clear explanation of the benefits.

14.11.2 Engaging Hearts and Minds

This requires a provision of sufficient inspiration and motivation to set the scene for change and provide a driving force toward that change which includes leadership "pull" and individual staff "push."

14.11.3 Aligning Enterprise and Personal Objectives to Define Perception of Change

A clear description of the objectives should be used to align enterprise objectives with group objectives and personal objectives. This is likely to require an interactive dialogue to support information sharing across these different sets of objectives. This should be the first step in moving the organization and individual staff toward a convergence on the vision.

14.11.4 Defining the Skills and Tools Available Today and Those Required for Tomorrow

A deliberate structured analysis be conducted that compares skills and tools available within the organization today, with those that will be required to support tomorrow's future vision. A good way to address this is through the use of the capability maturity model as described in the preceding case study from the United Arab Emirates. Building on the analysis, recommendations should be identified and implemented to develop the workforce and organization to align with the future vision. It will also be necessary to update the analysis and also the future vision at least on an annual basis.

14.11.5 Integration of the Change Management Process Across the Entire Organization to Optimize Efficiency

Change management should be more than a one-time deal on a short-term campaign, the change should be implemented in an integrated manner across the organization. Methods for maintaining and monitoring the effects of the changes should also be identified and implemented. Small organizational and process improvement groups with a focus on specific tasks within the organization have been shown to be extraordinarily successful in this regard.

14.11.6   How Long Should the Change Take?

The amount of time it takes to implement change videos according to the scope of the change:

- Individual procedures can take between one and three months to implement meaningful change.
- Entire business process changes can take between 3 and 12 months, depending on the number of processes involved. Note that simplification of business processes and reduction in the number of processes would be an important activity within this change.
- The business unit can take between one and three years for complete change to be implemented.
- An entire organization can take between three and six years depending on the size and scope of the existing organization.

These times can be shortened through the use of centralized management, the adoption of agile techniques, the application of innovation management techniques, and strong abilities to motivate people. Change management is a specialist subject that may require the enlistment of appropriate specialist resources.

14.12   Summary

This chapter describes the purpose of an organization and how it can support the smart mobility revolution. It describes an overview of the essential elements of a smart mobility organizational framework. It addresses performance management aspects of the organization, differentiating between performance, and productivity. An extensive case study is described that illustrates how a capability maturity model approach was taken to the analysis of current organizational capabilities and future requirements. This information should provide some good guidance on the development of a specific approach to organizational change for a smart mobility application. In particular, the Smart City Recommendations for Improvement Actions that were defined at the end of the assignment represent a rich source of strategies that can be implemented to tune and organization to the needs of smart mobility and Smart cities.

References

1 APQC process classification framework (PCF) – city government version. chrome-extension://efaid nbmnnnibpcajpcglclefindmkaj/https://www.apqc.org/system/files/City%20Government_ v721_020619.pdf, retrieved July 7, 2022, at 10:08 PM Central European Time.
2 Capability maturity model for software, version 1.1. Carnegie Mellon Software Engineering Institute, 1993.
3 Value chain and ministry of infrastructure development processes. Ministry of Infrastructure Development, received May 5, 2020.
4 UAE smart data framework part two implementation guide. Ministry of Infrastructure Development, received May 5, 2020.
5 The vision and future plan for open data. Ministry of Infrastructure Development, received May 5, 2020.
6 Why do projects fail? A resource for organizational learning focused on improving project success rates. http://calleam.com/WTPF/?Page_id=1445, retrieved July 10, 2022, at 8:29 PM Central European Time.
7 Daniel Locke consulting, change management process: the ultimate step-by-step guide. https:// daniellock.com/change-management-process/, retrieved July 10, 2022, at 20 PM Central European Time.

15

Smart Mobility Operational Management

15.1 Informational Objectives

After reading this chapter, you should be able to:

- Define operational management and the essential elements
- Explain the importance of operational management in smart mobility
- Describe current traffic operational management processes
- Define an operational management process that takes full advantage of expedience from other industries
- Explain the need for effective decision support in operational management
- Describe the roles of people, business models, and technology in operational management
- Explain the relationship between operational management and performance management
- Describe an example of the application of technology to decision support in traffic management
- List some appropriate advice and lessons learned

15.2 Introduction

Effective operational management is a critical success factor in smart mobility. As many of the initiatives in smart mobility are delivered within a framework of projects or programs, it is important to differentiate these from operational management. It could be considered that operational management is a project that has no end. This requires a different philosophical approach as many people in the world of smart mobility have a project or program management culture. They specialize in the project lifecycle from planning, through design to project delivery. In many cases, the specialists then move on to a new project and involvement in the initial project wanes. The authors have found during their experience advising public sector agencies on the successful implementation of smart mobility that many project management specialists evolve into an operational management role without really realizing it. Therefore, it is important to recognize and clearly define the boundary between the end of a project and the beginning of operations. While there may be an overlap between the two, it is important to define this overlap as a transition period, with a fixed duration.

Smart Mobility: Using Technology to Improve Transportation in Smart Cities, First Edition. Bob McQueen, Ammar Safi, and Shafia Alkheyaili.

It is the authors' experience that people who excel in project and program management do not necessarily have the skills for successful operational management. Project management requires a focus over a defined period and usually involves the management and creation of new implementations. On the other hand, operational management tends to be a much longer-term assignment requiring a distinct set of skills including customer service, within a more repetitive framework. In this chapter, the essential elements of operational management are defined and explored. An overview of the operational management processes currently used in traffic and transportation is provided before the introduction of an amended process based on lessons learned and experiences gained in industries beyond transportation. The role of operations management within the overall system lifecycle is also considered and the importance of operations management to the success of smart mobility initiatives is defined and discussed. While the focus of this chapter is on operational management, it should also be noted that the other elements of service delivery are equally as important. Operational management works within a larger framework of these elements as depicted in Figure 15.1.

This figure shows how operational management fits within the larger framework of service delivery elements. Variations on this figure have been used throughout the book to illustrate the range of activities involved in delivering smart mobility services.

The activities required to deliver smart mobility services could also be viewed from the system lifecycle perspective as depicted in Figure 15.2.

This typifies the activities in terms of an investment plan lifecycle curve. It shows that operations become important after planning, design, and implementation have been achieved.

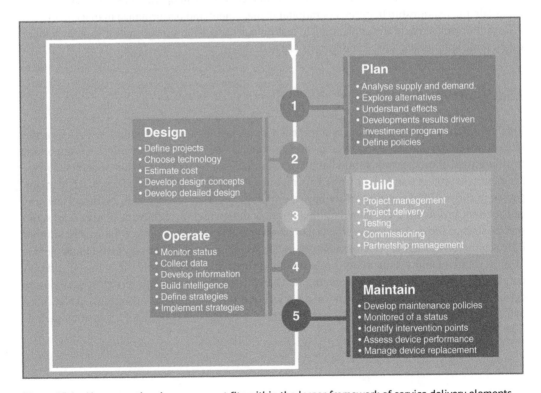

Figure 15.1 How operational management fits within the larger framework of service delivery elements.

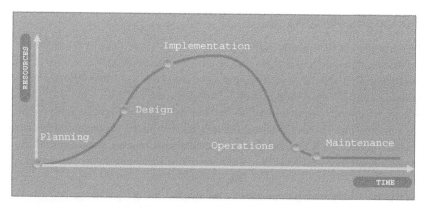

Figure 15.2   Operational management within a lifecycle perspective.

The figure is not to scale, but it also indicates that operations and maintenance will typically require somewhere between 15% and 20% of the capital invested in planning, design, and implementation. The resources required for operations and maintenance of smart mobility services typically exceed the figure of 10% of capital, which is the accepted wisdom for infrastructure-focused initiatives. Smart mobility services require a higher level of customer interaction and resources to support the continuing operation of the system. It is also interesting to note that it would be expected that most of the challenges related to the smart mobility services lifecycle would be toward the left-hand end of the diagram where large capital investment is required. In practice, the main challenges seem to be in finding resources for the right-hand side of the diagram, especially during operations and maintenance. There are many resource funds available for capital and implementation, with lesser resources reserved for operations and maintenance. This highlights the need for efficient operations and maintenance to ensure that funds are available and invested wisely and that the quality of the services is preserved.

15.3   Definition and Essential Elements of Operational Management

Operational management involves the management of available resources to achieve a predefined set of objectives, in the most efficient way. It can be defined as "the administration of business practices to create the highest level of efficiency possible within an organization" [1]. The steps taken to do this are as follows:

- Defining and agreeing objectives
- Managing processes efficiently
- The use of technology to improve processes and implementations
- Implementing strategies
- Collaboration on implementation and evaluation
- Evaluating the effects of the strategies
- Adapting operational management and performance measures in the light of the effects of the strategies

15.3.1 Defining and Agreeing Objectives

The objectives can be defined at both strategic and tactical levels. At the strategic level, these are commonly referred to as outcomes and can include one or more of the following:

Safety: reduction in the number of crashes within the zone of control of the Traffic Management Center. These can include fatalities, injury, and property damage crashes.

Efficiency: the difference between output and input. In other words, the difference between the value of the outcomes achieved and the investment expended in creating the management system enhanced user experience: focusing on the end user or traveler, this involves ensuring that the traveler has an enjoyable experience and feels that the investment in management systems has created a worthwhile effect.

Equity management: this is of growing importance and involves ensuring that the benefits of the transportation system are available to all types of users on an equitable basis.

Decarbonization: outcomes under this category include reduction in the amount of fuel used and reduction in emissions. This can be achieved by better traffic and transportation management and a switch from internal combustion to electric vehicles.

At a tactical level, the Traffic Management Center can be focused on achieving performance measure targets related to total delay on the network, number of stops, amount of acceleration, and deceleration. This is measured during both recurring and nonrecurring congestion. The latter include weather-related events, construction, and sporting or leisure events.

15.3.2 Managing Processes Efficiently

This involves the management of available resources to ensure that business processes supported by the Traffic Management Center in pursuit of agreed objectives are operated as efficiently as possible. Such resources can include the following:

- Hardware
- Software
- Technologies
- Data
- People

It is worth taking some time at this point to ensure that there is a shared perspective on the meaning of efficiency and how it compares to other terms used to measure operations management. A simplified definition of efficiency is the difference between the resources invested and the value of the outcomes. For example, if $50 million has been invested in the establishment of a Traffic Management Center which results in total benefits to the community of $300 million, then the efficiency would be $250 million, or 600%. The benefits received are six times the initial investment. This is not to be confused with the term "productivity," which is used to measure the amount of work that can be done in a given time.

At a tactical level, efficiency is typically not measured as a single point, but rather as a zone as depicted in Figure 15.3. This depicts the balance of efficiency between freeways, arterials, and urban surface streets, buses, light rail, freight, aviation, taxis, and transportation network companies. In summary, efficiency measures how well operational management addresses the defined objectives and the level of resources required to do so. Efficiency zone could also be defined on a temporal basis with efficiency variations by time of day.

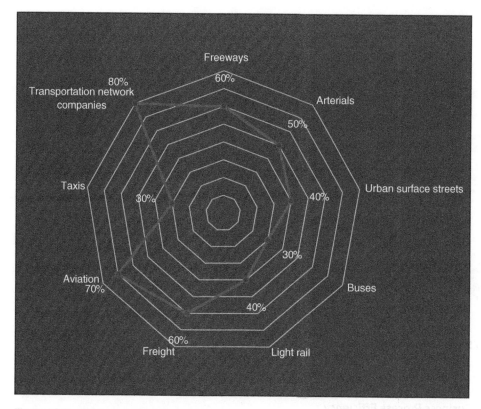

Figure 15.3 Efficiency as a zone.

15.3.3 The Use of Technology to Improve Processes and Implementations

A range of technologies can be applied to improve internal business processes and the strategies to be implemented. These include the full range of technologies as described in Chapter 8, which provides a detailed description of each technology. Figure 15.4 is reproduced from Chapter 8, Smart Mobility Technology Applications, as a reminder of the range of technologies that can be employed.

Sensing includes the collection of data from a variety of sources, telecommunications is the transmission of data from the point of collection to the Traffic Management Center. Data management involves technologies to convert data, to information, to insight, and strategies. Information and control technologies are used to deliver information to the traveler. Smart vehicles technologies include connected and automated vehicles.

15.3.4 Data Efficiency

Effective use of the maximum available data. This provides easy data access and optimizes the number of resources required to turn data into information. An efficient dataset management approach would enable any question to be answered any time and form part of a smart data management approach that minimizes data storage and processing costs while spanning silos and opening organizational cockpits. An efficient approach to data management would involve

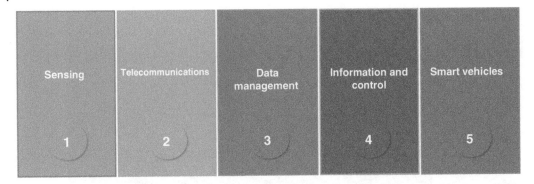

Figure 15.4 Range of available technologies.

Figure 15.5 Business process elements.

a centralized, coherent data storage capability that can ingest multiple traditional and non-traditional data sources. This would support the application of advanced analytics and enable such analytics to be shared in the Traffic Management Center and across regional stakeholders. The best approach would also support the valuation of the data being processed and ways to monetize data [2].

15.3.5 Business Process Efficiency

Business process elements are depicted in simple terms in Figure 15.5.

The figure shows the essential elements of a business process map: input, output, and activity. The activity is defined as the resources or effort required to convert input into output. A whole chain of these elements usually comprises a business process map that reflects the current configuration and activity, setting the scene for business process improvement. Here again, efficiency would be defined as the difference between output and input in terms of resources invested and value delivered.

15.3.6 Implementing Strategies

The strategies can consist of traffic advisory and traffic control actions. Typically, these are achieved using traffic signals, dynamic message signs, and automated actuators to open and close barriers. Figure 15.6 provides a summary of the type of actions that can be included in a strategy or response.

The actions have been grouped into three categories: Information, Control, and Management. The information category includes several channels that can be used to deliver information to travelers. The control category addresses traffic control activities, and the management category addresses additional actions related to the overall management of the traffic network.

15.3.6.1 Information Category
15.3.6.1.1 Variable Message Signs or Dynamic Message Signs
These are larger roadside structures, connected to the Traffic Management Center by a communication network. This allows messages from the Traffic Management Center to be displayed at the roadside to passing vehicles [3].

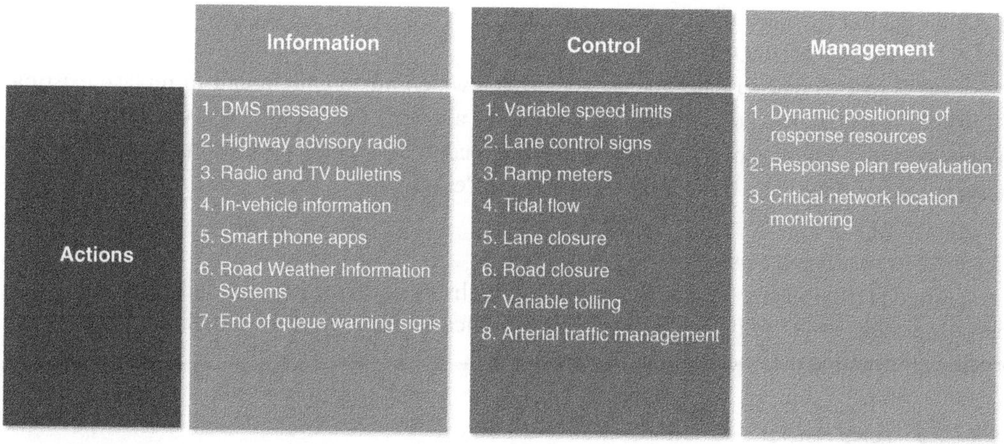
Information
Control
Management
Actions
1. DMS messages
2. Highway advisory radio
3. Radio and TV bulletins
4. In-vehicle information
5. Smart phone apps
6. Road Weather Information Systems
7. End of queue warning signs
1. Variable speed limits
2. Lane control signs
3. Ramp meters
4. Tidal flow
5. Lane closure
6. Road closure
7. Variable tolling
8. Arterial traffic management
1. Dynamic positioning of response resources
2. Response plan reevaluation
3. Critical network location monitoring

Figure 15.6 Possible strategy elements.

15.3.6.1.2 Highway Advisory Radio (HAR)

While this delivery channel has been replaced by dynamic message signs, highway advisory radio is still in use, especially in many parts of the United States. It involves the use of our specific sign with flashing lights informing drivers to tune in to a specific radio frequency to receive highway advisory information. The lights flash to inform drivers that new information is available [4].

15.3.6.1.3 Radio and TV Bulletins

The Traffic Management Center can have a direct feed to provide information to local radio and TV stations. The stations then broadcast traffic advisory information at regular intervals, via terrestrial TV, cable TV, radio, or radio data system [5].

15.3.6.1.4 In-vehicle Information Systems

This includes several different devices that can be installed in vehicles to provide information to drivers. The information can be provided using audio, visual information on flatscreen displays, or a combination of the two [6].

15.3.6.1.5 Smartphone Apps

In a comparable way to vehicle information systems, personal information can be delivered to cell phones by smartphone users using appropriate smartphone apps. There is some concern that the use of smartphones while driving can lead to distraction that will have an adverse impact on safety levels [7].

15.3.6.1.6 Road Weather Information Systems (RWIS)

Road Weather Information Systems have been specially developed to monitor weather conditions and provide suitable alerts and advisories to travelers [8].

15.3.6.1.7 End of Traffic Queue Warning Signs

These are special-purpose dynamic message signs that warn drivers when they are approaching the end of a queue. The signs are located just in advance of locations where queues are expected on a regular basis but are detected by sensors and CCTV in the Traffic Management Center [9].

15.3.6.2   Control Category

15.3.6.2.1   Variable Speed Limits

These are a special type of dynamic message signs which are small enough to replace existing speed limit signs along the highway. Where legally allowed, the variable speed limit signs can inform drivers of the prevailing speed limit on the section of road. The speed limit information is delivered to the signs from the Traffic Management Center [10].

15.3.6.2.2   Lane Control Signs

These are typically placed on gantries above the highway with one sign per lane. The signs can display variable speed limits as described above indicate when drivers should change lanes and provide information on lane closures in advance [11].

15.3.6.2.3   Ramp Meters

These are a specialized form of traffic signals installed on highway ramps, just before the ramp joins the main highway. The signs display either green or red aspects indicating that traffic in the lane controlled by the signals can proceed or must stop. This enables the floor vehicles onto the highway to be metered to fit into gaps in the mainline traffic flow. Sensors are typically installed on both the ramp and mainline highway to enable effective metering. These are typically operated based on one vehicle per green signal [12].

15.3.6.2.4   Tidal Flow

Movable barriers can be used to make additional lanes available to vehicles in traveling one direction or the other. These are typically used during AM and PM peaks when there is a significant tidal element to the traffic. In other words, a disproportionate volume of traffic is traveling in one direction, compared to the other [13].

15.3.6.2.5   Lane Closure

Specific lanes of the highway can be closed making use of barriers or the lane control signals described above.

15.3.6.2.6   Road Closure

Specific sections of the highway can be closed, making use of barriers or dynamic message sig. ns. These are typically reinforced by police presence to ensure that drivers comply with the information.

15.3.6.2.7   Variable Tolling

In the case of toll roads, the current charge for use of the road can be varied using variable or dynamic tolling techniques. These require special-purpose dynamic message signs to inform the driver of the toll currently in force. Congestion pricing could also be applied using similar techniques [14].

15.3.6.2.8   Arterial Traffic Management

Traffic signals on arterials can be subject to revised signal timings based on information from the Traffic Management Center. In some cases, this form of traffic control is also extended to urban surface streets, depending on whether these are within the jurisdiction of the Traffic Management Center [15].

15.3.6.3 Management Category
15.3.6.3.1 Dynamic Positioning of Response Resources
In the light of information regarding current traffic conditions and the current strategies that have been implemented, staff and management at the Traffic Management Center may reposition response resources. This allows the management of response times by moving response resources closer to critical locations. These locations may change depending on the strategy being implemented, and the effects of the strategy and traffic patterns. Resources include emergency vehicles and tow trucks.

15.3.6.3.2 Critical Network Location Monitoring
Specific network locations which are become critical can be subject to additional monitoring including refocusing closed-circuit TV cameras and deploying additional resources to increase monitoring of these locations.

15.3.6.3.3 Response Plan Reevaluation
Depending on the observed effect of the strategies being implemented, the response plan may be reevaluated, either in real-time or on a summative basis at regular intervals.

15.3.7 Collaboration on Implementation and Evaluation

Typically, a Traffic Management Center will control traffic within the boundaries of any smart city and overlap with other jurisdictions. Stakeholders from these jurisdictions should also be involved in assisting with implementation and the subsequent evaluation of the strategy effects. Many Traffic Management Centers also play a regional role spanning multiple jurisdictions.

15.3.8 Evaluating the Effects of the Strategies

This is an essential element of the aspects of operating a Traffic Management Center. The observed effects of the strategies will be measured against a few key performance indicators to determine if the desired results were achieved. A comparison of intended results against actual results will then be developed and used to reveal elements of the strategy that need to be improved. Mathematical simulation modeling tools can be used to assist in the evaluation.

This high-level overview of operational management for traffic and transportation is intended to orientate the reader by providing an overview rather than a detailed exposition.

The typical operational management business processes that are currently in use for traffic and transit will be described later in this chapter.

15.4 The Importance of Operational Management to the Success of Smart Mobility

Operational management is a critical factor in the success of smart mobility within a smart city. Smart mobility is delivered as a range of services. The delivery of the services must be conducted in an effective and efficient manner. Efficient operational management will result in the following outcomes:

- Highest service quality
- Best return on investment for mobility services

- A higher level of traveler satisfaction
- Minimization of fuel consumption
- Minimization of emissions
- Improving the equity of service delivery

15.5 Current Transportation Operational Management Processes

A review was conducted by the authors regarding typical traffic operational management business processes currently in use around the world [16–25].

Transportation operational management has a wider context including aviation, freight, and logistics. These are beyond the scope of this chapter and beyond the experience and expertise of the authors. The focus has been placed on the business processes associated with traffic and transit management. Figure 15.7 summarizes the typical business processes associated with traffic management.

15.5.1 Traffic Monitoring, Surveillance, and Detection

The use of roadside sensors and CCTV to monitor current traffic conditions and detect anomalies.

15.5.2 Data Management

The conversion of data into information, enabling the extraction of insight from the information and the development of appropriate response strategies,

15.5.3 Incident Management

Detecting anomalies due to traffic incidents, verification of the nature of the incident, dispatch of appropriate resources, and incident clearance. Incident management also includes continuous traffic management at both the regional and local levels during the incident.

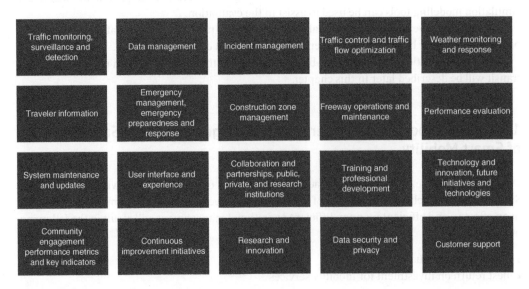

Figure 15.7 Typical traffic management business processes.

15.6 Traffic Control and Traffic Flow Optimization

The use of various traffic control techniques as described earlier to manage traffic and optimize traffic flow. The typical process for traffic control and traffic flow optimization on arterials is illustrated in Figure 15.8.

This depicts a process beginning with detection of current traffic conditions, the selection of an appropriate plan to be implemented, and consideration of wider corridor effects. Other regional stakeholders are then consulted on the proposed plan to be implemented, leading to the implementation of a plan appropriate to the nature and effects of the incident. This process can also be applied to urban surface streets.

15.6.1 Weather Monitoring and Response

The use of weather sensors and information from external systems to determine current weather patterns and provide input to strategy responses.

15.6.2 Traveler Information

Broadcasting information to drivers and other travelers using multiple delivery channels.

15.6.3 Emergency Management, Emergency Preparedness, and Response

Working in coordination with emergency management services, preparing plans, preparing for predefined emergencies, and implementing appropriate responses.

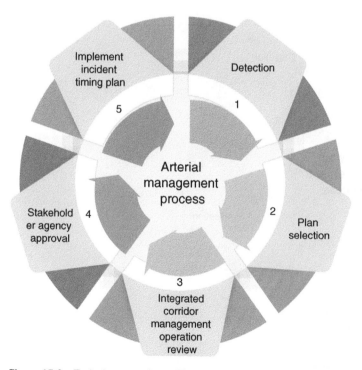

Figure 15.8 Typical process for traffic control and traffic flow optimization on arterials.

15.6.4   Construction Zone Management

Giving special attention to construction zone locations including controlling the speed and monitoring of traffic flow through the zone. This can also include the repositioning of tow trucks to enable the clearance of incidents quickly within critical sections of the construction zone

15.6.5   Freeway Operations and Maintenance

Operation and management of limited access highways, including routine operation, operation during special events, incidents, and managing the maintenance of such facilities.

Figure 15.9 illustrates a typical freeway operations and management process. Incidents are detected and then verified to confirm the responses required and ascertain the nature of the response resources required. A predefined plan is then selected and implemented. This leads to incident clearance, including local and regional traffic management as appropriate.

15.6.6   Performance Evaluation

Formative and summative performance evaluation of the operational management of the Traffic Management Center and the resources involved.

15.6.7   System Maintenance and Updates

Identifying, designing, and applying system updates and managing the system maintenance process to ensure optimal operation of TMC systems and assets.

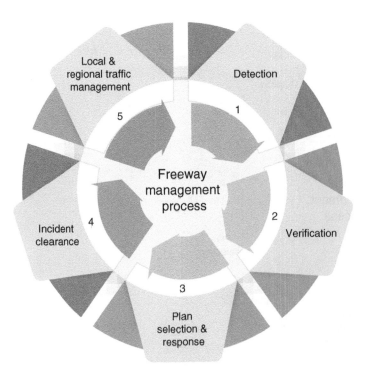

Figure 15.9   Freeway management process.

15.6.8   User Interface and Experience

The management of the interface between people and systems. This includes managers and operators in the Traffic Management Center and the interface with the end user, the traveler.

15.6.9   Collaboration and Partnerships, Public, Private, and Research Institutions

Alignment of Traffic Management Center objectives and operations to take account of resources available within the public sector, the private sector, and research institutions. In some cases, certain business processes within the Traffic Management Center may be outsourced to the private sector.

15.6.10   Training and Professional Development

Ongoing identification of training and professional development needs and creation of the appropriate programs.

15.6.11   Technology and Innovation, Future Initiatives, and Technologies

Scanning for innovative technologies and novel approaches, identifying opportunities to improve operations through technology, and studying the use of new practices and procedures. This activity can result in proposals for improvements to data collection, information processing, and the interface with TMC operators and travelers.

15.6.12   Community Engagement, Performance Metrics, and Key Indicators

The communication of predefined performance metrics and key indicators to the public. This will also include reporting to nontechnical decision-makers and regular performance evaluation.

15.6.13   Continuous Improvement Initiatives

This activity involves the application of a continuous improvement methodology. Typically, this will involve regular meetings of TMC operators and management to review the operation of the business processes that are supported and the identification of business process improvement opportunities. This may also be conducted under the auspices of a formal business process improvement program.

15.6.14   Research and Innovation

Continuous research to identify possibilities for improving operational management through innovation and research. This is typically done in collaboration with local research institutions and private-sector enterprises.

15.6.15   Data Security and Privacy

Defining and maintaining data security and privacy arrangements including data governance planning and data security arrangements. This can also include the definition and implementation of data governance arrangements.

15.6.16 Customer Support

Supporting the interface between the Traffic Management Center and the end user, traveler, or customer. This typically involves a call-center-style approach to managing requests for information from the traveling public.

The research review also identified typical business processes for transit management which are configured to the specific needs of transit. The transit business processes are depicted in Figure 15.10 and explained in the following section.

15.6.17 Service Planning and Scheduling

The development of service plans that define frequency and location of transit routes. Detailed schedules will then be developed to support the management of the service implementation.

15.6.18 Fleet Management

The vehicles that comprise the transit fleet are subject to management including the monitoring of vehicle location, vehicle condition, and scheduling of appropriate maintenance.

15.6.19 Crew Management

The management of transit operators includes assignment of vehicles, management of shifts, and overall personnel management.

15.6.20 Real-time Operations Control

Typically using automated vehicle location technology, real-time operations are controlled and managed from a centralized transit management center.

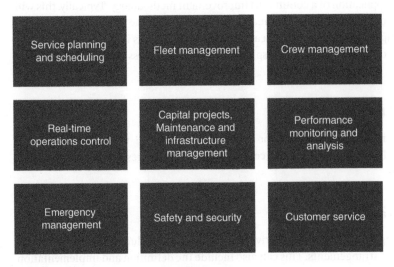

Figure 15.10 Typical transit business processes.

15.6.21 Capital Projects, Maintenance, and Infrastructure Management

The transit management center will also be responsible for the definition of required capital projects, the need for maintenance, and ongoing infrastructure management.

15.6.22 Performance Monitoring and Analysis

Together with real-time operations control, a performance management approach will be implemented to compare service performance against the intended. Analytics will be used to determine a detailed analysis of service **performance, leading to the definition of improvement proposals.**

15.7 Emergency Management

Management of crew and fleet in the event of nontypical operating conditions. This can involve vehicle rewriting and adjustment to crew schedules.

15.7.1 Safety and Security

Ongoing monitoring of the safety and security of the services being provided. Safety and security levels are compared against target levels and appropriate improvements are proposed. Note that this may also include the management of perception of safety

15.7.2 Customer Service

Travelers making use of the transit system are provided with customer support services including information to make best use of the services and responses to specific information requests.

15.8 A Complete Operational Management Approach Incorporating Lessons Learned from Other Industries

The typical business processes explained in the preceding section describe current practice in Traffic Management Centers around the world. The practice has grown organically and evolved in line with the evolution of the Traffic Management Center. While the processes represent best practice as currently applied, there is a significant benefit in taking a new look at the business processes associated with operational management and a TMC. To do this, the authors conducted a research review of operational management best practices beyond transportation. Industry studied included manufacturing and financial services.

The resulting business processes can be referred to as "industrial grade" operational management as they span across industries and different service delivery approaches. Some of the resulting operational management processes overlap with those defined for traffic and transit. Other processes introduce new insight into the requirements for effective operational management. The various processes are defined in a more generic language that make it easier to take a step back and understand requirements from a holistic perspective, rather than an empirical or traditional approach.

Figure 15.11 illustrates the complete operational management framework that was defined as a result of a study of other industries and other service delivery operations [26–35].

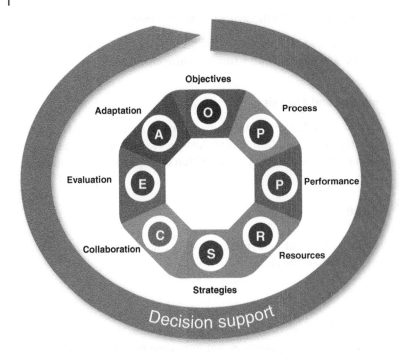

Figure 15.11 Complete operational management framework.

The authors believe that this forms a useful model and checklist for reviewing current operational management activities within Traffic Management Centers. It represents a streamlined approach that clearly identifies the need for decision support on a continuous basis across all processes. There are eight steps in the overall process described as follows.

15.8.1 Objectives

Clear definition of the objectives relating to operational management within the organization. This includes defining the scope of the operational management activity, including identification of the departments, functions, and processes being addressed. Strategic objectives can include safety, efficiency, user experience, equity attainment, and decarbonization. Examples of tactical objectives include response time to incidents, clearance time for incidents, the volume of traffic that moves through the network daily, the number of stops, and the amount of delay.

15.8.2 Process

This involves a definition of business processes to be supported to achieve the objectives and outcomes. It also involves the definition of "who does what," ensuring that every activity has an input, an output, and an honor. Existing processes and workflow should be mapped out to identify bottlenecks, redundancies, and opportunities for improvement. This can form the basis for a process redesign to eliminate inefficiencies, standardized procedures, and enhance productivity. From a data perspective, the process map will enable the identification of parallel or duplicate data inputs and provide the basis for clustering activities to improve efficiency. The resulting maps of both the existing process and the proposed redesign process should be clearly documented and communicated to all relevant stakeholders and those involved in the implementation of the processes.

15.8.3 Performance

This requires a definition of key performance indicators (KPIs) that are of direct relevance to the operational management and the successful support of the business processes. These KPIs should align with the art organization strategic goals and objectives, as defined in step one. KPIs can include metrics such as productivity, quality, customer satisfaction, efficiency, and employee performance. Based on the KPIs performance targets should be agreed and then monitored for achievement of goals. This aspect of performance management will be addressed in more detail in Chapter 11, Smart Mobility Performance Management, which addresses performance management for smart mobility, in more detail.

15.8.4 Resources

Involves the assessment of resource requirements to support the business processes and enable effective operations. The resources will include human, technology, equipment, and facilities. In addition, resources can also include hardware, software, data, and of course people. Effective management of resources requires an up-to-date perspective on emerging technologies and trends relevant to operational management best practices.

Existing resources should be evaluated to determine the effect they have in enhancing operational efficiency. New resources should be properly integrated within existing business processes and professional development training provided as appropriate.

15.8.5 Strategies

The development of strategies to optimize processes, streamline workflows, and enhance overall operational efficiency. The use of lean management principles and Six Sigma methodologies are other relevant frameworks that can be useful.

15.8.6 Collaboration

Fostering collaboration and effective communication between different departments, teams, and organizations involved in operational management. Cross-functional cooperation should be encouraged to build solutions to inter-departmental challenges and promote the attainment of shared goals. This can be further reinforced by using appropriate communication channels to ensure timely and accurate information flow across the organization and among regional stakeholders. This communication should be conducted with all those who affect overall operating conditions within the region and should include coordination of strategies and alignment of objectives. The communication should form part of our coherent and deliberate program to engage key stakeholders including end customers, those providing support to the TMC, partners, and others involved in regional transportation management. The overall goal of collaboration should be to build strong relationships based on trust, transparency, and mutual understanding that helps to align goals and strategies and optimize traffic management across the region. Continuous improvement should also be conducted in a collaborative manner.

15.8.7 Evaluation

This includes short-term changes to strategies because of effectiveness monitoring. This is based on formative evaluation which delivers feedback in a timely manner to enable ongoing course corrections. This can include regular staff meetings to evaluate past performance and suggest changes. This should include the establishment of monitoring mechanisms to track KPIs, evaluate performance, and identify areas of concern. A control system should be implemented to ensure inherence to standard processes, policies, and quality standards. The effectiveness of evaluation can be improved by recognizing and rewarding individuals and teams for exceptional performance and contributions to operational excellence. Feedback from the evaluation process can be used to identify areas for improvement and drive future decision-making.

15.8.8 Adaptation

This involves longer-term changes to the library of strategies, responses, and operational management of the business processes. It is based on summative, retrospective evaluation on a longer cycle than evaluation. Again, it can involve regular staff meetings to evaluate par performance and suggest changes. Adaptation should fit within a culture of continuous improvement within the organization and encourage feedback and suggestions from staff to identify potential areas for optimization. As adaptation takes place and changes to business processes and operational procedures are suggested, it is important to ensure that staff are equipped with the necessary tools and resources to perform their roles effectively within the changed environment. Successful adaptation will also involve continuous monitoring of market trends, customer demands, and industry developments to ensure that operational management strategies remain relevant and well aligned with changing conditions. Adaptation should also include the definition of arrangements to scale TMC operations to cover a larger future network or other modes of transportation. Periodic reviews and audits of operational management processes should also be conducted to ensure compliance with established procedures and attainment of the agreed objectives and outcomes.

15.8.9 Decision Support and Operational Management

The last step in the complete operational management cycle continuously operates in parallel with the other processes. This step provides effective decision support to each of the individual processes. In Traffic Management Centers, staff and management are required to make fast decisions that can have regional consequences. Therefore, it is important that they have the best available decision support system to take better decisions based on better information and advice. Many Traffic Management Centers around the world lack formal decision support leaving it to staff and management experience and opinions.

In the next section of this chapter, an example based on work conducted for two US transportation agencies will be described. The specific agencies will not be named to protect confidentiality. Since the objective is to illustrate operational management principles, and the example is a product of information from both agencies, the specific names of the agencies are not relevant anyway.

15.9 Application Example: Decision Support for Traffic Management

15.9.1 Background

The information in the example is based on work conducted with two agencies. The transportation agencies in question have a long history in the successful operation of Traffic Management Centers that address traffic management in dense urban areas and rural areas. The regions feature special event activities in the form of sporting events, concerts, and theme parks. The Traffic Management Centers manage freeways and arterials. One of the agencies recently implemented variable tolling on a section of freeway and has significant expertise in freeway, incident, and arterial traffic management.

15.9.2 Challenges

Operational management for a Traffic Management Center addresses many challenges; however, two specific challenges were identified as being of particular concern. Data quality and appropriate plan selection.

15.9.2.1 Data Quality

The data on which decision-making and plan selection is made comes from a variety of sources of variable quality. Sometimes it is difficult to understand the quality of the data from the disparate sources. The data sources also do not provide complete geographic coverage across the region.

15.9.2.2 Decision Support for Plan Selection

The appropriate plan for prevailing traffic conditions is selected on a subjective basis based on the experience of senior managers and staff in the Traffic Management Center. This puts managers and staff under considerable pressure to make rapid decisions that have region-wide significance. Effective decision support is vital.

15.9.3 Approach

A private sector solution provider was engaged to develop an approach that would apply artificial intelligence and machine learning to address the two problems identified above. The approach involves the use of software as a service to deliver the solution, avoiding the need for a lengthy period of procurement of hardware and commissioning. The concept was to supplement the existing Advanced Traffic Management System, rather than replace it, by focusing on improving the business processes related to data collection, data fusion, analysis, strategy selection, implementation, and presentation of results.

Figure 15.12 illustrates the intended integration of the existing advanced traffic management system, with a Software as A Service, AI power decision support system.

The figure shows a range of data sources been brought together into a unified dataset. Within the unified dataset activity, data quality management is supported by comparing the same data from various sources using orthogonal sensing techniques. The unified dataset is a foundation for delivering increased situational awareness and providing the fuel required for simulation and emulation. It also enables the maximum value to be extracted from all available data and sets the scene for data exchange. Advanced analytics drive dashboards to convey the information visually and are used to inform the selection of appropriate strategies. TMC operations include support for the business processes defined earlier, the output from which are traffic

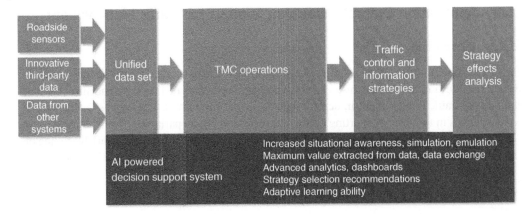

Figure 15.12 Integration of an existing advanced traffic management system with an AI power decision support system.

control and information actions that comprise a strategy. The effects of the strategy are assessed by the decision support system, with results used as feedback to a library of traffic control and information strategies.

The simulation and emulation conducted by the decision support system supports two learning cycles – **Network** and **Response** as depicted in Figure 15.13.

The "network" state learning cycle involves developing an understanding of network geometry, network contexts, and any updates made to the network over the operational period. The "response" learning cycle involves the measurement of response effects, the evaluation of response plans, and the revision of response plans.

The approach harnesses a mathematical traffic simulation model in combination with a decision support system engine. The relationship between the simulation model and the decision support system engine is depicted in Figure 15.14.

This shows that the simulation model is used to accept data from the data quality management system along with data regarding target objectives and plan tactics. The simulation model takes

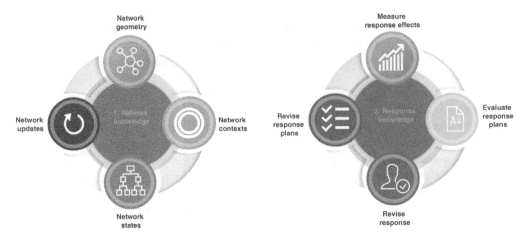

Figure 15.13 Decision support system learning cycles.

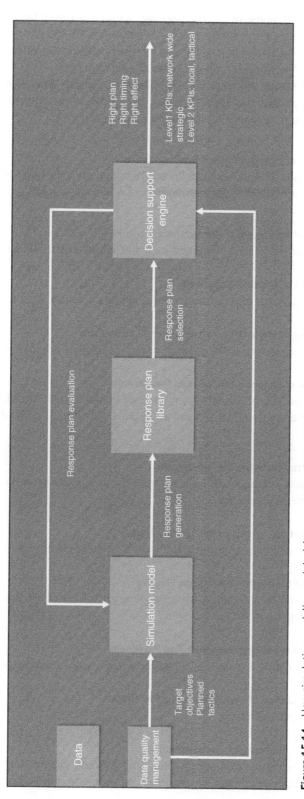

Figure 15.14 How simulation modeling and decision support work together?

this input and generates response plans. The response plans are accepted by the decision support system which is responsible for plan selection given current traffic and network conditions. The decision support system makes use of strategic and local key performance indicators to ensure that the right plan is selected at the right time and generates the right effect.

The decision support system was implemented using the following Work Breakdown Structure. This provides some insight into the type of activities that will be delivered by the solution provider and which will also require support from the public client.

15.9.3.1 Development of Data Quality and Business Cases

Initial assessment is made of available data, the business case for acquisition of third-party data, and the exploration of arrangements for data supply.

15.9.3.2 Definition and Agreement of Requirements

Requirements outcomes are defined, communicated, and agreed with the client.

15.9.3.3 Establishment of Dynamic Data Feeds

Dynamic data feeds required as input to the decision support system are acquired and established.

15.9.3.4 Establishment of Static Data such as GIS and Census Data

Static data feeds required as input to the decision support system are acquired and established.

15.9.3.5 Development of Output Interfaces and Dashboards

The user interfaces and dashboards used to communicate output from the decision support system to management and operating staff are defined and agreed. An example of the dashboard is illustrated in Figure 15.15.

15.9.3.6 Data Fusion to Create the Unified Dataset

Sophisticated data fusion techniques are used to create the unified dataset from all available data, suitably verified, and made compatible.

15.9.3.7 Localization of the System to the Specific Agency

The system implementation is customized to the specific needs of the agency.

15.9.3.8 Calibration of the Model and Analytics Engine

Through initial operations, models, and analytics engines are calibrated by monitoring output and comparing to the expected.

15.9.3.9 Digitalization of Existing Response Plan Playbooks

A digital response plan library is created by digitalization of existing response plans.

15.9.3.10 Operationalization of the Decision Support System

The system is brought into operation.

15.9.3.11 Interaction, Collaboration, Socialization

The system is used as an interaction collaboration and learning tool for staff and management within the Traffic Management Center and regional transportation stakeholders.

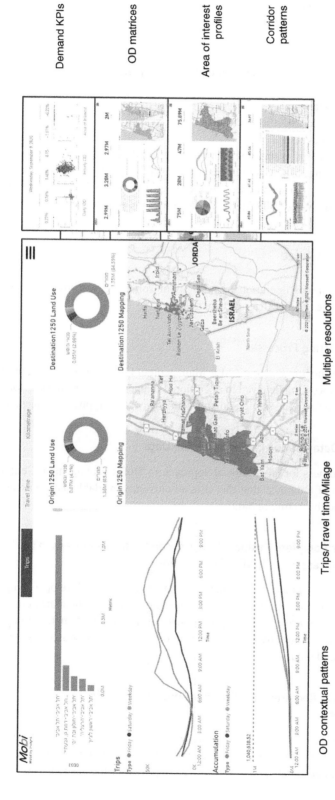

Demand KPIs

OD matrices

Area of interest profiles

Corridor patterns

OD contextual patterns   Trips/Travel time/Mileage   Multiple resolutions

Figure 15.15 Example dashboards from a decision support system. *Source:* Copyright Mobility Insight Ltd, used by permission.

15.9.3.12 Solution Delivery

The solution is delivered and handed over to the agency.

15.9.3.13 Evaluation

The solution is subject to a period of evaluation by an independent assessor.

15.9.3.14 Next Stage Planning

Plans to scale up and expand the decision support system including enhanced features and wide-area coverage are developed.

The authors hope that this example will bring operational management concepts to life and in particular highlight the importance of an effective decision support system. The work breakdown structure provided can also give some insight into how such a decision support system might be implemented within an existing Traffic Management Center and advanced traffic management system.

15.10 Practical Advice

To summarize this chapter on operational management, some practical advice is offered on how to apply the contents of the chapter in a practical situation.

15.10.1 Review Existing Processes

In addition to considering how existing Traffic Management Centers around the globe operate, also review the business processing involved in operational management for things beyond transportation. Examples include financial services, manufacturing, water supply, and telecommunications.

15.10.2 Create a Data Catalog Listing Available Data

As a step toward readiness for an enhanced approach to Traffic Management Center operations, consider the creation of a data catalog that lists all currently available data sources and provide some information and the nature of data.

15.10.3 Use of the Complete Business Process Map to Highlight Gaps in the Existing Business Processes for Traffic Management

The review of current best practices around the world can be used to develop a complete business process map to illustrate the vision of the innovative approach to traffic management operations. This can then be compared to a business process map of the existing situation to identify gaps in terms of inputs, outputs, and activities. The gaps may also address resource gaps such as people, hardware, software, and data.

15.10.4 Define Data Requirements for Next Generation Traffic Management Center

Based on the business process map for the desired future traffic management operations approach, it is possible to define the data required to support the desired activities. A gap analysis can also be conducted to identify additional data that would be required for the desired vision.

15.10.5 Establish Data Fusion and Data Quality Arrangements

An essential part of the implementation of the decision support system is the establishment of data fusion and data quality arrangements. The example shown shows the establishment of a unified dataset that enables the extraction of the maximum value from data and provides a platform for data fusion and quality management.

15.10.6 Make Sure the Organizational Arrangements Including Job Objectives and Job Descriptions Aligned to the Innovative Approach

The need to ensure that the appropriate resources are available to support the desired future approach to traffic operational management should put a particular emphasis on people resources. It is imperative the people involved have the appropriate skills and expertise to manage existing processes and new ones. This may involve the need for training in innovative technologies as well as new practices.

References

1 Hayes, A. Investopedia, March 28, 2023, Operations management: understanding and using it. https://www.investopedia.com/terms/o/operations-management.asp, retrieved July 03, 2023, at 12:53 PM Central European Time.
2 Talend. Data monetization: a complete guide. https://www.talend.com/resources/data-monetization/, retrieved July 03, 2023, at 12:54 PM Central European Time.
3 Washington State Department of Transportation. Intelligent transportation systems, variable message signs. https://tsmowa.org/category/intelligent-transportation-systems/variable-message-signs, retrieved July 03, 2023, at 12:05 PM Central European Time.
4 Virginia Department of Transportation, Highway Advisory Radio. https://www.virginiadot.org/travel/highway_advisory_radio.asp, retrieved July 03, 2023, at 12:07 PM Central European Time.
5 Wikipedia, Traffic Announcement (Radio Data Systems), Wikipedia, Traffic Announcement (Radio DataSystems). https://en.wikipedia.org/wiki/Traffic_announcement_(radio_data_systems), retrieved July 03, 2023, at 1:02 PM Central European Time.
6 Starkey, N.J.; Charlton, S.G.August 2020, Frontiers in sustainable cities. https://www.frontiersin.org/articles/10.3389/frsc.2020.00039/full, retrieved July 03, 2023, at 12:10 PM Central European Time.
7 Hanna, K.T.; Wigmore, I. Whatis.com, Mobile App. https://www.techtarget.com/whatis/definition/mobile-app#:~:text=Other%20examples%20include%20games%2C%20socialof%20programming%20languages%20and%20frameworks, retrieved July 03, 2023, at 1:02 PM Central European Time.
8 Federal Highway Administration. 21st-Century operations using 21st-century technologies, road weather information system. https://ops.fhwa.dot.gov/weather/faq.htm#:~:text=A%20Road%20Weather%20Information%20System%20(RWIS)%20is%20comprised%20of%20Environmental, and%2For%20water%20level%20conditions, retrieved July 03, 2023, at 12:08 PM Central European Time.
9 Washington State Department of Transportation. Queue warnings. https://tsmowa.org/category/intelligent-transportation-systems/queue-warnings, retrieved July 03, 2023, at 12:01 PM Central European Time.
10 National Highways England. Variable speed limits. https://nationalhighways.co.uk/road-safety/variable-speed-limits/, retrieved July 03, 2023, at 12:02 PM Central European Time.

11 Federal Highway Administration. Manual on uniform traffic control devices. https://mutcd.fhwa. dot.gov/htm/2003r1/part4/part4j.htm, retrieved July 03, 2023, at 12:04 PM Central European Time.

12 Federal Highway Administration. Ramp metering: a proven, cost-effective operational strategy – a primer. https://ops.fhwa.dot.gov/publications/fhwahop14020/sec1.htm, retrieved July 03, 2023, at 12:05 PM Central European Time.

13 New South Wales Government. Transport for New South Wales, lane management system (TidalFlow). https://www.transport.nsw.gov.au/projects/b-line-bus/b-line-stops/lane-management-system-tidal-flow, retrieved July 03, 2023, at 12:15 PM Central European Time.

14 Federal Highway Administration. Tolling and pricing program, congestion pricing – primer: overview. https://ops.fhwa.dot.gov/publications/fhwahop08039/cp_prim1_03.htm, retrieved July 03, 2023, at 12:16 PM Central European Time.

15 Federal Highway Administration. Arterial management program, traffic signal program management. https://ops.fhwa.dot.gov/arterial_mgmt/traffic_sig.htm, retrieved July 03, 2023, at 12:17 PM Central European Time.

16 Federal Highway Administration. Office of operations, 21st-century operations using 21st-century technologies. https://ops.fhwa.dot.gov/index.htm, retrieved July 03, 2023, at 12:18 PM Central European Time.

17 United States Department of Transportation, Office of the Assistant's Secretary for Research and Technology, Intelligent Transportation Systems Joint Program Office, Smart Community Resource Center. https://www.its.dot.gov/index.htm, retrieved July 03, 2023, at 11:04 AM Central European Time.

18 Institute of Transportation Engineers. Traffic engineering handbook, 7e. https://www.ite.org/ traffic-engineering-handbook/, retrieved July 03, 2023, at 11:05 AM Central European Time.

19 Runar. En guide for utforming av vegtrafikklys, July 27, 2020, http://ntoctalks.com/, retrieved July 03, 2023, at 11:07 AM Central European Time.

20 US Department of Transportation. Newsroom, https://www.transportation.gov/, retrieved July 03, 2023, at 11:09 AM Central European Time.

21 European Commission. Mobility and transport, highlights. https://ec.europa.eu/transport/ home_en, retrieved July 03, 2023, at 11:10 AM Central European Time.

22 ERTICO. European commission releases C – ITS platform phase 2 final report, October 10, 2017. https://erticonetwork.com/european-its-platform/, retrieved July 03, 2023, 11:15 AM Central European Time.

23 Intelligent Transport. 2018, Traffic management best practices in Europe. https://www. intelligenttransport.com/?s=Traffic+management+best+practices+in+Europe, retrieved July 03, 2023, at 11:16 AM Central European Time.

24 The CIVITAS Initiative. Sustainable urban mobility planning, https://civitas.eu/, retrieved July 3, 2023 11:18 AM Central European Time.

25 The European Conference of Transport Research Institutes (ECTRI). http://www.ectri.org/, retrieved July 03, 2023, at 11:20 AM Central European Time.

26 Womack, J.P.; Jones, D.T.; Roos, D. 1990, The machine that changed the world: the story of lean production. https://www.goodreads.com/book/show/289467.Lean_Thinking?ref=nav_sb:ss_1_65, retrieved July 03, 2023, at 1:10 PM Central European Time.

27 Womack, J.P.; Jones, D.T. 1996, Lean thinking: banish waste and create wealth in your corporation. https://www.goodreads.com/book/show/289467.Lean_Thinking?ref=nav_sb:ss_1_13, retrieved July 03, 2023, at 1:15 PM Central European Time.

28 Goetsch, D.L.; Davis, S.B. 2008, Quality management for organizational excellence: introduction to total quality. https://www.goodreads.com/book/

show/6384456-quality-management-for-organizational-excellence?ref=nav_sb:ss_1_81, retrieved July 03, 2023, at 1:37 PM Central European Time.

29 Edwards Deming, W.; Orsini, J.; Deming Cahill, D. (Eds.) 2012, The essential Deming: leadership principles from the father of total quality management. https://www.goodreads.com/book/show/15854713-the-essential-deming?ref=nav_sb:ss_1_21, retrieved July 03, 2023, at 1:35 PM Central European Time.

30 The Cancel for Six Sigma Certification. 2018, Six Sigma: a complete step-by-step guide: a complete training and reference guide for white belts, yellow belts, green belts, and black belts. https://www.goodreads.com/book/show/40971064-six-sigma?ref=nav_sb:ss_1_9, retrieved July 03, 2023, at 1:35 PM Central European Time.

31 George, M.L. 2003, Lean Six Sigma for service: how to use lean speed in Six Sigma quality to improve services and transactions. https://www.goodreads.com/book/show/283453.Lean_Six_Sigma, retrieved July 03, 2023, at 1:44 PM Central European Time.

32 Ries, E. 2011, The lean startup. https://www.goodreads.com/book/show/10127019-the-lean-startup, retrieved July 03, 2023, at 1:45 PM Central European Time.

33 Sutherland, J. 2014, Scrum: the art of doing twice the work in half the time. https://www.goodreads.com/book/show/19288230-scrum?ref=nav_sb:ss_1_5, retrieved July 03, 2023, at 1:53 PM Central European Time.

34 Simchi-Levi, D. 1999, Designing and managing the supply chain. https://www.goodreads.com/book/show/249764.Designing_and_Managing_the_Supply_Chain?ref=nav_sb:ss_1_39, retrieved July 03, 2023, at 1:55 PM Central European Time.

35 Handfield, R.B.; Nichols, E.L. 1998, Introduction to supply chain management. https://www.goodreads.com/book/show/681104.Introduction_to_Supply_Chain_Management_Technologies, retrieved July 03, 2023, at 1:57 PM Central European Time.

16

Summary

This is a book about mobility, about how effective application of technology can significantly increase mobility and accessibility options to all travelers while providing new possibilities for avoiding or managing undesirable side effects. The smart mobility revolution is underway that involves sudden and dramatic change as emerging technologies make new things possible. Smart mobility lies in the nexus between technology and business, therefore, this book deals with both.

Technology is addressed and described at a level of sufficient depth that enables the reader to understand capabilities and constraints, enabling effective choices. Business approaches and models are also important elements in this book as it is important to focus beyond technology to discover the best way to apply it. The authors have witnessed the evolution of smart mobility with a significant shift from an emphasis on making technology work to identifying and applying the best way to use it.

At the end of each chapter, somebody has been provided with the essential points in chapter; therefore, it will not be repeated here. Instead, the essential practical lessons learned are collected into the summary chapter.

16.1 Cities

Although there are multiple dimensions to smart mobility applications, including rural, interurban, freight, maritime, and aviation applications, this book deals primarily with the application of smart mobility in urban areas. This does not mean that the authors consider the other application areas to be unimportant, rather it indicates the particular importance of the application of smart mobility in urban areas. More than half the people on the planet live in urban areas and there has been considerable activity in applying technology to make cities smarter. There are smart city programs in various stages of implementation. In most major cities in the world, these go beyond mobility and address the application of technology to all city services including healthcare, water, energy, and education. Set smart mobility within the appropriate context, the chapter was devoted to smart city applications, describing how smart city services are being revolutionized through the application of technology.

An interesting lesson learned with respect to smart cities is the fact that there seems to be no universally agreed definition of a smart city. It means different things in different cities and to provide a focus, a working definition has been provided. A useful lesson for smart mobility comes from smart cities focusing on the needs of people. There are occasions when the needs of the user are eclipsed by fascination with smart mobility technologies. The smart cities approach reminders of the importance

Smart Mobility: Using Technology to Improve Transportation in Smart Cities, First Edition. Bob McQueen, Ammar Safi, and Shafia Alkheyaili.
© 2024 The Institute of Electrical and Electronics Engineers, Inc. Published 2024 by John Wiley & Sons, Inc.

and the incredible value of understanding the needs, preferences, and characteristics of the end user. While many smart city programs use similar technology applications to smart mobility, it is this latter area that is the primary subject of this book. The authors hope that by maintaining a focus on smart mobility in urban areas, it has been possible to treat this subject in sufficient depth to be of value.

16.2 Lessons

The overall objective of the book is to provide a reference guide for transportation professionals, those teaching, and those studying the subject. To be comprehensive, requires a balanced approach be taken to the content of the book. The intention is not to deliver a detailed how-to guide with respect to technology but to provide sufficient detail to enable a working understanding of the technology and all the other topics for which some knowledge is required in order to take full advantage of the smart mobility revolution. It is a rich subject area that requires careful treatment. One of the primary motivations in authoring the book lies in the ability of the authors to share practical experiences and lessons learned from more than eighty collective years of experience in both the public and private sector. It is that experience in the accumulated first-hand knowledge of the power of smart mobility that is the motivation to pass it on.

There are a great many lessons that can be applied, and nuggets of information that can be of considerable value in assisting public, private, and academic practitioners to extract the full value from the smart mobility revolution.

16.3 Value

Global experience indicates that there is significant value to be achieved and the application of smart mobility applications technologies to urban areas. Significant advances in safety, efficiency, and enhanced user experience have all been attained. It has also been understood that smart mobility has a significant impact on decarbonization and improving equity related to transportation.

To take a structured approach to the subject of smart mobility, it is necessary to define and understand the problems being addressed. A concise definition of the desired outcomes provide strong guidance for technology deployment. While technology can be fascinating, there is no obvious value unless there are significant outcomes and real results. One of the most powerful aspects of smart mobility is the ability to deliver these outcomes while avoiding or managing any undesirable side effects. Mobility is often a question of balance between efficiency, accessibility, and economic well-being on the one hand, against energy consumption, emissions, and damage to the environment, on the other hand.

It is also important to understand the difference between features, benefits, and values. Features are the characteristics of a product or a service that make it interesting to the buyer. Benefits refer to the tangible improvements that the product or service will bring to the buyers' life. Values are the end results are outcomes both for the individual and for society as a whole.

Benefits can include an improvement in safety, an increase in efficiency, enhanced user experiences, more equitable and accessible transportation, and decarbonization. It is vital to be able to quantify these benefits and others to justify current investment in smart mobility and set expectations for future investments. This is currently a challenge, as many smart mobility implementations have insufficient data regarding the before and after situations. The lack of coherent standards for smart mobility evaluation also means that even when such data has been collected, it may not be suitable for comparison. This is a big area of opportunity in smart mobility.

Appropriate data forms the basis for scientific analysis of the effects of smart mobility. Key performance indicators and analytics derived from such data form the basis for a solid business case for smart mobility investment. It is important to have a scientific approach supported by good documentation in order to take rational investment decisions. The vacuum created by the lack of this ability can lead to unduly biased and heavily politicized investment decisions. Smart mobility has the inherent advantage that implementations designed to deliver specific services such as electronic payment and advanced traffic management, also have the ability to collect the quantity and quality of data required for solid evaluation approaches.

16.4   Challenges and Opportunities

It is readily apparent that the definition and agreement of objectives is an important factor in successful application of smart mobility. However, a successful approach should also be built around a clear understanding of the challenges and opportunities presented by the implementation. One of the best ways to define these challenges and opportunities is to establish a dialogue with prior implementers. Those practitioners who have implemented the same smart mobility initiatives, or something similar in the recent past. The authors have found that the most effective way to do this is to engage these prior implementers directly in a dialogue.

Preparation for establishing such a dialogue can take the form of literature review and scanning published information. However, this information source tends to focus on successes and achievements and does not provide detail on challenges and opportunities. Having conducted a scan for published information, it is essential to build on that knowledge with a direct dialogue with those practitioners responsible for the prior implementation.

The information exchange during the dialogue can include sharing of key performance indicators used, examples of benefit and cost evaluation, and results achieved. The approach methodology used for guiding the implementation should also be ascertained along with the value and constraints associated with the approach.

The objective in establishing and maintaining such a dialogue should be to develop sufficient understanding of the prior implementation to enable it to serve as a model of the one being planned. The value is in the model, not the recipe. No two cities are the same, resulting in different starting points and different approaches, one size does not fit all, but a good model and delivering incredible value.

16.5   Global Progress

It is obvious from research conducted by the authors that there is considerable momentum in the application of smart mobility initiatives worldwide. A small sample of ten leading cities was selected for detailed study. The study resulted in a good understanding of the applications that have been deployed worldwide and the lessons learned.

In fact, it is interesting that while each city had different requirements and slightly different approaches, a pattern emerged illustrating common themes and approaches. It is not surprising that the ten leading cities exhibited a high degree of commonality in the smart mobility applications implemented. Information from the 10 leading cities provides good examples that can be used to illustrate the characteristics and results that can be expected from smart mobility. This can set the scene for later, more detailed explanation of the technologies, products, and services that form the basis of the smart mobility initiative.

These examples can serve as an excellent communication tool when explaining the meaning of smart mobility, the importance of smart mobility, and the results that can be expected. The lessons learned from each city can also serve as the raw material for the development of a checklist for future smart mobility investments. Public and private sector cooperation, effective use of data, and the deliberate focus on understanding the characteristics and preferences of people, all featured heavily in the study.

The intense activity exhibited around the world also confirms that the smart mobility revolution is underway in the form of physical implementations and not just plans.

16.6   Effective Planning

It is also important to take a robust and effective approach to planning and smart mobility implementations. This does not mean that the planning process needs to take a long time. An accelerated approach to the establishment of a smart mobility plan has been explained in the book. There are sufficient examples to enable building blocks to be defined to accelerate the planning process. Building on practical experiences and lessons learned from the large number of prior implementations around the globe, it is possible to build on the alien experience of others to put together an effective plan for the implementation of smart mobility in a city. A structured framework approach is possible that bridges three milestones in the implementation timeline. The first milestone is the definition of the starting point. This includes an evaluation of prior investments and the unique characteristics of the transportation system for the city. It can also include a summary of specific needs, issues, problems, and objectives. It is also important to note that objectives define what is desired. It is equally important to define what is to be avoided.

The second milestone is the development of a future vision. This describes the intended outcome of the smart mobility implementation from user and the end-user perspectives. This defines the desired endpoint.

The third milestone is the development of a roadmap or transition strategy to get from the first milestone to the second one, from today's situation to the desired future vision. An important dimension in the roadmap will be the identification of implementation stages that while moving the initiative forward to the vision, also attain clear and significant value at each stage. These milestones will primarily guide the technical and business elements of the implementation. However, they will also play a significant role in establishing and maintaining effective communications with stakeholders including the general public.

16.7   Solutions and Business Models

Smart mobility solutions represent a collection of technologies designed to deliver specific services. It is likely that these services will be implemented in a staged manner over a period of time. This makes it particularly important that the overall architecture defining how the initiatives will fit together is clearly understood and agreed.

There are multiple elements that define smart mobility in an urban area. These include technology-driven solutions and services, existing transportation infrastructure, and business models that define the relative roles of the public and private sector. The technologies that drive smart mobility solutions are compelling and it would be easy to get distracted by the technology, with its features and functions. This makes it important to maintain a focus on the desired end result, on the outcomes to be achieved, and how each individual element fits into the bigger picture.

Business models and an important aspect of smart mobility initiatives. They define who will do what, who will invest, and who will obtain the return on investment. Such business models can also identify those smart mobility initiatives that can be supported by a pay-as-you-go approach, while the end user pays for the service. The business model can also identify those smart mobility initiatives that cannot be supported by the user-pays approach and are suitable cases for funding through public, tax-derived funds. The authors believe that a good business model can be the key to unlocking investment from both the public and the private sector, ensuring alignment of objectives and maximizing synergy.

16.8 Technology

This is a fundamental component in smart mobility and the choice of appropriate technology is critical to success. However, it must be recognized that technology is merely a building block used to create products and services. It is the services that deliver values, benefits, and outcomes. Technologies do not deliver value; they enable value creation through the delivery of services.

There are a wide variety of technologies used to enable smart mobility solutions. Most of these technologies were not specifically developed for smart mobility applications but were created for other areas of industry or commerce. Therefore, it is important to have a good understanding of the technologies that drive smart mobility solutions, their capabilities, their characteristics, and their limitations. In most cases, existing technologies and adapted and adopted for smart mobility purposes, sufficient knowledge of the technologies is vital to guide this adaptation and adoption.

A powerful and challenging feature of the technologies used in smart mobility initiatives is the speed at which they emerge. The cycle from invention to commercialization of technology has shortened dramatically, requiring that some effort be placed in understanding emerging as well as existing technologies. This also requires a certain degree of flexibility to be integrated into smart mobility initiatives to enable the adoption of powerful new technologies as they emerge. This is one of the common lessons learned from the global review of progress.

The major technology groups used in smart mobility, including sensors, telecommunications, information delivery, data management, and in-vehicle technologies, present significant opportunities, and challenges. A multidisciplinary approach is required to smart mobility initiatives to bring a detailed understanding of these technologies to bear on the initiative. It is also important that transportation practitioners expand their horizons to include a solid understanding of technological capabilities and limitations. The speed of emergence of technologies also requires the continuous scanning to take place for new technologies representing new opportunities.

16.9 Opportunities and Challenges

Based on a clear definition of the problems to be addressed, it is possible to navigate the challenges and take full advantage of the opportunities offered by smart mobility. A unique characteristic of smart mobility has the potential to attend the benefits while avoiding or mitigating any undesirable side effects. Since smart mobility involves the application of technology, there are a number of lessons learned from the information technology industry that can be of great value to smart mobility. The specialized nature of information technology planning and application, along with the complexity, can create opportunities for cost overruns and delays. A clear understanding of the challenges and practical lessons learned from the past can go a long way to the successful management of information technology.

16.10 Accelerating Success

There is a need to accelerate progress in the implementation of smart mobility. It is clear that the outcomes that can be achieved are worth making a significant effort to realize. The longer it takes to implement, the longer it will be before the benefits are truly realized. It would be a fair assessment of the current situation to say that we have "islands of progress." Certain cities and certain parts of cities demonstrate considerable progress toward implementation. At the moment, it is difficult to see how these islands will expand and amalgamate to deliver a consistent, coherent benefit delivery. Of necessity, progress is spotty. This has enabled lessons to be learned through the adoption of an incremental approach that enables knowledge gain and practical experience, before scaling up and widescale deployment.

It is time to build on the early focused success by adopting larger objectives. Entire cities, entire corridors, and complete transportation networks could be the focus of the application of smart mobility. The ability to understand opportunities and challenges and accurately quantify values and benefits suggests that we are at a point of maturity in smart mobility and have the capability to go to the next level. There is enough experience and sufficient expertise has been developed. The industry can now move quickly and boldly, while confident of achieving new higher-level targets.

Planning for smart mobility is now well understood, and the development and implementation of strategies is well defined. New knowledge has been acquired on how to fit technology-driven solutions together while maintaining a focus on outcomes. There is great experience in defining effective technical, organizational, and commercial arrangements for smart mobility. It is time to apply this new knowledge and experience and start making some mighty leaps.

16.11 Managing Performance

One of the great things about smart mobility is the new opportunities that are delivered for measuring, understanding, and managing performance. A large volume of new data is created that, with the right data management approach, can be used to drive effective performance. A higher level of situational awareness can be achieved that results in a greater understanding and insight into variations in the demand for transportation and operating conditions. In the past, practitioners relied on judgment and making the best of what little data was available. To some extent, it was like working in the dark. Now with the availability of data, the lights are on, and a clear understanding of performance can be achieved. The information technology applications that form the basis for smart mobility solutions can also be used to improve the effectiveness of project management by managing the data to extract insight and understanding. This in turn moves the focus away from making sure technology works, to making sure that we get the best from it.

16.12   People Focus

With all the possible distractions of technology and its characteristics, it can be easy to forget one of the most important elements of smart mobility – people. We have unparalleled opportunities to understand in great detail the characteristics and preferences of the traveler and all other people who benefit from effective mobility. Remembering that people who do not travel benefit from smart mobility also, in the form of better air quality, better living standards, higher safety, and lower cost of transported goods. Given that new capability, there is the possibility to take a much better, more effective approach to understanding people and integrating that understanding into the planning and design process for smart mobility.

Smart mobility offers the ultimate solution for equity in transportation. In the past, the population has been divided into groups with similar characteristics. This was done either because the data was not available or was too large and complex to be analyzed. Now there is a new possibility to plan and design smart mobility services for the individual. Surely this is the ultimate and equitable transportation.

Smart mobility enables researchers to study and understand the people involved in transportation on a direct basis. It obviates the need for experts to function as surrogate users and replaces them with actual user data. This does not replace the expert, but merely better informs the experts' opinions and design choices.

There are also some great possibilities to establish a continuous dialogue between users, operators, planners, and designers. It is an ability to establish a continuous feedback loop, enabling a detailed understanding of the user experience, how well smart mobility designs are working, and whether the scope for improvement. The user can now be treated as an informed provider of scientific data regarding operational effectiveness. This can be achieved on a complete trip basis from original origin to ultimate destination.

In an interesting way, applying smart mobility to transportation can actually humanize the process, through the ability to communicate and understand. However, it will be necessary for us to reassess our current procedures and develop better ways to take full advantage of the data flow and the insight obtained.

16.13   Policies and Strategies

Smart mobility provides challenges and opportunities with respect to policies and strategies. The primary challenge relates to ensuring that policies and strategies keep pace with the implementation of technology. This is particularly important when the private sector have moved into a new, more prominent leadership role in the application of smart mobility in cities. When policy and strategy lag behind technology implementation then it becomes difficult to avoid undesirable side effects. When policy and strategy lead technology implementation ,it is vital to ensure that while the public interest is safeguarded, the sufficient growth space for new technology and innovation. The latest thinking in policy is that the new data available enables the end user to play a more active role. Future policymaking, strategies, and standards will be driven from a local level based on end-user characteristics and needs.

16.14   Organizing for Success

The need for technological and commercial coherence using business models has been identified. It is also essential that suitable organizational arrangements are defined and implemented to enable the smart mobility revolution. This will require a detailed assessment of organizational capabilities and the identification of skill gaps that the marriage when moving from traditional transportation processes to smart mobility. It is necessary to consider how performance management can be applied to the organization through the use of revised job descriptions and job objectives. A coherent plan is required on high new skills will be acquired and what shape the new organization should be to optimize the benefits of smart mobility. This should be recognized as a significant change management challenge, requiring the application of change management techniques and tools.

16.15 Managing Operations

In a similar way to performance management, smart mobility offers opportunities and challenges. Many smart mobility initiatives such as mobility as a service, and micromobility require higher level of operational management to succeed. In particular, current traffic or transportation management centers can benefit significantly from the new data possibilities offered by smart mobility. Based on artificial intelligence can also be used to turbocharge these existing traffic management centers. Effective operational management requires that rapid decisions be taken that can have citywide or even regional repercussions with respect to transportation. Effective operational management will need higher situational awareness and stronger abilities to analyze data and extract insight. This must be matched by the ability to convert the insight into appropriate strategies and responses, which in turn can be measured for effectiveness.

Smart mobility also offers new possibilities to bridge the distance between nontechnical decisionmakers and operational managers. The use of common data with customized dashboards designed for each role can ensure that political goals and objectives are well aligned with operational management strategies. This becomes of even greater importance when bold political promises with regard to safety, efficiency, and the environment are being made. It is crucial to understand how operational management is making progress to deliver on those promises.

16.16 Conclusions

To achieve the intended objective, this book has of necessity covered a wide range of material. It is not a business book and is not a technology book, it contains elements of both. This reflects the nature of smart mobility as an interface discipline between business and technology. The ability to run transportation as a business is one of the interesting capabilities offered by smart mobility.

To conclude, here are some of the most important lessons contained in the book.

16.16.1 Learn from the Past

While smart mobility is an emerging area, there are considerable lessons to be learned from the past. These include prior implementations of smart mobility applications. There are also lessons to be learned from other industries that the view should be both retrospective and wide. This should also include the establishment and operation of a feedback loop from current smart mobility initiatives to future ones.

16.16.2 Keep Looking for the Future

It is important to scan continuously for emerging technologies and for successful results in the application of new technologies. Pilots and demonstrations can also be used to assess the suitability of these technologies. Plans for smart mobility implementation should have sufficient flexibility to incorporate new technologies and be scalable.

16.16.3 Effectively Engage the Private Sector

Over the past 15 years, the emphasis shifted from public leadership of smart mobility to private sector. This makes it imperative the public sector effectively engages the private sector in smart mobility implementation. This can include a range of business models, with varying roles for both the public and private sector. In addition to alignment of resources and investment, objectives and motivation must also be aligned.

16.16.4 Make Better Use of Academia as the Independent Trusted Advisor

Academia represents an amazing resource available to smart mobility implementers. The sector can serve as the independent trusted advisors providing input to planners and designers on the capabilities and limitations of technology. Academia can also be the source of up-to-date advice on emerging technologies and practical lessons learned in prior implementations.

16.16.5 In Addition to Technologies from Other Industries, Look for Best Practices and Tools

The point was made earlier that technologies utilized in smart mobility typically originate from other industries. In addition to adapting and adopting technologies, smart mobility practitioners should also look for best practices, techniques, and tools from other industries that could be of value in smart mobility.

16.16.6 Understand the Underlying Processes

A good understanding of the underlying processes that are being supported through investments in activities can go a long way to maximize investment efficiency and align solutions understanding the business processes can also avoid the situation where good technology is applied to bad process, by identifying and eliminating the weaknesses in the process. In a world of rapidly accelerating technology, the process can also serve to maintain a focus on "what" needs to be done instead of "how."

16.16.7 Bridge the Spaces Between Stakeholders

There are numerous technical and nontechnical stakeholders involved in a typical smart mobility initiative. It is important to ensure that any space between stakeholders is bridged through the use of technology, business models, and organizational arrangements. This also includes the public, the end user who can sometimes be ignored in the planning, design, and implementation of smart mobility. This can be enabled by the new level of connectivity offered by smart mobility applications.

16.16.8 Move Quickly Through Planning

There are enough resources and enough insight at the industry's disposal to accelerate the planning process for smart mobility. Using lessons learned and experiences from prior deployments, taking advantage of the sophisticated communication tools available today, planning can be accelerated. This can be done while maintaining a high degree of confidence enabling bold products to be made.

16.16.9 World-class Communications

Because of the innovative nature of smart mobility, there is a strong need for world-class communications. This should extend from defining the intended outcomes to communicating progress toward the future vision. The communication should extend to all aspects of the trip and be designed for the individual. Communication should also be highly effective between the public and private sector and especially between nontechnical decision-makers and operational managers. World-class communications should be used to support clarity of purpose and well-aligned objectives across the city.

16.16.10 Effective Use of Data

There is a repeating theme throughout the research conducted by the authors and that is the value of the effective use of data. Smart mobility technologies include sophisticated data capture, data processing and information, exchange, and technologies. These must be implemented within a coherent framework that includes technological capabilities, organizational arrangements, and business models. It is vital that others commonality across the data being used by nontechnical decision-makers, planners, designers, implementers, operators, and maintainers.

New abilities to capture data and extract insight should be used to fully support maximum adaptability for the smart mobility implementations based on a higher level of monitorability. A higher level of situational awareness provides the basis for better decisions and the attainment of real results.

The authors hope that the reader has enjoyed reading this book and has been able to learn some things from our past experiences. It would be a considerable success if this book formed the basis or contributed to a more effective application of smart mobility in urban areas. Ideally, it would contribute to a bottom-line impact in urban areas around the world with respect to safety, efficiency, enhanced user experience, equity, and decarbonization the authors believe it is important to be equipped for the smart mobility revolution to take advantage of all opportunities offered.

Mobility is one of the essential ingredients to urban living. Smart mobility holds the promise of significantly enhanced levels of mobility for all people while balancing the cost and impact of achieving the new levels. Driverless vehicles and mobility as a service play a leading role in the smart mobility revolution, and it requires experience and expertise to get the best from these new technology applications.

The authors look forward to a future where these and other technology applications are used to the best effect to take our urban transportation systems to a new level of effectiveness and performance. The future is bright. If we get it right.

Index

Note: *Italicized* and **bold** page numbers refer to *figures* and tables, respectively.

Smart Mobility: Using Technology to Improve Transportation in Smart Cities, First Edition. Bob McQueen,
Ammar Safi, and Shafia Alkheyaili.
© 2024 The Institute of Electrical and Electronics Engineers, Inc. Published 2024 by John Wiley & Sons, Inc.

Printed and bound by CPI Group (UK) Ltd, Croydon, CR0 4YY

27/08/2024

14546154-0001